DRY CREEK

Peopling of the Americas Publications
Sponsored by the Center for the Study of the First Americans
GENERAL EDITORS: MICHAEL R. WATERS AND TED GOEBEL

PUBLICATION OF THIS BOOK WAS MADE POSSIBLE IN PART BY
FUNDING FROM THE NATIONAL SCIENCE FOUNDATION.

Dry Creek

ARCHAEOLOGY AND PALEOECOLOGY OF A
LATE PLEISTOCENE ALASKAN HUNTING CAMP

W. ROGER POWERS, R. DALE GUTHRIE,
AND JOHN F. HOFFECKER

Edited by Ted Goebel

TEXAS A&M UNIVERSITY PRESS
College Station

Copyright © 2017 by Center for the Study of the First Americans
All rights reserved
First edition

Portions of chapter 8 were previously published as "Dry Creek Revisited: New Excavations, Radiocarbon Dates, and Site Formation Inform on the Peopling of Eastern Beringia" in the October 2015 issue of *American Antiquity* (Volume 80, Number 4, pages 671–694)

This paper meets the requirements of ANSI/NISO Z39.48-1992
(Permanence of Paper). Binding materials have been chosen for durability.
Manufactured in China by Everbest Printing Co. through FCI Print Group
∞

Library of Congress Cataloging-in-Publication Data

Names: Powers, William Roger, author. | Guthrie, R. Dale, 1936– author. | Hoffecker, John F., author. | Goebel, Ted, editor.
Title: Dry Creek: archaeology and paleoecology of a Late Pleistocene Alaskan hunting camp / W. Roger Powers, R. Dale Guthrie, and John F. Hoffecker; edited by Ted Goebel.
Description: First edition. | College Station: Texas A&M University Press, [2017] | "Peopling of the Americas publications." | Includes bibliographical references and index.
Identifiers: LCCN 2016046610 (print) | LCCN 2016048214 (ebook) |
ISBN 9781623495381 (printed case: alk. paper) | ISBN 9781623495398 (ebook)
Subjects: LCSH: Dry Creek Site (Alaska) | Excavations (Archaeology)—Alaska—Denali Borough. | Denali Borough (Alaska)—Antiquities. | Antiquities, Prehistoric. | Paleo-Indians—Alaska—Dry Creek Site. | Archaeological geology—Alaska—Dry Creek Site. | Paleoecology—Alaska—Dry Creek Site. | Geology, Stratigraphic—Pleistocene.
Classification: LCC F912.D79 P69 2017 (print) | LCC F912.D79 (ebook) | DDC 979.8/6—dc23
LC record available at https://lccn.loc.gov/2016046610

In Memoriam
William Roger Powers
1942–2003

CONTENTS

Preface   ix
Ted Goebel

PART 1. THE DRY CREEK REPORT

Chapter One. Introduction   3
R. Dale Guthrie and W. Roger Powers

Chapter Two. The Dry Creek Site: A History and Description   9
W. Roger Powers

Chapter Three. The Geology of the Dry Creek Site   25
W. Roger Powers

Chapter Four. Lithic Technology of the Dry Creek Site   37
W. Roger Powers

Chapter Five. The Occupation Floors at the Dry Creek Site   107
John F. Hoffecker

Chapter Six. Paleoecology of the Dry Creek Site and Its Implications for Early Hunters   153
R. Dale Guthrie

Chapter Seven. Dry Creek and Its Place in the Early Archaeology of the North   193
W. Roger Powers and R. Dale Guthrie

Appendix A. Component IV at the Dry Creek Site   203
W. Roger Powers

PART 2. DRY CREEK UPDATE

Chapter Eight. New Geoarchaeology and Geochronology at Dry Creek   219
Kelly E. Graf, Lyndsay M. DiPietro, Kathryn Krasinski, Brendan J. Culleton, Douglas J. Kennett, Angela K. Gore, and Heather L. Smith

Chapter Nine. A Dry Creek Retrospective   261
Ted Goebel and John F. Hoffecker

References Cited   289
Contributors   321
Index   323

PREFACE

TED GOEBEL

Historically, Dry Creek is one of the most important archaeological sites of Beringia, the Ice Age gateway to the Americas. To understand Dry Creek's significance, consider the state of Alaskan archaeology in 1974, when Roger Powers, professor of anthropology at the University of Alaska Fairbanks, initiated its full-scale excavation. Geomorphologists understood that the Bering Land Bridge persisted from about 35,000 to 11,000 radiocarbon years ago (^{14}C yr BP, or 40,000 to 13,000 calendar years ago [cal yr BP]),[1] but no evidence of humans dating to this time had been found in eastern Beringia to demonstrate its use as a migration route from Asia to America. Likewise, paleoecologists knew that late Pleistocene Alaska was home to a rich fauna of now-extinct megamammals—predominantly mammoth, horse, and steppe bison—but unlike in northern Eurasia and temperate North America, there was still no evidence that early Alaskans encountered, let alone hunted, these animals. South of the Canadian ice sheets, archaeologists had established the widespread presence of the Clovis Culture by about 11,000 ^{14}C yr BP (13,000 cal yr BP), but north of the ice sheets, no such complex—as old as Clovis or older—had been found. Simply put, despite years of searching (e.g., Anderson 1968a; Müller Beck 1967; West 1967), even as late as 1974, Alaskan archaeologists had not directly contributed to the scientific investigation of the peopling of the Americas.

All this changed with the discovery and excavation of the Dry Creek site. Excavated from 1973 through 1977, Dry Creek became the first place in Alaska where a series of late Pleistocene cultural occupations was unearthed in a well-stratified context. Initial radiocarbon dates suggested the presence of humans by 11,120 ± 85 ^{14}C yr BP (12,970 cal yr BP), the first uncontested claim that humans lived in Alaska before the flooding of the Bering Land Bridge. Faunal remains from the site's late Pleistocene cultural components included steppe bison (*Bison priscus*), suggesting the Dry Creek site's early occupants hunted at least one member of Beringia's now-extinct complex of megafauna. Moreover, at 11,120 ^{14}C yr BP (12,970 cal yr BP), Dry Creek's earliest occupation was contemporaneous to Clovis, and large-scale excavations provided an important chronologically constrained sample of tools, cores, and debitage, providing a tantalizing first glimpse of what Clovis-aged lithic technology in Alaska looked like. For some, it provided the long-sought-after link between the

Paleoindian archaeology of temperate North America and the early archaeology of Beringia. At the same time, the Dry Creek excavations demonstrated that even within the late Pleistocene, there was much variability in Beringian lithic technologies and tools, variability that we still do not completely understand. Without question, the Dry Creek excavations provided the framework through which Alaskan archaeologists would work for many years to come.

The tragedy of Dry Creek, of course, is that its final report was never published, so the site's potential significance was never felt by the American, Canadian, and Siberian archaeological communities. Instead, the site's record was presented only in a short series of journal articles and edited-book chapters, and these focused more on its geology than archaeology. In 1983, Powers and his team submitted a complete, detailed report of the site's excavation to the US Department of the Interior's National Park Service (the agency that provided much of the field project's funding), and there were plans for the report's publication as a book. Within a couple of years, Powers submitted a revised and updated version to the Alaska Anthropological Association for potential publication in their new Aurora book series, and although peer reviews were generally positive, there were enough editorial changes requested to keep him from quickly finishing the revisions and updating his team's interpretations. Other activities started to get in the way. By 1985, Powers and his students had turned their attention to a series of new archaeological sites in the Nenana Valley—Moose Creek, Panguingue Creek, and Walker Road—so that dissemination of the full Dry Creek findings gradually became overshadowed by the new work. Powers, however, never entirely gave up on the idea of publishing the Dry Creek report, and even upon retiring in 2003, he was still talking about significantly updating and finally publishing it. However, he was destined to never finish the project, because of his untimely death just months after leaving his faculty position at the University of Alaska.

A decade later, while John Hoffecker, Owen Mason, and I were flying across the Seward Peninsula on a winter consulting trip, our talk inevitably turned to Roger and the work in the Nenana Valley we had been involved in as students during the 1970s and 1980s. At some point in the conversation, I said that we owed it to Roger to finish the book for him, and John and Owen quickly challenged me to do so. Later that year, Kelly Graf and I visited Roger's widow, Alicia Powers, in Pittsburgh, and Alicia gladly gave us Roger's documents relating to Dry Creek, among them the last version of the report's text as well as a full set of the report's figures, although many of the figures were water damaged and nonreproducible. With these materials in hand, we obtained a small grant from the Arctic Social Sciences Program at the National Science Foundation to create a new electronic version of the manuscript's text and all its figures and to help cover the cost of the book's production and printing.

We immediately got to work on the manuscript. Our objectives were twofold. First, we sought to present the final version of the book that Powers, Guthrie, and Hoffecker had created by 1986–87. Second, we planned to provide an update of work accomplished at the site and with its collections since 1983, putting the record from Dry Creek squarely in the context of Beringian archaeology and paleoecology today, the mid-2010s. Heather Smith, then a PhD student at Texas A&M University, served as a research assistant for the project, retyping the manuscript and re-creating as many of the original artifact plates as possible. This required a trip to the University of Alaska Museum of the North in Fairbanks, where she was able to access the Dry Creek collection and digitally photograph its still-available artifacts. Unfortunately, not all of the tools and cores were found, so we could not re-create all of the original artifact plates in their entirety. For providing Heather access to the collection, we are especially grateful to the museum's staff, including Jeff Rasic, Scott Shirar, and Sam Coffman. Heather and Josh Keene, another PhD student at Texas A&M, also redrafted the report's maps, stratigraphic profiles, and other line art using a current version of the drawing program Adobe Illustrator.

Simultaneous to these editing activities, Kelly Graf earned funding from the Elfrieda Frank Foundation to revisit Dry Creek and address some lingering issues regarding the site's geoarchaeology and geochronology. In 2011, she reopened Powers's original excavation, exposed its stratigraphic profiles, and excavated an additional 10 m^2 of the site, establishing the stratigraphic relationship of its two late Pleistocene cultural components and obtaining the first accelerator radiocarbon dates on hearth features preserved within them. Given the significance of these results, we deemed it important to include a full description of Graf's activities as part of the report's update. Although presented originally in an article published in the journal *American Antiquity* (Graf et al. 2015), in chapter 8 we present the full results of her team's excavations, expanding their report to include details on the new archaeological assemblages recovered, not just the geoarchaeology and geochronology of the Dry Creek site.

Hence we offer *Dry Creek: Archaeology and Paleoecology of a Late Pleistocene Alaskan Hunting Camp*. Part 1 presents the original Dry Creek report much the way Powers, Guthrie, and Hoffecker envisioned it in the mid-1980s. Most of the chapters have been only minimally revised since the 1983 report, except that Hoffecker's chapter 5 was extensively revised, for the last time in 1988. My editorial changes have been minimal, correcting typographic errors, inserting calendar ages where appropriate, and inserting occasional notes referring readers to updated discussions in the final chapter. Part 2 presents an update on the site, with a first chapter fully reporting Graf's 2011 geoarchaeological investigations and a second chapter providing a retrospective of the research conducted at the site and on its collections since 1984. We intend to continue this program by fully reporting other large-scale excavation projects that have been carried out in the Nenana Valley, next focusing on the Owl Ridge and Walker Road sites, both of which have been the subjects of reinvestigations in recent years.

Finally, the authors and editor of the book wish to express our sincere gratitude to Alicia Powers, who graciously opened Roger's personal archives to us, so we could retrieve the related Dry Creek documents and assemble the book. Also, we wish to acknowledge the support of the Arctic Social Sciences Program at the National Science Foundation. Without this the Dry Creek monograph may never have been finished.

Note

1. When Dry Creek was excavated and originally reported, radiocarbon-calibration curves had not yet been developed for the late Pleistocene–early Holocene period. For consistency, dates and ages are presented as radiocarbon years ago (^{14}C yr BP) as well as calendar years ago (cal yr BP) throughout the volume, with calendar dates being editorially added to Part 1. Calibrations were done with Calib 7.1 (http://calib.qub.ac.uk/calib; Reimer et al. 2013), with median calendar dates being presented in the text.

PART 1
The Dry Creek Report

CHAPTER ONE

Introduction

R. DALE GUTHRIE AND W. ROGER POWERS

Research Philosophy: A Multidisciplinary Approach

There seem to be at least two ways in which sites are currently dug. One is to obtain the artifactual material in order to produce a point in time and space with typological identity so that a cumulative pattern might emerge of the "phylogenetic" distribution and chronology of people and their traditions. Studies of palynology, paleontology, and geology all become ancillary subdisciplines directed toward this end.

Another way is to look at an archaeological site as an important and possibly unique occasion to focus on that point in human prehistory and all the natural forces that were affecting and molding it. It is an opportunity for natural historians from several disciplines to live together in the field and share laboratories, arguing over the various meanings of Quaternary events, using the site almost as an informal symposium for interdisciplinary discussion. As such, it becomes an occasion to pursue the interrelationship between an environment and a people, with the underlying assumption that there are causal connections. Approaching it from this angle, the presence of *Chloridae* phytoliths in fire hearths and their rarity in the rest of the sediments, indicating the use of buffalo chips for fuel (Lewis 1978), might be as important to our general understanding as whether the occupants of a site used Cody knives. Likewise, information about sediments in the site dating from when there is *no* archaeological evidence at all might sometimes tell us as much about humans as the artifact-laden levels. People are beginning to argue that there are long segments of time during which there is no evidence of human occupancy in areas occupied by humans both before and after (Reher 1974). These absences might be the result of environmental restrictions.

The bothersome thing about the "natural history" approach of looking at the ecological setting of an archaeological site, past and present, as opposed to the "prehistory" approach, is that no one person can be trained to have—or even gain through a lifetime of experience—the necessary skills and angles of perception to gather the information. As difficult as it is to coordinate specialists with varying interests and disperse the responsibility for a watermark site, it seems obvious that an interdisciplinary approach is by far the most productive. By this, we do not mean that a site should be dug and the specimens sent to respective specialists for identification; rather, there should be a cooperative focus by various Quaternary researchers using, in addition to their rote expertise, a creative eye for identifying and resolving new questions. Excavations are best carried out with a multidisciplinary team of researchers working together to investigate Quaternary biota, climate, and people, each member having knowledge of the others' specialties and all interested in a common focus.

Such a synthesis at the Dry Creek site admittedly began after excavations were initiated but provided the fuel for many hours of discussion among authors and their colleagues. Some approaches to collecting and analyzing the data went unbelievably well; others we would now do differently. The sediments were unexpectedly devoid of small mammals and invertebrates. The ground squirrel burrows offered great potential, but the expected fossil nests never occurred within the site. Also, the sediments contained a poor pollen record, and pollen cores in nearby lakes were not sufficiently old.

In retrospect, we could probably have benefited from a person interested in Quaternary soils. Also, the geoarchaeological and bioarchaeological studies were disjunct, the former having been conducted during the early part of the project, with the latter following during the main thrust of the excavations and analyses. The interdisciplinary character of the research could have been greatly strengthened had these phases of the research run concurrently. Despite our disappointments and misjudgments, the site has been the origin of much new information and many new ideas.

Research Hypothesis

The Dry Creek site, like most Early Man sites, was an accidental discovery, but one that was to be the first occurrence in the northern part of North America of a multicomponent site containing many thousands of lithic artifacts and identifiable large-mammal remains that span a good part of the eleventh millennium ^{14}C yr BP (thirteenth millennium cal yr BP). Because of its discovery, it was possible to develop a research strategy that can predict the occurrence of similar sites in a comparable topography. This twofold quality—its importance for understanding the lifeways of early hunters at the site itself and its ability to open the door to the discovery of early sites—underscores the fundamental importance of Dry Creek for northern archaeology.

The site was excavated over three summers by several Quaternary researchers with multidisciplinary approaches in geology, archaeology, and paleobiology. As tests were conducted and excavations expanded, the importance of the site began to emerge, and it became quickly obvious that Dry Creek possessed several features that, in combination, made it ideal for further investigations. First of all, the site was deeply stratified, at least by Alaskan standards, and the radiocarbon date from near the base of the section, in what was later to be called Component II, indicated a probable terminal Pleistocene age. Second, the site was multicomponent, with two occupation horizons near the base of the section containing preserved fauna and a third nearer the top. In addition, the archaeological remains occurred in clusters, and they contained variable compositions of artifacts.

These features of the site allowed us to pose several hypotheses that could be tested by further excavation:

1. The two stratigraphically separate early components could provide us with new information on temporally separated lithic and faunal assemblages.
2. The clustering of artifacts would allow us to isolate and define sets of tools, waste materials, and associated fauna within each of the early components.
3. Identification of age, sex, species, and seasonality of the fauna would allow us to reconstruct some aspects of the paleoecology of the site—the Nenana Valley specifically and the northern Alaska Range foothills in general.
4. The associated fauna and artifact clusters could allow us to reconstruct, in part, the procurement and processing techniques of the game species.
5. The preservation of bone raised the possibility that the articulation of microblades with bi- or unilaterally grooved bone/antler points would provide information on the presumed existence of the composite inset technique and that the manufacture and maintenance of these points could be described.
6. Given a comparable quality of information from both of the early components, we could examine some aspects of subsistence activity and possibly study the problem of either stability or change in the patterns of site use.
7. A familiarity with the surrounding area and its seasonal climatic variations would permit us to study how these affected large mammal distributions. Combined with information from the archaeological site, this would allow us to reconstruct the reasons for the site's particular location and what kinds of activities would have likely occurred there.

Research strategies, the kinds of paleoecological data sought, and our ideas about what we were seeing underwent a series of changes as more and more data came in. These ideas continued to change back at the university, where careful analysis revealed many things that were not obvious in the field.

Though other analyses will be conducted with the Dry Creek material over the years, and our ideas and interpretations will no doubt continue to change and evolve, the data from the site do provide important new information pertinent to some of the major questions in the study of Early Man in North America.

General Conclusions

1. SETTLEMENT PATTERN. The Dry Creek site was occupied several times as a temporary hunting camp, or "spike camp," by early peoples. The site is located on a prominence too exposed for long-term winter camping comfort, but it is ideal as an observation point. The artifacts indicate extensive weapon repair and manufacture and some large mammal processing. Judging from these observations, one can conclude that the Dry Creek site was primarily a hunting camp and not a main habitation. Most of the moraine-top Denali Complex sites in the Tangle Lakes area (West 1981) and ridge prominences (such as the Campus site) seem to fall into this general category of hunting and processing stations.

Dry Creek was evidently a site where people camped briefly when out looking for game, repairing and manufacturing new weapon tips while they watched. Meat and hides already obtained were probably rough processed and dried for transport to the main camp.

This pattern of spike camps could indicate an orbital exploitation strategy in which small groups on hunting forays radiated out from a central, more permanent campsite. Such an orbital pattern of resource use would exploit a large area and thus allow a critical group size of 25–100 people to be maintained without the necessity of constant movement due to overhunting.

Although they are poorly preserved, the bones found at the site do corroborate Paleoindian and Paleolithic evidence of a reliance on large mammals for food. Mountain sheep (*Ovis*), wapiti (*Cervus*), and bison (*Bison*) were used at various times at the Dry Creek site.

Reconstruction of the ecology of range use by large ungulates suggests a fall-winter concentration in the area, because the wind from the pass keeps the rangeland free of snow and greatly increases the ungulates' access to critical winter forage.

Rather than positing that nomadic groups moved over the landscape, cropping game as they went, this orbital model suggests a more believable hunting strategy in which new ground at the periphery of the "wheel" would be explored using a mobile-hunting focus, as game concentrations varied within a large patrolled area.

Such central base camps undoubtedly occurred in quite different areas than the spike campsites, which are the major sites known thus far. The former are probably in areas not necessarily conducive to hunting efficiently but next to open water in winter and away from strong down-valley winds. Field surveys in search of base campsites might need a quite different search image than that associated with the spike campsites.

2. **HUNTING OF EXTINCT FAUNA.** In Alaska, a substantial body of data about Pleistocene fauna and their paleoecology had accumulated over the years, but little was known about the rates and patterns of their extinctions and redistributions. One could say that there was a complex Pleistocene fauna and Holocene fauna and that they were remarkably dissimilar and represented adaptations to very different environments. From the archaeological perspective, we knew next to nothing about the species hunted by early Alaskan peoples simply because lithic assemblages had not yet occurred with faunal assemblages. Hence ideas about early hunting activities were, at worst, speculations and, at best, logical constructions.

Judging from the few fossils at Dry Creek and from other radiocarbon dates on the Alaskan megafauna, the lower levels of the Dry Creek site date to a time when many of the extinctions had just occurred. Mammoths, for example, have not been dated into the lower 11,000s ^{14}C yr BP (13,000s cal yr BP) in Siberia or Alaska. Our surveys of Quaternary deposits in the vicinity of Dry Creek found mammoth remains dating between 12,340 ± 205 ^{14}C yr BP (GX-6284; 14,450 cal yr BP) and 12,240 ± 180 ^{14}C yr BP (I-10,532; 14,260 cal yr BP; Ten Brink and Waythomas 1985). These data fit into a large body of information relating to the demise of the mammoth steppe throughout northern Europe, Asia, and North America. Although horses, mammoths, camels, saiga, lions, and others might have already become extinct in Alaska at the time when the lower two levels of Dry Creek were occupied, other grazers such as wapiti and bison had not. Neither wapiti nor bison are native to Alaska today.

Thus the lower Dry Creek components document the remnants of a grassland environment that was once the dominant Pleistocene habitat in Alaska. The Dry Creek people were still hunting relict Pleistocene fauna (the grazing ungulates) and might not have shifted to the more mesic-nivian adapted caribou and moose. Thus the Dry Creek site dates to an important time and is located in an area of North America important to the Quaternary paleoecological interests, especially those concerned with the pattern of large-mammal extinctions that occurred during the glacial/postglacial transition. But just as interesting and important is the question of Clovis technology—its origins, spread, and adaption to a doomed fauna.

3. **CLOVIS ORIGINS**. One of the arguments for the Clovis projectile point not being derived from an Alaskan precursor has been the presumed use in Alaska of inset microblades for projectile points. There is a hiatus between the technologies involved in the manufacture of microblade insets and Clovis points. Microblades and the characteristic wedge-shaped cores do not occur in Clovis sites.

Component I at Dry Creek, dated at around 11,100 ^{14}C yr BP (13,000 cal yr BP), lacks microblades but does contain broken and complete triangular bifaces (chapter 4). These are basally thinned and potentially could have been related to the ancestral line that produced the Clovis Tradition. Also within Component II, dating around 10,500 ^{14}C yr BP (12,500 cal yr BP), there are several clusters of artifacts (chapter 5) that, instead of microblades, contain broken bifaces possibly similar to Hell Gap points of the Great Plains. These dichotomous clusters in Component II suggest either that there were two different groups occupying the area at the same time or that the people who produced the characteristic Denali Complex had another activity that depended on biface projectile points in addition to their composite antler-microblade points.

Either of these different interpretations would argue for a strong biface tradition in the north that could have given rise to Clovis points at the time of the southward colonization in North America prior to 11,500 ^{14}C yr BP (13,350 cal yr BP).

4. **DRY CREEK AND THE DIUKTAI CULTURE**. The Dry Creek stoneworkers strongly emphasized the manufacture of microblades, wedge-shaped cores, and a specialized burin technique, all of which are well represented in northeast Siberian sites of an earlier age. These techniques are probably derived from a technology that has been called the Diuktai Culture (Mochanov 1977). While there is reason to be cautious about the dates from the older Diuktai sites (Abramova 1979), there seems little doubt that this tradition, characterized by both microblade and bifacial technologies, became widespread in northeastern Siberia and spread to Alaska during the terminal Pleistocene (West 1981) and is represented in part at Dry Creek Component II. Whether the bifacial point technology mentioned previously is derived from the Diuktai Culture or from some area within America is a problem that cannot presently be solved. These types of bifaces at Dry Creek that we have called projectile points are unknown at Diuktai sites, although some specimens at Dry Creek, which we feel confident were used as knives, have been called points at Ushki Lake (Dikov 1977). While the Diuktai Culture contains bifacial pieces that could represent the technological base for bifacial projectile points at Dry Creek, this aspect of the technology would appear to have its origins in either eastern Beringia or elsewhere in North America. Hence there is an interface between Siberian and North American lithic techniques that we can see at Dry Creek and other slightly younger Alaskan sites, and it seems reasonable that if local antecedent developments gave rise to this situation, we should see evidence of it in the archaeological record as more new sites are discovered.

CHAPTER TWO

The Dry Creek Site
A History and Description

W. ROGER POWERS

Location

The Dry Creek site (HEA-005) lies in the Nenana River valley of central Alaska and is located about 180 km southwest of Fairbanks near the town of Healy (figure 2.1). The site proper is situated on a prominent bluff that lies on the north side of the bed of Dry Creek and is about 0.5 km upstream from the Parks Highway bridge over Dry Creek (figures 2.2, 2.3).

The prominent southeast-facing bluff on which the site is situated was formed by the downcutting of Dry Creek through a glacial outwash plain of Healy Age (Illinoian/early Wisconsinan [Wahrhaftig 1958]). This incision formed the Healy terrace along Dry Creek, which is composed of glaciofluvial sediments and is mantled by a continuous eolian formation of sands and loesses.

Regional Setting

The Nenana River valley transects two major physiographic provinces—the Pacific Mountain System and the Intermontane Plateau (Wahrhaftig 1965). The former comprises the imposing mountainous massif generally called the Alaska Range. Technically, the mountains visible from the Dry Creek site (figure 2.4) are referred to as the Outer Range, while the Inner Range ("Alaska Range" proper) lies farther to the south. The Intermontane Plateau in this region is a zone of foothills about 50 km wide that stretches north from the Outer Range until it merges with the Tanana-Kuskokwim lowland, a vast interior region composed of piedmont alluvial fans that begin about 50 km north of the Dry Creek site and continue northward until they interdigitate with Tanana River floodplain deposits.

The Nenana River, which originates in the Nenana Glacier on the south side of the Alaska Range, cuts north across this range and emerges from the mountains at Healy. From this

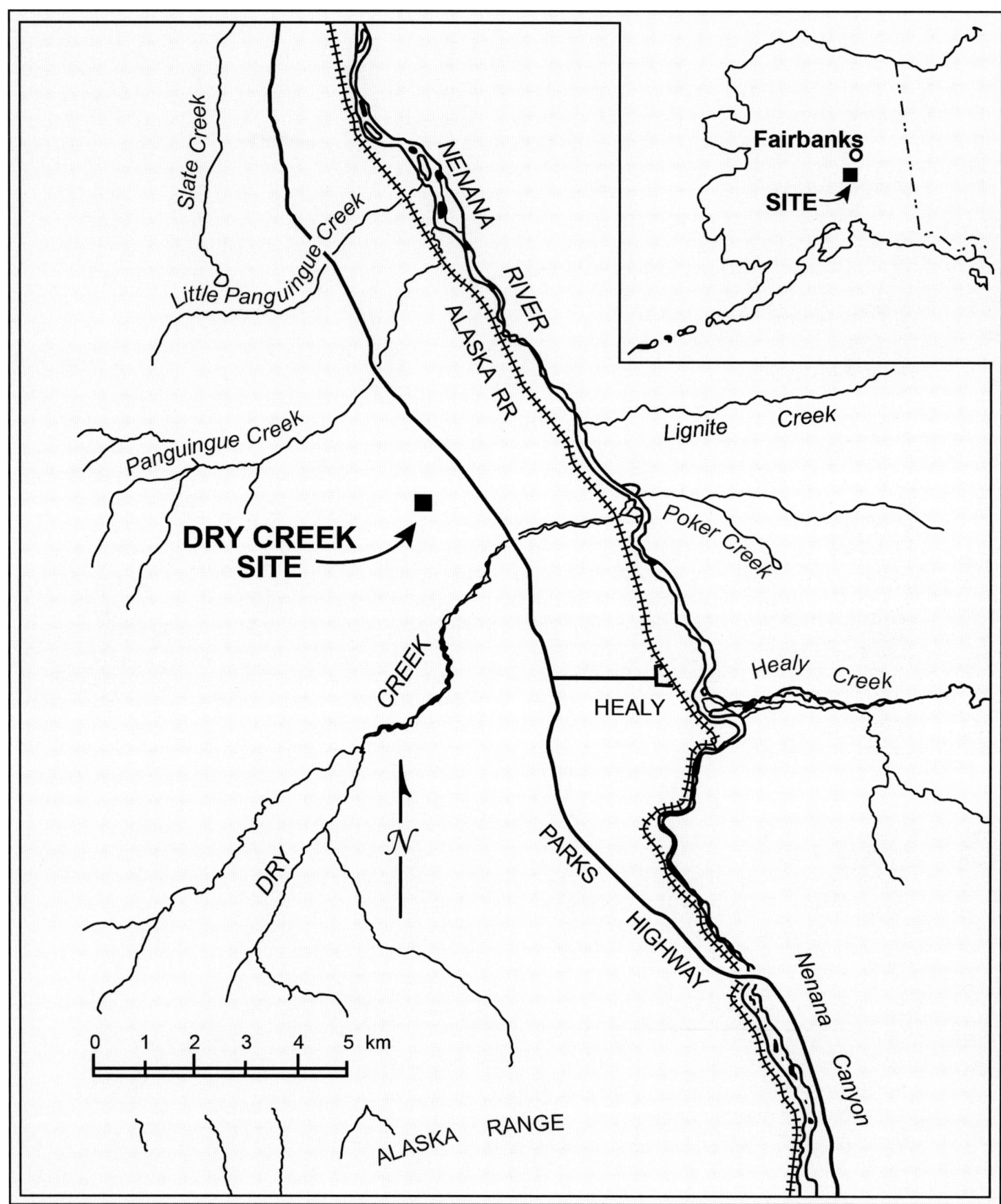

FIGURE 2.1. Location map of central Alaska and the study area.

Figure 2.2. View of Dry Creek. Parks Highway is in low center and the Dry Creek bluff is at the center of picture. The view is upstream to the southwest. The Outer Range of the central Alaska Range is in the distance.

Figure 2.3. The Dry Creek site. The view is downstream to the northeast, showing topography of the Nenana Valley.

FIGURE 2.4. View of Dry Creek and the Riley Creek terrace surface with the Outer Range in the background.

point, it flows through the foothill zone and emerges onto the Tanana-Kuskokwim lowlands, which it crosses, and empties into the Tanana River at the village of Nenana.

The valley of the Nenana River between the Outer Range and the lowlands possesses a somewhat subdued topography relative to the majesty of the surrounding mountains and is dominated by suites of terraces standing as erosional remnants of the ancient floodplains of the Nenana River and its tributaries. Downcutting by the Nenana and its tributaries has dissected and eroded these terraces so that the topographically lower portion of the valley is dominated by a steplike appearance as these terraces progressively drop to the modern floodplain of the Nenana. This erosional activity has produced a topography composed of broad, sweeping, relatively even terrace surfaces that end abruptly in steep bluffs (terrace risers) as much as 60 m in height, which separate the different terrace surfaces. Further incision has occurred through the action of tributary streams, such as Dry Creek, which have cut valleys through the terraces. As a result of these various erosional activities, the terrace systems are heavily dissected and present a very striking contrast between broad, even areas and precipitous terrace edges and stream valleys (figures 2.2, 2.3).

While absolute elevations are seldom great, relative topographic relief is quite imposing. The Nenana Valley is 394 m above mean sea level at Healy, which is situated at the base of the Outer Range. This range in turn rises to 1,356 m at Sugar Loaf Mountain and 1,742 m at Mount Healy. Thus in the immediate Dry Creek area, there is about 1,350 m of relief.

Above the terrace systems, the topography of the valley margins is less striking and is characterized by higher, more evenly rounded hills. Several prominent mountains, termed "domes" in this area, rise to 1,200–1,300 m and provide some scenic relief to a rather monotonous landscape. The upper reaches of the Nenana's eastern tributary streams originate in these hills, and in a few localities—for example, Lignite Creek—some spectacular badlands topography is present.

At the present time, the Dry Creek site lies roughly on an ecotone between the alpine, open, herbaceous tundra and the lowland forests, and the major vegetation communities of the Nenana Valley and neighboring hinterlands are simply an intricate interplay of these two ecosystems. In the lower elevations of the valley (below 600 m), the open, poorly drained terrace surfaces are dominated by shrub and herbaceous communities with scattered stands of black spruce (muskeg) and isolated copses of deciduous trees. Mixed deciduous and coniferous associations (balsam poplar, aspen, willow, alder, and spruce) are commonly found along stream valley margins and sometimes on the south-facing slopes of the hills and stream valleys. North-facing slopes are generally covered with a dense black spruce forest and sphagnum moss undercover. South-facing slopes with a well-drained substrate typically bear aspen, poplar, and willow stands with an herbaceous ground cover. Terrace edges can be quite grassy but more generally have a dense mixed deciduous and coniferous tree cover.

Above 600 m as a rule, tundra associations (meadows and bogs) are dominant and grade from shrub and cotton grass bogs at lower elevations to the drier heaths on the higher mountain slopes.

At the present time, the only large mammals that occupy the lowland forests are moose and black bears, while mountain sheep, caribou, and brown bears can be found on the higher mountain slopes.

History of Archaeological Investigations at Dry Creek
1973
In May 1973, Charles Holmes located displaced artifacts at the base of the loess mantle and on the debris slope at the Dry Creek bluff. Further investigations revealed that the artifacts were eroding from a cultural horizon lying about 1.3 m below the present surface. A charcoal sample that lay at the same depth as the artifacts was collected from a presumed hearth by Thomas D. Hamilton. During the summer of 1973, Robert Stuckenrath of the Smithsonian Institution determined the age of this sample to be $10{,}690 \pm 250$ ^{14}C yr BP (12,540 cal yr BP). It appeared that a site with a deeply buried microblade technology had been dated to the terminal Pleistocene.

The surface of the Dry Creek site is a relatively flat terrace covered with a dense stand of black and white spruce and a considerable amount of deadfall. An herbaceous plant cover established near the bluff edge, composed mainly of grasses, extends sporadically into the wood (figure 2.5). The debris slope below the site supports scattered clumps of aspen and willow and a sparse herbaceous cover including some xeric components such as *Artemisia*. A hundred meters upstream, the terrace riser is stable and wooded, but because the site lies at the outside of a meander loop of Dry Creek that is undercutting the entire terrace edge, this slope is actively eroding. At the time of the discovery, the edge of the loess mantle blanketing the terrace was suffering from heavy wind erosion and was undergoing considerable destruction from slumping (figure 2.6).

Initially, cultural remains were encountered along the eroded loess bluff for a distance of about 50 m. However, the center of density of the flakes and tool parts lay at the highest point on the terrace surface (figure 2.7). It was for this reason that test excavations were begun here. Three test pits were excavated at this time along the bluff edge, and these further confirmed the presence of in situ cultural material in the loess (cf. figures 2.8, 2.9).

With this information at hand, a five-year program of archaeological and paleoecological (geology, paleontology, paleobotany, palynology) studies was initiated at the Dry Creek site.

Figure 2.5. Spruce woods on the Dry Creek site prior to clearing for excavation.

Figure 2.6. Wind destruction of exposed 1973 test pit at the edge of the bluff.

Figure 2.7. View of the central site area prior to the 1976 excavations.

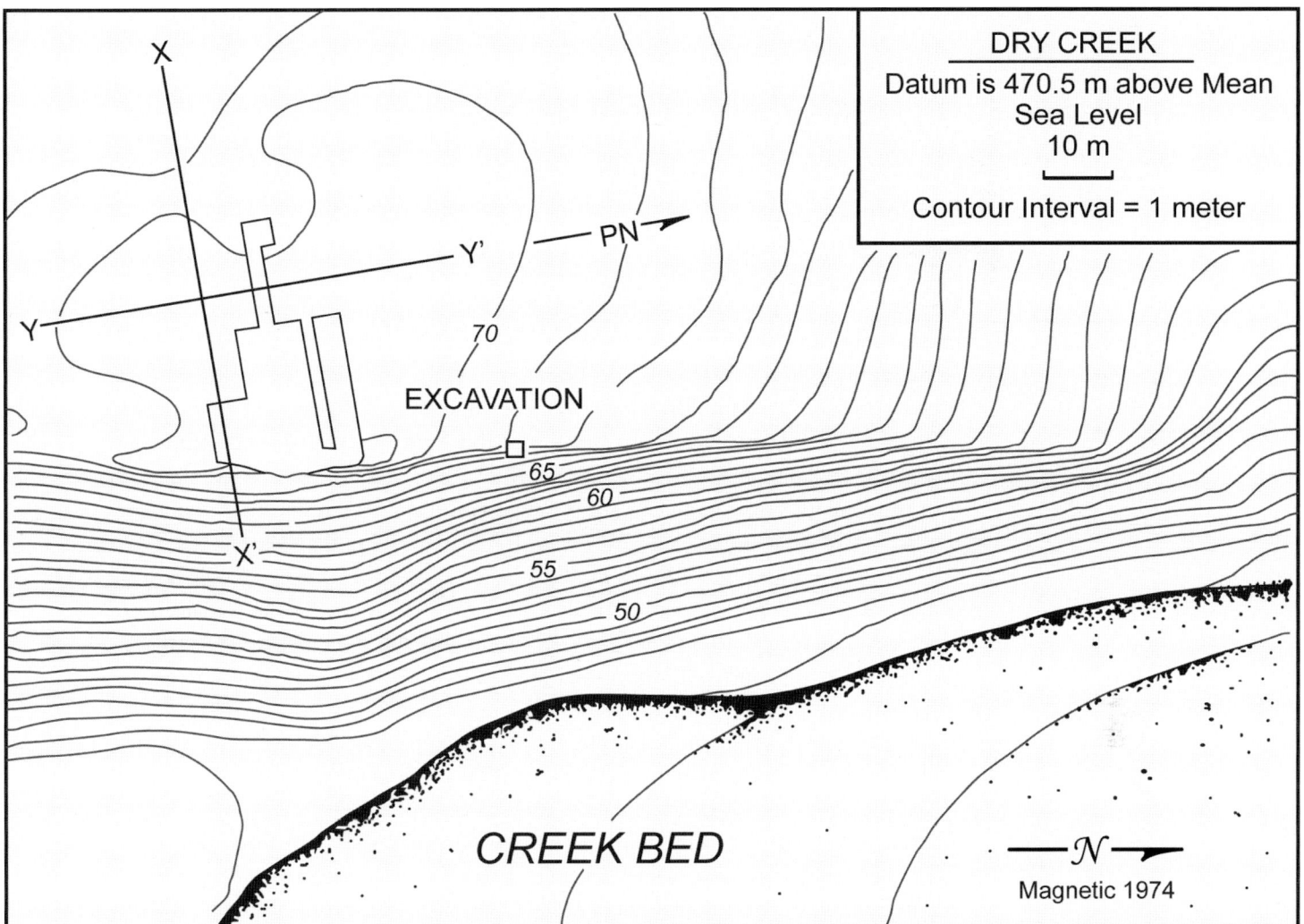

FIGURE 2.8. Topographic map of the Dry Creek site showing grid axes and completed excavation area.

1974

After the initial test excavations, conducted in the late summer of 1973, plans were made for a full-scale testing program that reached fruition in May and June of 1974 with a National Science Foundation institutional grant from the University of Alaska. It was during this time that the site emerged as one worthy of extensive study.

The site was mapped, a provenience datum was established about 25 m west of the bluff edge, and a metric Union Grid was laid out over the area in which excavation was anticipated to occur. The north-south axis (Y-Y') was oriented parallel to the bluff edge so that the east-west axis (X-X') would intersect the stratigraphic section at approximately 90°. This was done to avoid odd angles in the excavation units along the bluff edge. As a result, project north (PN) is 20° from true north (figure 2.8).

A 2 × 15 m test trench excavated at approximately 90° of the bluff edge was linked to planimetric excavations at the bluff edge (figure 2.9). The results of these test excavations are summarized as follows:

1. The north wall of the test trench was mapped, and the geological units within the section were defined and related to the regional geology (Thorson and Hamilton 1977).

Additional radiocarbon samples revealed that the loess section spanned the last 11,000 ^{14}C yr BP (13,000 cal yr BP). These dates also provided control for the stratigraphic units of loesses, sands, and paleosols, all of which record changing environmental conditions during the terminal phase of the Pleistocene and throughout the entire Holocene (cf. figure 2.10).

2. Three (and possibly four) cultural components were found in the Dry Creek loess mantle. The components are definable on geological grounds, and even in those rare cases where the lower two components come vertically close, there is still a clear stratigraphic separation (Powers and Hamilton 1978). This phase of the excavation produced 2,827 artifacts. The oldest of the components (I) contained flakes, flake cores, retouched flakes, a triangular biface, possible burins, a chopper, and end scrapers. No microblades, microblade cores, or any by-products of microblade production were recovered. Component II overlies Component I, and it is from this horizon that the radiocarbon date of 10,690 ± 250 ^{14}C yr BP (12,540 cal yr BP) was secured. This component produced microblades, wedge-shaped microcores, burins, and a variety of bifacially flaked pieces, most of which appear to have been knives. In addition, there are large choppers, biface blanks, anvil stones, hammerstones, unworked stones, and pebbles. Component III was noted in the 1974 excavations. It was composed of 573 waste flakes, 1 bladelike flake, 3 blades, and 1 biface fragment. It appeared to be similar to Component II, although the undiagnostic character of the material prohibited further refinement or identification. The uppermost horizon, Component IV, produced 2 side-notched point bases and flakes (Appendix A).

3. Fragments of dentition were found in a poorly preserved state. However, they were sufficient to allow preliminary identification by R. Dale Guthrie at the University of Alaska in 1974, who was not yet a part of the project. The majority of the remains were of *Bison* sp. It was also thought that one fragmentary specimen of a hyposodont molar might be from a horse (*Equus* sp.), and some large oval stains occasionally seen in the excavation might be the remains of mammoth tooth plates (*Mammuthus* sp.). Unfortunately, further study failed to confirm the presence of either *Equus* or *Mammuthus*.

4. Tools and flakes were clustered in definable areas of high concentration with intervening areas of low artifact frequency. While only one obvious hearth was located, there was evidence of burning in areas of high artifact concentration.

As a result of this information—that is, the stratification, distinct archaeological units, extinct fauna, and horizontal concentrations of cultural material—more extensive excavations at the Dry Creek site were planned.

A problem was encountered during this phase of the work that would affect all future excavation strategies. Within a closed trench, the thawing loess could not dry out. By necessity, the work was performed in a very mucky environment that compelled the crew to use winter clothing. Also, the walls would not stand, and shoring became a constant problem. In spite of our best efforts, portions of the walls collapsed after the excavation was completed.

Experience required that broad open areas should be excavated concurrently so that both sunlight and wind could reach the thawing loess. This approach proved to be very successful, and the excavations of 1976 and 1977 suffered very little from collapsing walls.

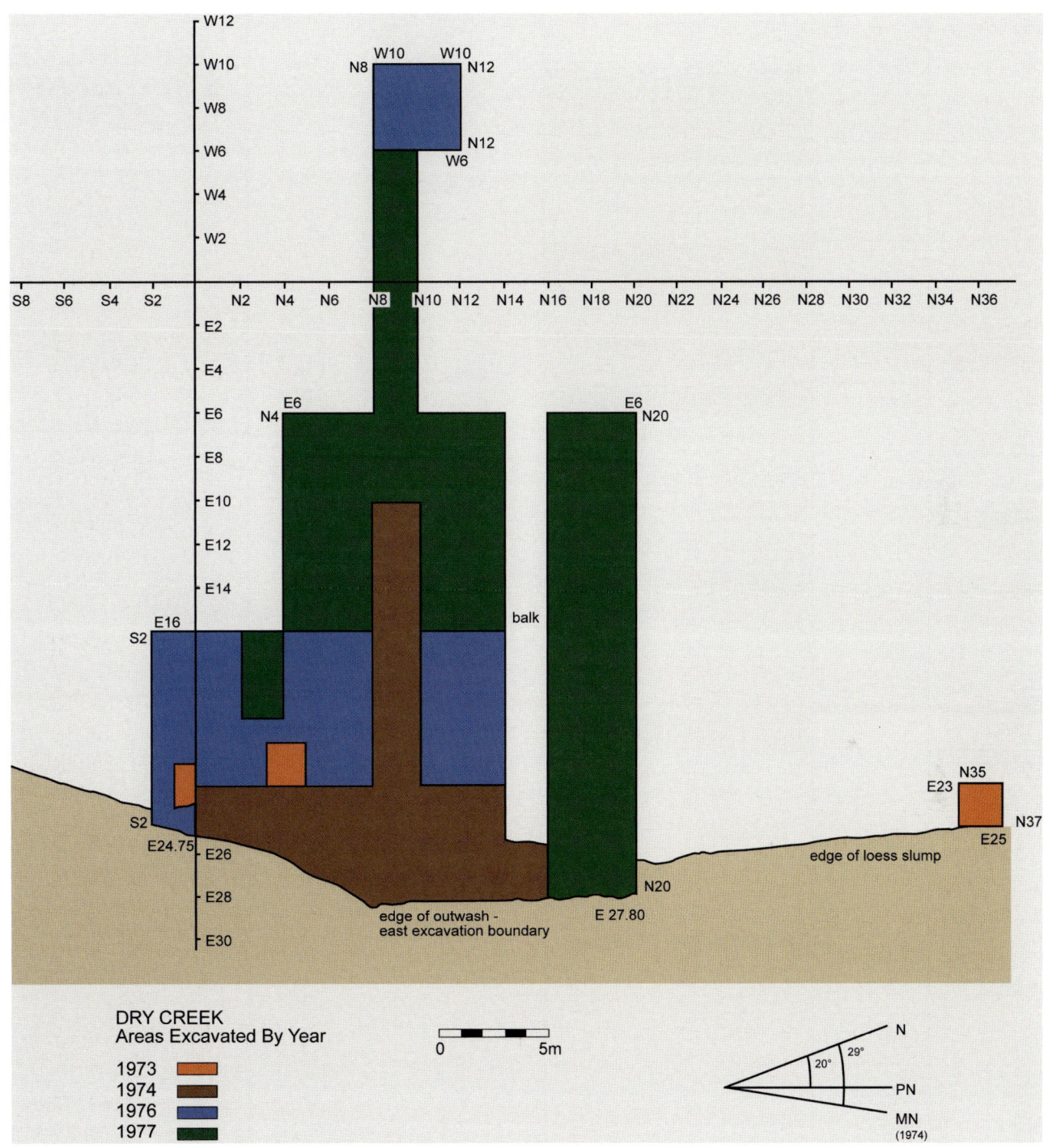

FIGURE 2.9. Grid system of the Dry Creek site showing excavation areas by year.

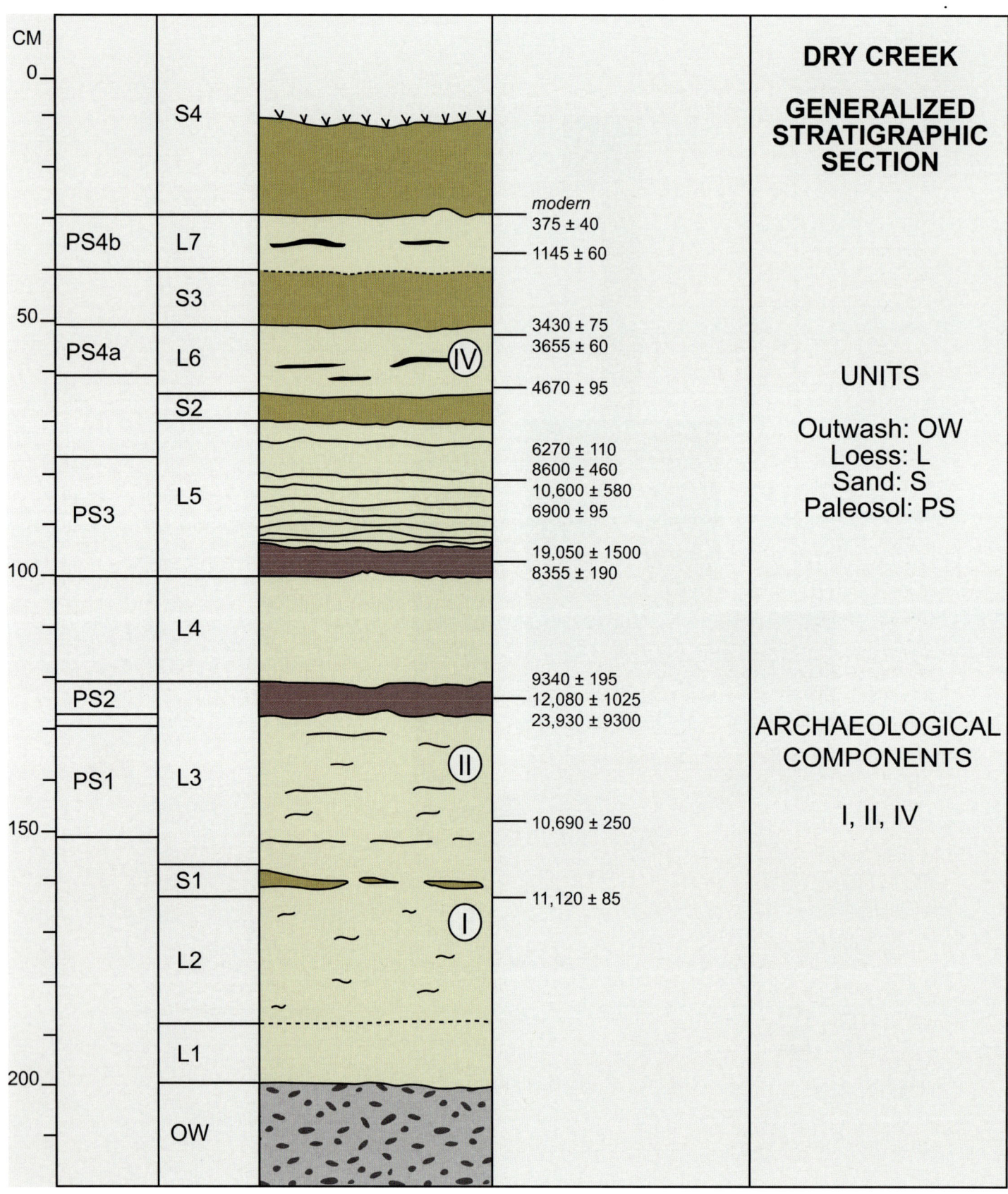

FIGURE 2.10. Generalized stratigraphic section of the Dry Creek site. Radiocarbon dates are shown in ¹⁴C yr BP.

1976

The first broadscale excavations were conducted during the summer of 1976 with support from the National Science Foundation, the National Geographic Society, and the Division of Parks of the State of Alaska. Students in an archaeological field school worked the site through June and July, and the excavation continued through August and September with volunteer labor. Geological studies were continued by Thomas D. Hamilton of the United States Geological Survey (USGS) and James McCalpin, then of the University of Alaska.

During this season, the excavation was expanded considerably (figure 2.9). Two 4 × 7 m units (Areas A and B) were opened between the bluff edge and grid line E16 and between grid lines S2 and N8. A 2-m balk was left between these two areas, the eastern half of which was removed near the end of the season. Further extensions conformed to the configuration of the bluff and kept the excavation orderly. A third 4 × 10 m unit (Area C) was opened between the bluff edge and grid line E16 and between grid lines N10 and N14. As we still had little control over the extent of the site, and as random test pits were impractical in the frozen loess, we opened another 4 × 4 m unit (Area D) between the grid lines W5 and W10 and N8 and N12.

An additional 60 m of stratigraphic profiles were taken, sediment and pollen samples were removed for analysis, and several radiocarbon samples were collected that provided our first dates from Components I and IV: 11,120 ± 85 ^{14}C yr BP (12,970 cal yr BP) and 4,670 ± 95 to 3,430 ± 75 ^{14}C yr BP (3,690 cal yr BP), respectively (figure 2.10). The excavation produced 12,951 cataloged specimens. While the same categories of artifacts were recovered from the components, it was clear that the vast majority were being found in Component II. Component I still contained flakes, side scrapers, end scrapers, and another triangular point or blank, but no microblade technology. Component II produced more microcore parts, burins, spalls, scrapers, and a variety of bifacial forms. Component IV yielded two more side-notched points, some end scrapers, and more flakes. No evidence of Component III could be detected. This prompted a thorough reexamination of the stratigraphic position of this component. It was realized that, in fact, this component was really the uppermost part of Component II. This called for a new numeration of the components, and it was decided that Component II should encompass Component III. Hence we were left with an odd system of numeration for the archaeological components—that is, I, II, and IV. It was thought that this system would create less confusion than renumbering Component IV to III, since the uppermost component (IV) had already been referred to in the literature (Powers and Hamilton 1978; Thorson and Hamilton 1977). However, it should be noted that the system of numeration in which Component IV is called III has been used elsewhere (Smith 1977).

The artifacts of all components were still occurring in concentrations of activity areas, and these were usually related directly to a burned area. Also, lithic raw materials were differentially distributed in the artifact concentrations.

All identifiable fragments of bone from the 1976 excavation are mandibular tooth fragments of *Bison*. Based on tooth morphology, our specimens are *Bison priscus*, the large Eurasian steppe bison.

Analysis of pollen samples taken during this season revealed poor pollen preservation in the highly oxidized loesses of Components I and II. Of those grains identified, fern spores are predominant. Some disturbance plants were noted, and arboreal pollen (*Betula*, *Alnus*, and *Picea*) was very weakly represented.

While conducting sedimentological analyses at the USGS labs in Menlo Park, California, James McCalpin discovered the presence of opaline phytoliths in the loess from the site. While little has been done with this potentially potent paleoecological tool in Alaska, it was realized that with the development of a phytolith key for northern vegetation, this might well develop into another source of ecological information. McCalpin did notice that the phytoliths from Loesses 2 and 3 (Components I and II) differed morphologically from those in the upper forest soils (Paleosols 4a and b; Component IV). Further analysis by our lab technician, Mary Calmes, revealed phytoliths from festucoid grasses. Thus opaline phytoliths could possibly help distinguish predominately herbaceous from sylvan landscapes and could conceivably offer further refinements.

The 1976 excavation failed to augment the sample of diagnostic artifacts in the oldest component, and this became a major concern. Was this really a nonmicroblade horizon, or was there a possible sampling error? It had always seemed possible that if Component I lacked a microblade technology, its antiquity could be much greater than that indicated by our single radiocarbon date. Muller-Beck (1967) had postulated the presence of a nonblade Mousteroid technology as a Clovis ancestor in North America, and such a technology would be succeeded by an Aurignacoid blade technology. This general situation appeared to be present at Dry Creek.

1977

The final excavation season at Dry Creek covered June and July of 1977 with full support from the National Geographic Society and the National Park Service as part of the Early Man in Alaska Program. The major objectives of this season were to excavate as large an area as possible, supplement the sample from Component I, and collect more radiocarbon samples from this horizon. As a result of these expanded excavations, the samples from Component II were also increased substantially.

Even larger areas were open during the summer of 1977, employing the same planimetric excavation strategy (figures 2.9, 2.11). Areas B and C of 1976 were extended to grid line E6 (figures 2.9, 2.12, 2.13), and the 1974 test trench was extended to connect Area D with the main excavation. Another 4 × 20 m unit (Area E) was also excavated at this time (figure 2.14).

During this season, an additional 172 m² were excavated to the surface of the Healy outwash, bringing the total Dry Creek excavation to 347 m². An additional 160 m of stratigraphic profiles were taken, and another 19,033 cataloged specimens were recovered, increasing the total sample size for all components at Dry Creek to 34,811. This figure does not include specimens collected from the surface and those for which provenience is incomplete or ambiguous.

Table 2.1 illustrates the numerical breakdown of the Dry Creek assemblage by component and year.

We found that the relative percentages of artifact categories from component to component did not change significantly and that the actual composition of the components remained generally consistent, with only a few new classes and types appearing in the 1977 season. The 1977 excavations produced more faunal remains of bison and, in addition, mountain sheep

Figure 2.11. View of the Dry Creek bluff during the 1977 excavation.

Figure 2.12. View of the 1977 excavation showing an excavated portion of Area C.

Figure 2.13. View of the 1977 excavation showing the expansion of Areas B and C.

FIGURE 2.14. View of Area E during the 1977 excavation. The majority of the area is on Component II (Loess 3, Paleosol 1).

and elk. It should be noted that for all genera at Dry Creek (*Bison*, *Ovis*, and *Cervus*), measurements on dentition indicate significantly larger animals than are known in extant species of these genera. Furthermore, only mountain sheep are presently in the area.

The presence of bird gastroliths (gizzard stones) was noted for the first time during the 1977 season. These occurred in small clusters in the site and were seen as another possible clue to understanding the paleoecology of the site, especially the seasonal nature of the occupations.

At the end of the 1977 excavation season, the entire area was backfilled with the use of a bulldozer (figure 2.15). Today, the disturbed area is being recolonized by herbaceous vegetation dominated by grasses and a few small willows and aspens.

Summary

Now that this phase of full-scale excavations has been completed, it is possible to briefly summarize the salient features of the site.

TABLE 2.1. Numerical breakdown of artifacts recovered from Dry Creek by cultural component and year

	1974	1976	1977	Total
C IV	145	907	1,320	2,372
C II	2,370	10,749	15,762	28,881
C I	312	1,295	1,951	3,558
Total	2,827	12,951	19,033	34,811

FIGURE 2.15. Backfilling the 1977 excavation.

First of all, the oldest cultural horizon appears to date to about 11,100 ^{14}C yr BP (13,000 cal yr BP), and there is still no evidence that a microblade technology was part of the tool kit. Of the total inventory from this component, 90 percent of the finds are waste flakes; 7.4 percent are pebbles, rocks, and rock fragments; 1.1 percent are bone and tooth fragments; and only 2.2 percent are chipped stone tools. The latter category is composed of retouched flakes, utilized flakes, small bifaces, biface fragments, side scrapers, end scrapers, scraper fragments, choppers, cores, core fragments, split pebbles, and cobbles.

Component II dates to the mid-eleventh millennium ^{14}C yr BP (mid-thirteenth millennium cal yr BP) and accounts for 84 percent of the entire site assemblage (N = 28,881). Again, a very high percentage (95.1 percent) is unmodified waste (flakes and bladelike flakes), with the remainder of the inventory constituting pebbles, rocks, and rock fragments (7.4 percent); bone and tooth fragments (1.1 percent); and chipped stone tools (2.2 percent).

A notable aspect of this component is the microblade technology and the numerous by-products of core manufacture (core tablets and rejuvenation flakes). Next, there is a series of bifacial tools composed of acuminate bifaces (elliptical, lanceolate, ovate, and deltoid), oval bifaces, rectangular bifaces, and bifacial preforms. A series of bifaces stands out; these have the appearance of projectile points and have been called such elsewhere in Alaska. These are roughly spatulate or slightly stemmed pieces with a very narrow straight base, expanding lateral edges, and a short, broad tip; the greatest width is just below the tip, and the lateral edges are ground. Only one edge of the tip shows use wear. Also, the tips are slightly asymmetric. It would appear that these pieces have been used as knives. The remainder of the Component II artifacts is made up of side scrapers, retouched and utilized flakes, burins and burin spalls, flake cores and fragments, hammerstones, anvil stones, flakes, split and battered pebbles, and cobbles.

In Component IV, almost 99 percent of the inventory is waste material (flakes). Actual tools include a few end scrapers, a boulder spall tool, and five projectile points that have weakly formed side notches.

While the association of artifacts in high numbers with the remains of extinct fauna in separate stratigraphic units makes Dry Creek unique in Alaska, at least for the present, the horizontal distribution of these remains in concentrations adds a new dimension to the importance of the site. This situation permits us to examine the assemblage in terms of activity areas within the site, since, as indicated previously, the artifact classes cluster, to a high degree, within separate areas. Also, these areas are often situated around or near a burned area.

The importance of this horizontal patterning slowly became apparent during the 1974 excavations, even though the majority of this work was confined to a trench. Also, during the course of this work, particularly at the end of the season, some areas of the site were bulk sampled—that is, flakes and microblades were collected within 25 cm^2 units and bagged accordingly. During our present analysis, it became necessary to renumber and re-sort these samples in order to integrate them and maintain consistency with the remainder of the collection.

During the excavations of 1976 and 1977, point provenience was recorded for the vast majority of the collection. Under certain circumstances, flake concentrations (flakes in physical context with each other) were bulk sampled, with the dimensions and positions of the clusters being recorded. This allowed us to structure our analysis of the spatial distributions of the artifacts as follows:

1. Mapping the exact horizontal arrangement of all artifactual and paleoecological data
2. Mapping the density of these data by arbitrary upper and lower limiting amounts of data per 1 m^2 units
3. Developing articulation matrices (the spatial distribution of artifact parts that fit together)

These maps then allowed us to delineate both the spatial distribution and the artifactual content of the clusters (activity areas) and to examine the variability within and between the clusters. We could also plot the position of different types of flakes (biface reduction flakes, sharpening flakes, core rejuvenation flakes, spalls, etc.) in order to relate these kinds of activities to the site as a whole.

CHAPTER THREE

The Geology of the Dry Creek Site

W. ROGER POWERS

Introduction: Regional Geology

The Pleistocene geology of the Nenana Valley has been studied in detail by Wahrhaftig (1958), and the regional geology of the Dry Creek site and its relationship to the broader geological framework of the Nenana Valley has been discussed thoroughly (Thorson and Hamilton 1977). Further field investigations concentrating on the surficial Pleistocene geology of the Nenana region have been conducted by the North Alaska Range Early Man Project, with Norman Ten Brink in charge of geological investigations. As a final report on these studies is still in progress, our understanding of the geological history of the Nenana Valley must rest on previous research.

Four major glacial episodes have been defined in the Nenana Valley, and three of these were responsible for the formation of the major terrace systems previously described. The Browne Glaciation is the oldest and might be of early Pleistocene or possibly late Pliocene age. No outwash terraces have been linked to this glaciation in the Nenana Valley. The next youngest glacial episode is the Dry Creek Glaciation, and it is probably of middle Pleistocene age. The terrace surfaces attributable to this glaciation are localized in the Nenana Valley, where they constitute the highest outwash terrace systems presently known. Following the Dry Creek episode, a major ice advance deposited morainal material about 1 km south of the Dry Creek site. This event, termed the Healy Glaciation, also deposited extensive glacial outwash sheets that now constitute the Healy terrace. It is at the edge of this terrace that the Dry Creek site is situated at the north side of Dry Creek. Outwash of this glaciation forms the substrate underlying the eolian deposits at the site. The Riley Creek Glaciation is the last major glacial episode and comprises several advances (Riley Creek I and II and the Carlo Readvance) that built outwash plains represented today by several terrace surfaces that lie below the Healy terrace, throughout the valley. The age of the Riley Creek Glaciation is considered to be late Wisconsinan, and it was during the waning phases of this event that the occupation of the Dry Creek site began.

Geology of the Dry Creek Site

As previously mentioned, the geology of the Dry Creek site has been studied in detail and published elsewhere (Thorson and Hamilton 1977). This description of the site geology is based entirely on this excellent work and supplemented only by field notes and by the laboratory report by James McCalpin, who continued field investigations in 1976. Subsequent observations were made by the site investigators and Norman Ten Brink in 1977.

With few exceptions, the stratigraphy of the Dry Creek site remained constant throughout the history of research. Because of this, the published report by Thorson and Hamilton (1977) still stands as the definitive study of the site's geology.

As introduced in the previous sections, the cultural remains at Dry Creek were contained in an eolian mantle overlying outwash deposits of the Healy Glaciation. This mantle is composed of sands and loesses that maintain a general thickness of 2.0 m. At the present time, the site lies within the zone of discontinuous permafrost, and the eolian deposits at Dry Creek remain thoroughly frozen except during summer months, when they thaw to a depth of about 0.5 m from the surface and 1.0 m from the face of the bluff.

The geological studies published by Thorson and Hamilton (1977) were based on natural bluff exposures, the stratigraphy of four test pits, and a 15 × 2 m exploratory trench that was excavated perpendicular to the bluff edge (cf. chapter 2). The most detailed stratigraphic profile was taken on the north wall of this trench along grid line N10 between E10 and E26 (Thorson and Hamilton 1977: figure 5). This profile formed the basis for the interpretation of the stratigraphy and the model against which future stratigraphic sections would be compared and monitored.

The stratigraphy of the Dry Creek site is represented by the vertical accumulation of eolian loesses (7 units) and sands (4 units), which was interrupted by five episodes of soil development (Paleosols 1–4b). The following general description of the lithological units provides more detailed information. It is taken verbatim from Thorson and Hamilton (1977: table 1), presenting stratigraphy from the top of the profile downward. All units are composed primarily of quartz, muscovite, and rock fragments; hence, mineralogy is not described individually for each unit. Munsell colors are on field-moist material.

Sand 4   Sand with minor silt and clay; light-yellowish-brown (10YR 6/4), very poorly sorted angular to subangular grains; peaty texture, with partially decomposed wood near base. Living spruce trees rooted at sharp lower contact.

Loess 7   Sandy silt with clay; poorly sorted angular grains, commonly with clay and oxide coatings; well-developed reddish-brown (5YR 5/4) buried soil (Paleosol 4b) with charcoal fragments. Gradational lower contact.

Sand 3   Silty sand with minor clay; yellowish-brown (10YR 5/6), poorly sorted angular to subangular grains; thickness variable. Sharp lower contact.

Loess 6   Sandy silt with minor clay; yellowish-brown (10YR 5/4), poorly sorted angular grains, commonly with clay and oxide coatings; contains archeologic Component IV; well-developed reddish-brown (5YR 5/4) buried soil (Paleosol 4a) with charcoal lenses and root casts. Gradual lower contact.

Sand 2   Sand with minor silt; brownish-yellow (10YR 6/6), very poorly sorted angular to subangular grains. Sharp lower contact.

Loess 5   Sandy silt with minor clay, mottled strong brown (7.5YR 5/6) to light olive-gray (5Y 6/2); poorly sorted angular grains, strongly folded and faulted, slightly to strongly deformed by creep and solifluction; altering dark organic, light olive-gray (5YR 6/2), and yellowish-brown (10YR 5/6) horizons (Paleosol 3). Sharp lower contact.

Loess 4   Sandy silt, mottled strong brown (7.5YR 5/6) to light olive-gray (5Y 6/2); poorly sorted angular grains; contains archeologic Component III. Gradational lower contact.

Loess 3   Sandy silt with minor clay, mottled strong brown (7.5YR 5/6) to light olive-gray (5Y 6/2); poorly sorted angular grains; contains archeologic Component II and decomposed bone fragments; nearly continuous dark organic horizons at top of unit (Paleosol 2), discontinuous dark organic horizons occur throughout unit (Paleosol 1). Sharp lower contact.

Sand 1   Medium sand; yellowish-brown (10YR 5/4), discontinuous sand lenses, with very well-sorted subrounded to subangular grains; weakly developed pitted texture on large quartz grains. Sharp lower contact.

Loess 2   Sandy silt with minor clay, mottled yellowish-brown (10YR 5/6) to light olive-gray (5Y 6/2); poorly sorted angular grains which coarsen upward; burrow casts common; contains archeologic Component I. Gradual lower contact.

Loess 1   Silt with minor fine sand; olive (5Y 5/3), very poorly sorted rounded to subrounded clasts of schist and other metamorphic and plutonic rocks; clasts wind polished at upper contact, and frost cracked, stained, and carbonate encrusted to 30–40 cm depth.

As can be seen from the foregoing description, schematic profile (figure 2.10), stratigraphic profiles, and photographs (figures 3.1–3.6), the oldest eolian deposit at the site is Loess 1. As this unit coarsens upward, the boundary with Loess 2 is gradual. Loess 1 was probably derived from the floodplain of the Nenana River. Loesses 2 and 3 are especially important because they contain archaeological Components I and II, respectively. These loess units were derived from the floodplain of Dry Creek. Sand 1, although discontinuous, separates Loesses 1 and 2 throughout most of the site and constitutes a clear stratigraphic break between Components I and II. It is thought to have been a sand sheet moving over the site when especially strong, active surface winds derived coarser-grained material from the front of the bluff.

Deposition was sporadically interrupted during the accumulation of Loess 3 by the formation of Paleosol units 1 and 2. Paleosol 1 occurs within Loess 3 as a series of discontinuous soil stringers (dark organic A horizons). They are more scattered and less predictable near the bluff edge but become more continuous and better developed away from the bluff and along the northern periphery of the site. This soil complex is thought to represent a series of immature tundra soils (Cryepts; Thorson and Hamilton 1977). The artifacts from Component II were found within and between these soil stringers.

Paleosol 2, which formed at the top of Loess 3, is thicker and nearly continuous throughout the excavation area. It is very often a single soil unit (dark organic A horizon) but locally bifurcates or even separates into a series of discontinuous soil stringers. Like Paleosol 1,

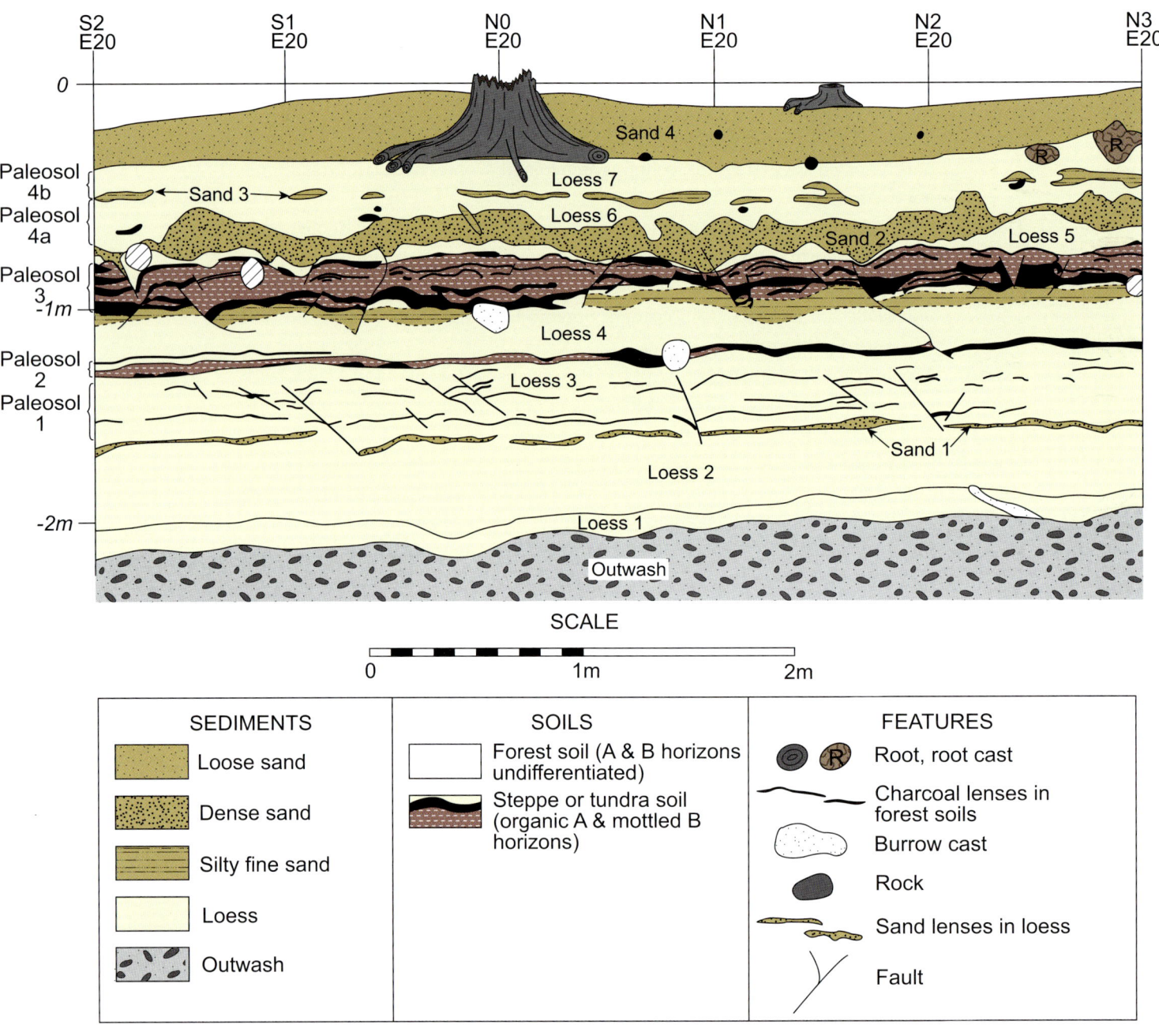

FIGURE 3.1. North-south stratigraphic profile along grid line E20 between S2 and N8. This section is parallel to the bluff edge.

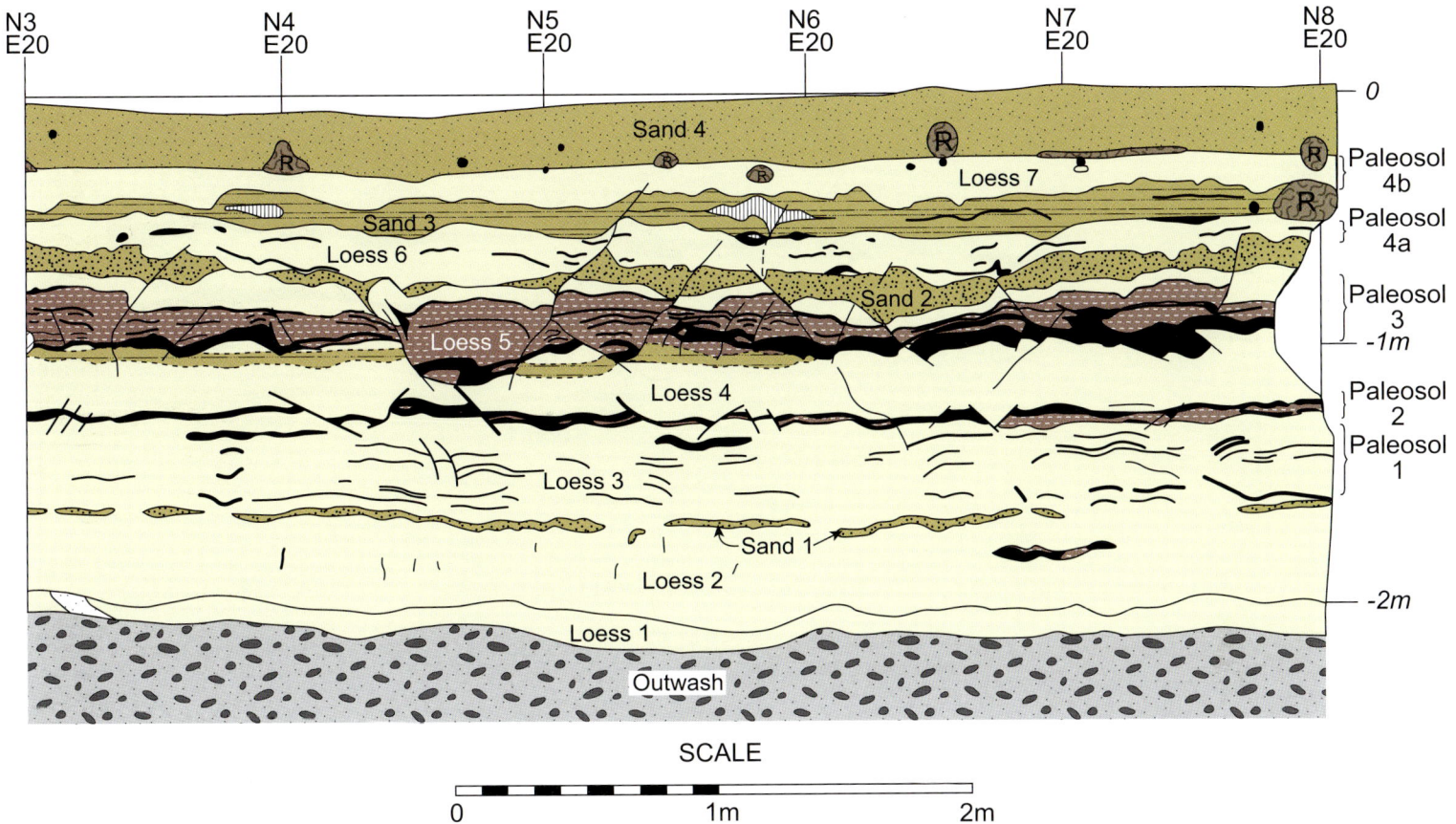

FIGURE 3.1. (cont.)

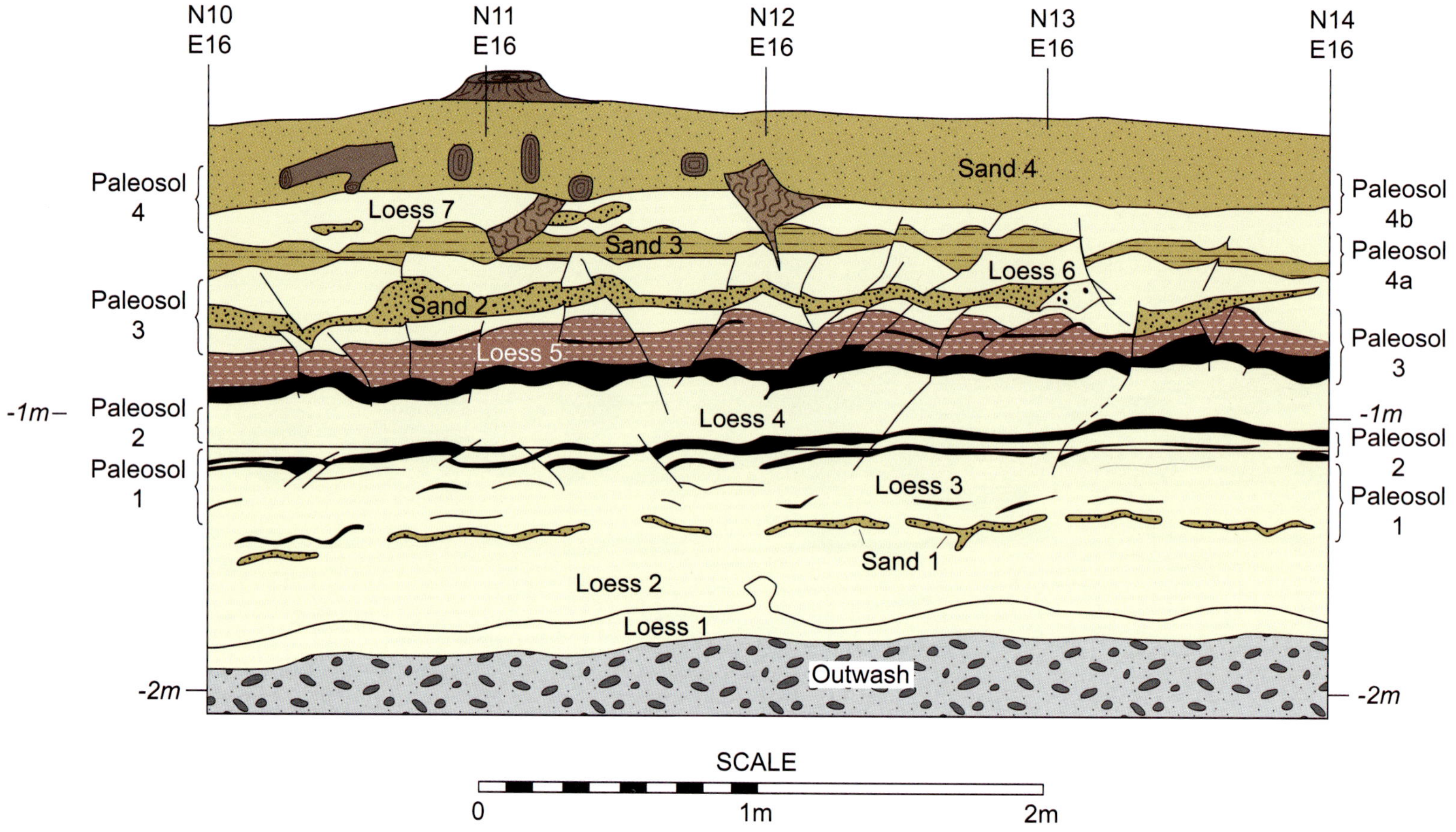

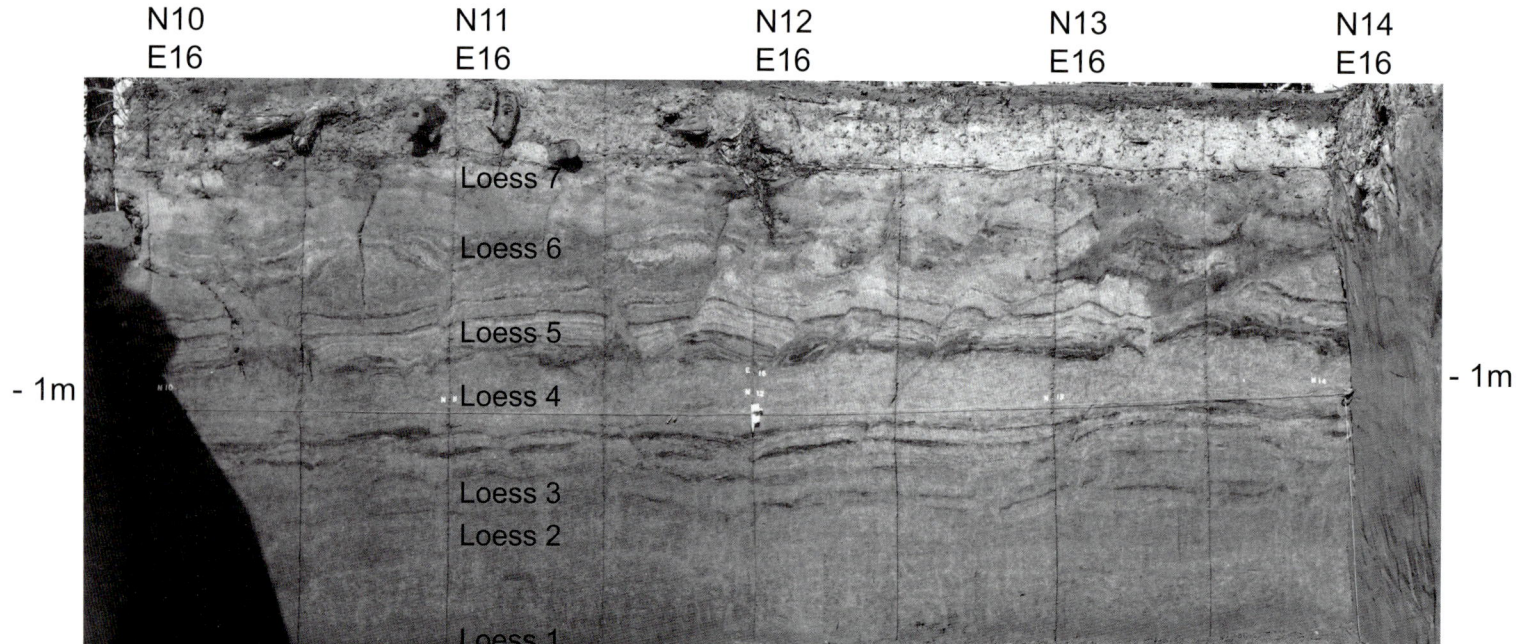

FIGURE 3.2. North-south stratigraphic profile along grid line E16 between N10 and N14. The section is parallel to the bluff edge. See figure 3.3 for a photograph of this section.

FIGURE 3.3. Photograph of the stratigraphy along grid line E16 between N10 and N14. This section is parallel to the bluff edge. Figure 3.2 is the stratigraphic profile of this section. The vertical lines are 0.5 m apart. The horizontal line is 1.0 m below subdatum.

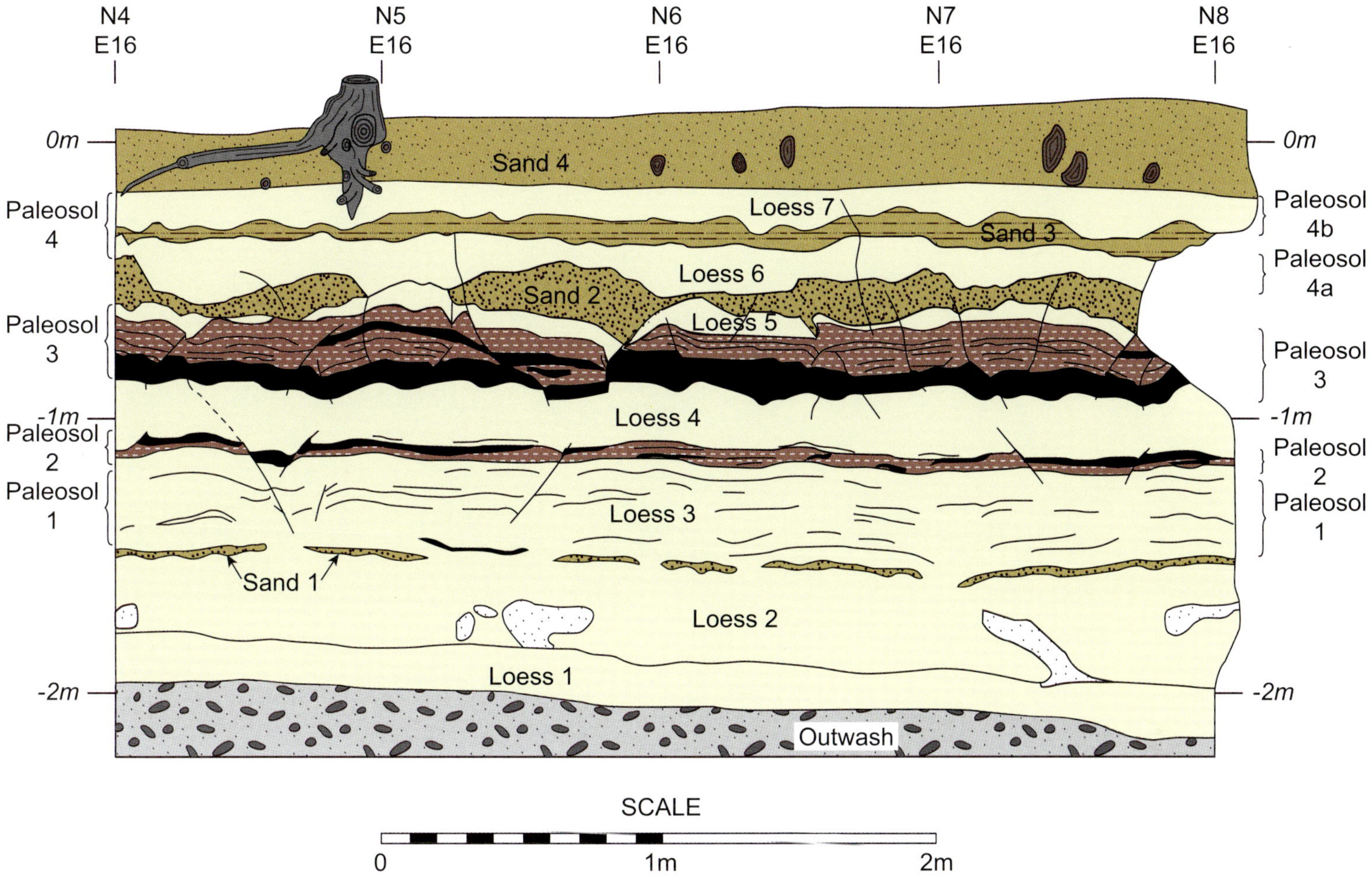

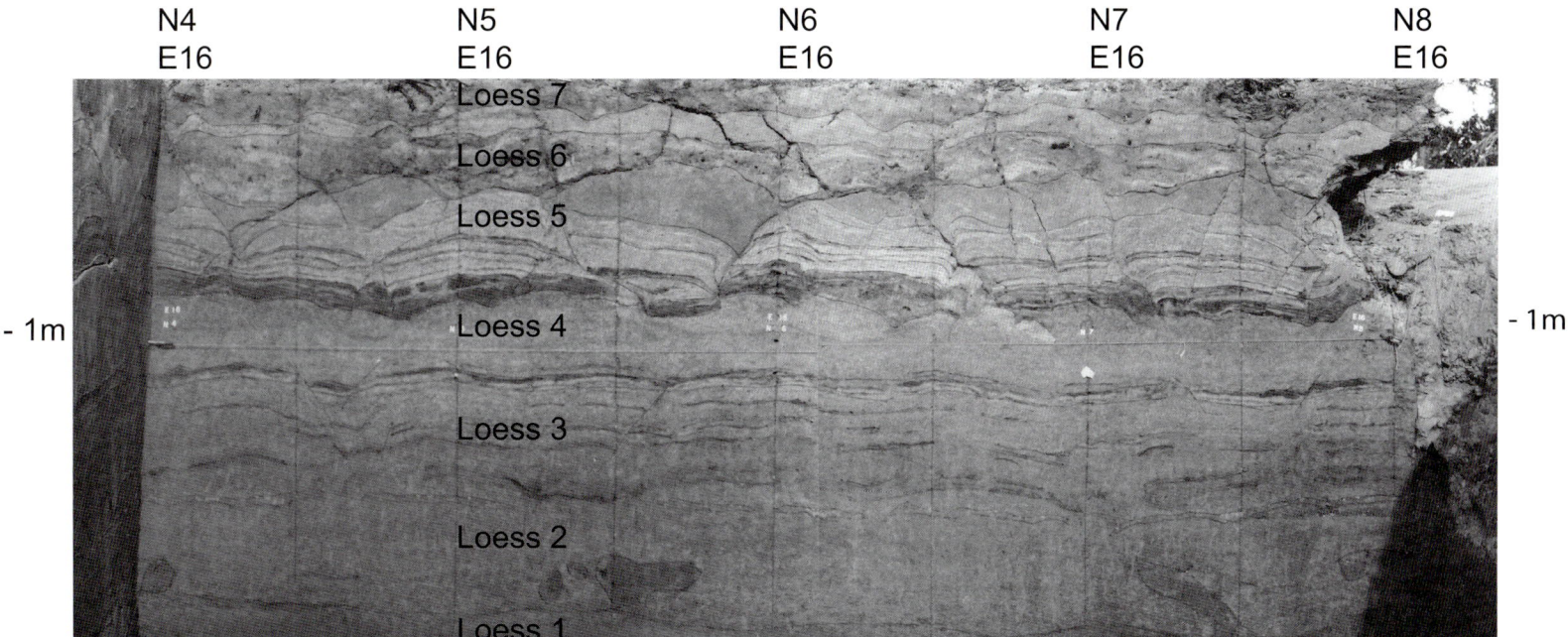

FIGURE 3.4. North-south stratigraphic profile along grid line E16 between N4.04 and N8. The irregular wall at the right-hand side of the picture is the collapsed south wall of the 1974 test trench. See figure 3.5 for a photograph of this section.

FIGURE 3.5. Photograph of the stratigraphy along grid line E16 between N4.04 and N8. The vertical lines are 0.5 m apart. The horizontal line is 1.0 m below subdatum.

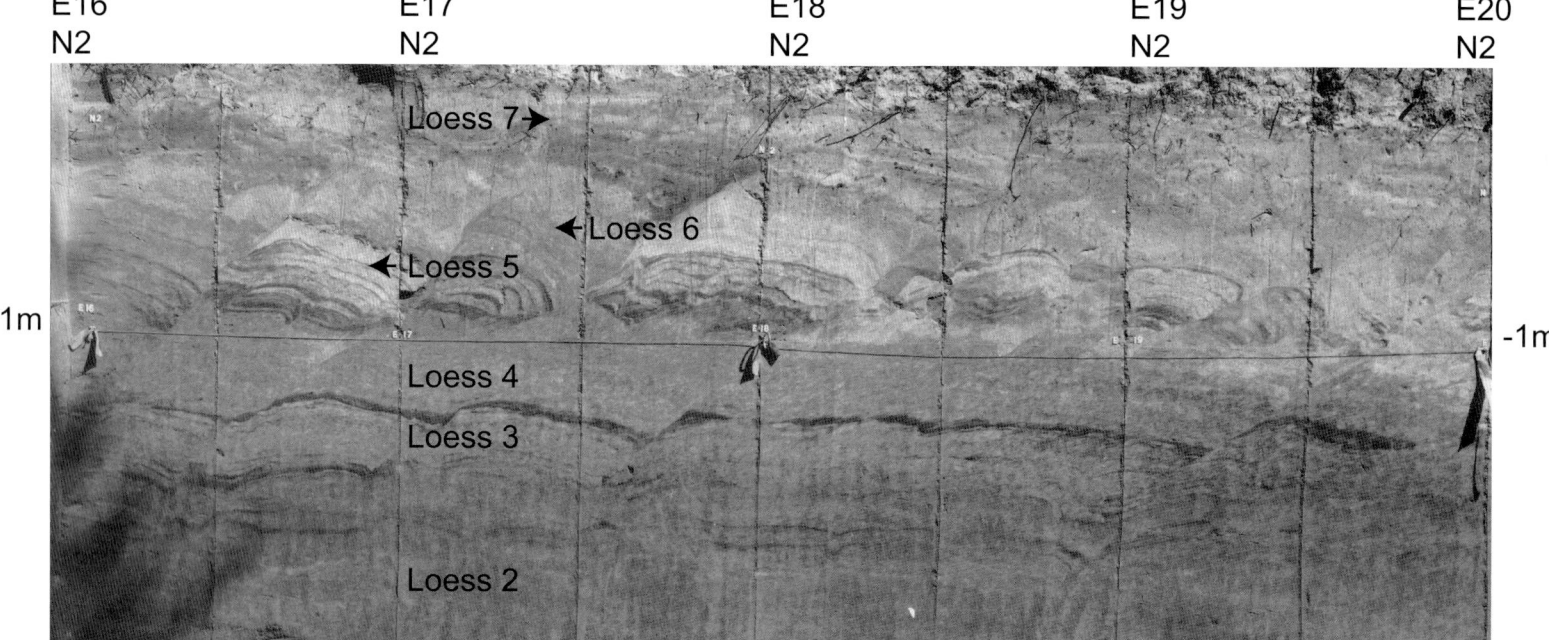

FIGURE 3.6. Photograph of the stratigraphy along grid line N2 between E16 and E20. This section is perpendicular to the bluff edge. The vertical lines are 0.5 m apart and the horizontal line is 1.0 m below the subdatum.

Thorson and Hamilton (1977) interpret this unit to be an immature Cryept. It is entirely sterile of cultural material.

Loess 4 represents renewed accumulation following the formation of Paleosol 2. The source area for this loess appears to be the Nenana River floodplain. This unit is better sorted and shows little thickness or textural variation over the excavation area.

There has been some confusion over the presence of artifacts in this loess unit. It appears that during the 1974 excavation period, a localized accumulation of flakes, microblades, and a few tools was associated with this unit. While depth measurements indicate the possible presence of a Component III, subsequent examination showed that in this area of the 1974 trench, the top of Loess 3 was higher than in most areas, and Paleosol 2 was often little more than a series of discontinuous soil stringers. These facts, plus structural deformation (see the following), created a situation where such a mistake was highly probable.

Loess 4 is followed by the accumulation of Loess 5 and the development of a thick set of soil units—Paleosol 3. Like Loess 4, this unit is also better sorted and is fairly consistent with respect to texture and thickness. Its source area is also thought to be the floodplain of the Nenana River (Thorson and Hamilton 1977). Loess 5 was continually interrupted by the formation of a series of soil units (Paleosol 3). The lower soil is the most strongly developed, and over much of the site, at least eight more organic horizons are present. These soils are more continuous and are represented by prominent, dark organic A horizons that are separated by light-gray and yellowish-brown zones. This series of soils, like Paleosols 1 and 2, is interpreted by Thorson and Hamilton (1977) as being immature Cryepts. This entire unit is continuous throughout the main excavation area. The only disturbance that has affected development has been structural (see the following).

During excavation, a few scattered flakes and one microblade core were found in Loess 5. Rather than a wedge-shaped microblade core, such as those recovered from Component II, this specimen is similar to a Tuktu core (cf. Campbell 1961). Even in this comparison, we hasten to add that it is aberrant. It is not only entirely clear that these scattered remains are in situ; it is possible that they have moved upward along one of the many cracks that run through this unit as a result of structural deformation. Again, since the situation is unclear, we have chosen to exclude these few remains from consideration.

Loess 5 is covered by Sand 2. It is very coarse grained and poorly sorted and becomes thinner and finer grained away from the bluff. It is thought to be derived from the Dry Creek floodplain and the exposed slope in front of the site when strong, gusting winds were sweeping the area.

Loess 6 overlies Sand 2 and, like Loesses 4 and 5, is better sorted and exhibits little thickness or textural variation. Again, the source area is thought to be the Nenana River floodplain.

Paleosol 4a occupies nearly the upper two-thirds of Loess 6. This soil differs considerably from the lower paleosol units. According to Thorson and Hamilton (1977), this paleosol is a subarctic brown forest soil (Ochrepts and Orthods) that is commonly found developing under the modern taiga of interior Alaska. This soil is thick and continuous and contains a prominent reddish-brown oxidized B horizon. Charred root casts indicate both development of forests during this episode and periodic burning. Local drainage conditions on the site area were variable. The B horizon of this soil exhibits oxidation near the bluff edge (better drainage), but at roughly 15 m from the bluff edge, it changes to a composite profile more similar to low-humic gley soils (forming in situations of poorer drainage; Thorson and Hamilton 1977). Archaeological Component IV occurs within this soil unit. It is an assemblage that differs considerably from the underlying components and probably represents the appearance of the Northern Archaic in the Nenana Valley.

Sand 3 overlies Loess 6 and, in contrast to Sands 2 and 4, is comparatively finer grained and exhibits better sorting. The floodplain of Dry Creek might have been farther from the site, or the exposed outwash surface of the bluff was more protected. This unit pinches out along the bluff edge and results in the vertical coalescence of Loesses 6 and 7, and hence Paleosols 4a and 4b.

Loess 7 overlies Sand 3 throughout most of the site. Again, this is a better-sorted loess bed and shows little thickness or textural change. It is also thought by Thorson and Hamilton (1977) to have derived from the floodplain of the Nenana River. This entire loess unit is occupied by Paleosol 4b, which is nearly identical to Paleosol 4a. As mentioned previously, this soil merges with Paleosol 4a near the bluff edge. Forest development and periodic burning are also suggested for this soil. It appears to have been better drained than Paleosol 4a, as it does not exhibit the transition to a gleyed condition away from the bluff (Thorson and Hamilton 1977).

The entire sequence is capped by Sand 4, which is presently building up along the bluff edge. This unit thins away from the bluff and is presently burying the trunks of the existing trees. It is derived from the exposed bluff edge that is presently being undercut by Dry Creek (Thorson and Hamilton 1977).

An examination of the stratigraphic profiles from Dry Creek reveals a section that has undergone striking deformation. The numerous cracks that run through the section are thought to be normal faults resulting in displacements of up to 50 cm. They run about 30° near the bluff but become progressively more vertical away from the bluff edge. The strike of most of these faults is parallel to the bluff edge, and many of these small fault blocks appear to have rotated counterclockwise as the entire mantle was expanding toward the bluff. This has resulted in the lateral stretching and minor vertical deformation of the lower part of the section (Loesses 1–4). However, the loess mantle underwent a major episode of stretching and faulting roughly concurrent with the deposition of Sand 3 (near the top of the section). It has been suggested that this might be the result of a strong local earthquake (Thorson and Hamilton 1977).

Evidence of solifluction disturbance is noticeable in the bluff edge, particularly in Paleosol 3. The movement appears to have been in a northeasterly direction along the edge of the terrace. Solifluction also affected Paleosol 3 within the excavation area where this unit slopes toward the bluff edge (Thorson and Hamilton 1977).

Because of the deformation by faulting that affected all the archaeological components in the Dry Creek site (Component I the least and Component IV the most), extreme care was necessary in determining the stratigraphic unit in which one was working and in monitoring the fault system both vertically and horizontally. As a result, there could be little doubt about the stratigraphic position of most finds, and as the mapping of the finds in Components I and II illustrates (chapter 5), the horizontal displacement of artifacts in these components was minimal.

Dating

Like other aspects of the geology of the Dry Creek site, the radiocarbon dates have been presented in detail and discussed by Thorson and Hamilton (1977), and the geochronology of the Dry Creek site has also been treated briefly by Powers and Hamilton (1978).[1]

Eighteen samples were dated by Dr. Robert Stuckenrath of the Smithsonian Institution. All samples except SI-1933b (375 ± 40 ^{14}C yr BP [430 cal yr BP]; Paleosol 4b, peat and roots) were identified as charcoal.

Of the 18 dates from the site (table 3.1), all but 3 are believed to represent the correct chronological succession of the site (cf. figure 2.10). The questionable samples have exceedingly high counting errors and are anomalously old considering the present general understanding of the late Pleistocene geological history of the Nenana Valley. Those samples that are considered incorrect are 23,930 ± 9,300 (SI-1938), 12,080 ± 1,025 (SI-1936), and 19,050 ± 1,500 (SI-1544) ^{14}C yr BP (45,020–7,180, 17,180–11,410, and 26,200–19,400 cal yr BP). Sample SI-1935A (10,600 ± 580 ^{14}C yr BP [12,270 cal yr BP]) from the top of Paleosol 3 (Loess 5) should also be treated with suspicion, as it is clearly not in line with the other dates from this unit. However, the incongruity is not of the magnitude represented by the aforementioned samples. One other sample, SI-2328 (7,985 ± 105 ^{14}C yr BP [8,840 cal yr BP]), was taken from the 4 × 4 m test pit excavated at the western extremity of the site in 1976. The stratigraphy here is so badly deformed by convolutions that little correlation can be made with the main excavation area.

The anomalous dates are difficult to explain. Thorson and Hamilton (1977) have discussed the possibility that during episodes of paleosol development, airborne coal, lignite, or

TABLE 3.1. Radiocarbon dates for the Dry Creek site[a]

Lab No.[b]	¹⁴C yr BP	Stratigraphic and Archaeological Units
SI-1933A	Modern	Paleosol 4b
SI-1933B	375 ± 40	Paleosol 4b
SI-2333	1,145 ± 60	Paleosol 4b
SI-2332	3,430 ± 75	Paleosol 4a Component IV
SI-1934	3,655 ± 60	Paleosol 4a Component IV
SI-1937	4,670 ± 95	Paleosol 4a Component IV
SI-2331	6,270 ± 110	Paleosol 3
SI-1935C	6,900 ± 95	Paleosol 3
SI-1935B	8,355 ± 190	Paleosol 3
SI-2115	8,600 ± 460	Paleosol 3
SI-1935A	10,600 ± 580	Paleosol 3
SI-1544	19,050 ± 1,500	Paleosol 3
SI-2329	9,340 ± 195	Paleosol 3
SI-2328	7,985 ± 105	Paleosol 2? (test pit)
SI-1936	12,080 ± 1,025	Paleosol 2
SI-1938	23,930 ± 9,300	Paleosol 2
SI-1561	10,690 ± 250	Paleosol I Component II
SI-2880	11,120 ± 85	Loess 2 Component I

[a]Modified after Thorson and Hamilton (1977: table 4).
[b]Smithsonian Institution.

ash from burning seams of the same might have been deposited at the site as part of the normal loess fallout. They also point out that the samples identified as charcoal actually yielded low amounts of residual carbon after nitration pretreatment, indicating that the original samples apparently contained high percentages of humic material. It is further noted that "if windblown particles of dead carbon had been transported to the site at a nearly constant rate, then those samples that yielded the smallest insoluble residue of charcoal after pre-treatment would show the greatest age anomalies. The largest samples with smallest counting errors, should yield the most nearly accurate dates" (Thorson and Hamilton 1977:167).

Research should be conducted to determine the presence or absence of airborne coal or lignite in the Nenana Valley because, in all likelihood, more of these seams are presently exposed than ever before due to modern mining activity and the burning of fossil fuels at the nearby Golden Valley Electrical Association coal-fired power plant at Healy. This would, at least, give us a point of departure for dealing with this potential problem in the future.

It should also be noted that wind direction is a critical factor. To derive airborne coal dust from the presently exposed coal seams would require wind direction from the north and east, neither of which are common occurrences in the area today. The effective winds (in this case, katabatic) are predominantly from the south. However, the Healy area is topographically complex, with several tributary valleys joining the Nenana, and this results in localized up- and down-valley drafts that could conceivably waft the dust over the whole region.

Until further attempts are made to solve this perplexing matter, we should consider it a very likely, albeit speculative, possibility that airborne coal dust had contaminated some of the Dry Creek radiocarbon dates, specifically those we have excluded from the site chronology.

Of the samples remaining, 11,120 ± 85 ^{14}C yr BP (SI-2880; 12,970 cal yr BP) provides a reasonable date for the upper part of Loess 2 and hence an upward limiting date for Component I. The date of 10,690 ± 250 ^{14}C yr BP (SI-1561; 12,540 cal yr BP), from a hearth in Component II, provides an age for the middle of Loess 3 (Paleosol 1). This is followed by the date of 9,340 ± 195 ^{14}C yr BP (SI-2329; 10,590 cal yr BP) from Paleosol 2 and applies to the top of Loess 3. Dates 8,355 ± 190 ^{14}C yr BP (SI-1935B; 9,310 cal yr BP) and 6,270 ± 110 ^{14}C yr BP (SI-2331; 7,180 cal yr BP) nicely bracket Loess 5 (Paleosol 3), with the date of 6,900 ± 95 ^{14}C yr BP (SI-1935C; 7,750 cal yr BP) constituting an average date for this unit. Two other dates from the top of Loess 5, 8,600 ± 460 (SI-2115) and 10,600 ± 580 ^{14}C yr BP (SI-1935A; 9,650 and 12,270 cal yr BP, respectively), show increased counting errors and are anomalous. Loess 6 (Paleosol 4a), which contains Component IV, is bracketed by 4,670 ± 95 ^{14}C yr BP (SI-1937) for the bottom and 3,655 ± 60 (SI-1934) and 3,430 ± 75 ^{14}C yr BP (SI-2332) for the top (5,400, 3,980, and 3,690 cal yr BP, respectively). Loess 7 (Paleosol 4b) has three samples taken from near the top of this unit that have provided dates of 1,145 ± 60 ^{14}C yr BP (SI-2333; 1,065 cal yr BP), 375 ± 40 ^{14}C yr BP (SI-1933B; 430 cal yr BP), and Modern (SI-1933A).

The only major weaknesses in the geochronology of the Dry Creek site are the number of single dates for the lower components and the possibility of contamination. In an attempt to overcome these obstacles, we have used only those dates with the lowest counting errors that, as fortune would have it, provide a consistent vertical series. The age of Component II is not excessive and is roughly synchronous with other early microblade sites in the Alaskan interior. Also, the nature of the fauna discovered in Components I and II increases our confidence that the Dry Creek loess cap has been correctly dated.

In addition to providing dates on the archaeological occupations at the Dry Creek site, the chronology provides estimates on major climatic changes in this part of the Nenana Valley. During the early history of the deposition at Dry Creek (Loess 1, Loess 2, Sand 1, and Loess 3), eolian sediments accumulated in what was probably an open herbaceous landscape concurrent with the major glacial-alluvial activity of the Riley Creek II Glaciation or the Carlo Readvance of the Riley Creek glacier. Our ideas relating to the nature of this environment are treated in greater detail in chapter 6. However, this major glacial activity probably ended by 8,500 ^{14}C yr BP (9,500 cal yr BP), and the subsequent vegetation history of the site area up to about 5,000 ^{14}C yr BP (5,700 cal yr BP) can be characterized by a slow deterioration of the dryer herbaceous cover, the development of tundra associations, poorer drainage conditions, a rising permafrost table in response to a more insulating vegetation cover, and finally, the appearance of the modern boreal forest shortly after 5,000 ^{14}C yr BP (5,700 cal yr BP), which has continued to the present day (Thorson and Hamilton 1977).

Note

1. Since preparation of the Dry Creek report, additional radiocarbon dates have been obtained (e.g., Bigelow and Powers 1994; Graf et al. 2015), as discussed in detail in chapters 8 and 9.

CHAPTER FOUR

Lithic Technology of the Dry Creek Site

W. ROGER POWERS

Introduction

The Dry Creek stoneworkers selected a wide range of raw materials for the production of their tools: most commonly used were rhyolite (light, dark, and banded), degraded quartzite (often referred to as just quartzite in the report),[1] grey chert, and chalcedony (including jasper). Of these, the rhyolite and quartzite are locally available, the former occurring east of the Nenana River and the latter in the bed load of Dry Creek, past and present. The cryptocrystalline rocks—gray chert and chalcedony—probably occur locally, considering their abundance, but the exact source is not presently known. Brown chert is available in the gravel bed of Dry Creek and is derived from the Tertiary gravels upstream. Less commonly encountered are pumice, diabase, sandstone, slate, argillite, schist, and quartz, all of which are available locally in the Tertiary-age Nenana gravels and the Pleistocene outwash and alluvial gravels of the Nenana River and its tributaries (e.g., Dry Creek). Several fine-grained and cryptocrystalline rocks were used, but uncommonly, and either occur sporadically in the local gravels or have distant source areas. These include green, black, and ferruginous cherts. Obsidian and devitrified volcanic glass are not presently known in the immediate area. There is a possibility that the obsidian might be derived from the Indian Mountain (Batza Téna) area on the Koyukuk River about 330 km by air to the northwest of Dry Creek (Holmes, personal communication).[2]

The Dry Creek stoneworkers were careful to reserve the fine-grained and cryptocrystalline rocks for tools in the finer end of the technology—that is, small bifaces, microblade cores, and small scrapers. There was a definite tendency to use these raw materials to the fullest extent. This could mean that most or all of these rocks were relatively rare in the immediate hinterlands or fairly difficult to come by. The coarser stone, especially the quartzites and dark rhyolites with phenocrysts, were used extensively for making opportunistic implements.

The total site assemblage is composed of 34,811 cataloged specimens. Component I accounts for 3,558 specimens, Component II for 28,881, and Component IV for 2,372. It is the 32,439 specimens from the two early components that are discussed in this chapter.

The lithic remains in the two early components at Dry Creek are of chipped and battered stone tools that account for only 8.3 percent of the assemblage. The remainder of the artifacts (91.7 percent) includes flakes or waste material resulting from stone tool manufacture. It is the tools themselves that are discussed in this chapter, and the emphasis is on technology and function. The spatial patterning of both tools and flakes, plus an analysis of the flakes, is discussed in chapter 5.

Certain judgments about the relative importance of artifact categories have been necessary. Hence, in describing certain artifact classes, particularly those that have controllable variability, measurements and attributes are provided individually. In other cases, where variability is more diffuse, measurements are provided as ranges and means, and descriptions are formalized.

Observations on the functions of some artifact types from the Dry Creek site are based on morphology and technology—for example, shape, manufacturing technique, and evidence of repair. In addition, microscopic examination of use wear and striations was conducted by Dr. T. A. Del Bene and presented in detail in his doctoral dissertation at the University of Connecticut (Del Bene 1981). The results of his research, which have been incorporated into the discussion of the lithic technology at Dry Creek, are based on both the dissertation and the lab notes.

Measurements for all classes are given in millimeters in a linear sequence, such as 102 × 67 × 14 mm, where the first is length, the second width, and the last thickness. This is true of all classes except wedge-shaped cores, where the sequence is length, width, and height. The length refers to the distance from the fluted surface to the back of the core; the width is the distance between the two sides, or faces, of the core; and the height is the distance from the base to the top of the working platform.

Component I: Artifacts

Of the total assemblage of 3,517 artifacts from Component I, only 43, or roughly 1 percent, can be classified as shaped tools. The remaining 99 percent of this assemblage is composed of flakes. Within Component I (Loess 2), most of the tools and flakes occurred within discrete clusters. The analysis of the spatial distribution of the finds and the contents of the clusters is presented and discussed in chapter 5.

The tools from Component I are classified as bifacial tools (8), scrapers (18), and miscellaneous artifacts (17).

Bifacial Tools
The bifacial tools can be further subdivided as follows: projectile point (1), point bases (2), biface base (1), biface tip (1), and bifacial knives (3).

Projectile Point (1)
This specimen is the single complete example of a point from Component I (figure 4.1A). It is a small, black chert, isosceles triangular biface measuring 31 × 16 × 4 mm. The edges are

FIGURE 4.1. Bifacial artifacts from Component I (A, projectile point; B–C, projectile point bases; D, biface base; E, biface base and tip rearticulated).

slightly excurvate, and the base is straight except for a small spur at one corner. There is no evidence of use wear or edge grinding, although there are faint hints of hafting wear on the base. All edges are straight and finished, and the tip is well formed and very sharp. Its flatness and symmetry render the piece perfectly suitable for penetration.

Point Bases (2)

Two basal fragments of probable points were also found in Component I. On morphological and technological grounds, they appear to be bases of points very similar to that just described. They are both of gray chert and were finished with fine pressure retouch. The larger specimen (figure 4.1B) has a straight base and straight ascending edges, while the smaller piece (figure 4.1C) is more irregular and has constricting edges and a small spur at one corner of the base. In this regard, it is similar to the complete specimen of a point from this component. There is no evidence of use wear, edge grinding, or hafting wear. The measurements are 8 × 23 × 3 mm (figure 4.1B) and 7 × 19 × 3 mm (figure 4.1C).

Biface Base (1)
This piece (figure 4.1D) is the basal fragment of a brown chert biface either broken during manufacture or discarded unfinished. Attempted fabrication essentially ruined one of its edges. It measures 31 × 60 × 20 mm.

Biface Tip (1)
This flake of ferruginous chert (figure 4.1E) has a triangular form and a lenticular cross section. There is unifacial retouch along one edge and bifacial retouch along the other. The bifacial edge and the tip are unfinished. This situation might be related to a snap that broke the base away from the piece. The fragment measures 32+ × 30+ × 5+ mm.[3]

Bifacial Knives (3)
Two of these knives were found in fragments that fit together to form complete tools. They have triangular outlines with slightly excurvate edges and convex bases. The cross sections are lenticular.

Of these, one (figure 4.2) measures 55 × 38 × 9 mm and was made on a flake of devitrified volcanic glass. It was broken during manufacture into three pieces: one entire edge, a portion of the tip, and the main body of the knife. The three fragments of the knife lay directly in the flaking debris resulting from its manufacture. The knife was broken when the worker applied excessive pressure while attempting to remove a knob at one edge. Crushing and the deep hinge fractures along the edge indicate that several attempts were made to remove

FIGURE 4.2. Bifacial knife from Component I.

the obstacle. This knife was in the pressure-flaking stage of bifacial reduction, and only some thinning along the edges and the tip remained before completion.

The other complete knife measures 64 × 36 × 8 mm.[4] It was made on a flake of ferruginous chert and was probably complete and in the process of use when it was broken. All final retouch has occurred along the edges, and there is isolated evidence of use wear. No clear evidence of hafting could be detected, but it seems possible that the specimen broke in half when either a blow or excessive pressure was applied, and a diagonal fracture developed that ran from high on one edge to the opposite corner of the base.

The third specimen possesses the same formal features, but it is a very crudely triangular rhyolite flake on which bifacial thinning and shaping was attempted. The flaking apparently miscarried and the piece was abandoned. It measures 45 × 32 × 10 mm.

Scrapers

The Dry Creek Component I scrapers are classified as follows: transverse scrapers (3), side scrapers (2), end scrapers (11), double end scraper (1), and end scraper/burin (1).

Transverse Scrapers (3)

These specimens are characterized by a unifacially flaked edge that is transverse to the long axis of a piece of dacite. On one specimen (figure 4.3A), this edge is straight and traverses the distal end of a flake, which retains cortex on the dorsal surface. The working edge lies at an angle of 70° to the ventral surface and appears to have been applied to a scrape- or cut-resistant material. It also appears to have been hand held. This specimen measures 110 × 61 × 10 mm.

The second transverse scraper (figure 4.3B) was made on a rectangular, waterworn slab. It has a rectangular cross section, and the unifacial working edge, formed by steep flaking, runs across one end of the slab. It lies at an angle of 70°–90° to the ventral surface of the tool. The edge is straight and fairly crude. Some additional finer retouch was applied, but most of the smaller flaking is edge damage that sporadically extends onto the ventral surface. It appears to have been hand held and used to pound on a resistant material. This piece measures 179 × 107 × 33 mm.

The third piece in this group (figure 4.3C) was made on a section of a longitudinally split oblong cobble. The dorsal surface is composed mainly of the facet of a previous flake removal that is bordered by cobble cortex. The unifacially flaked working edge is slightly convex. It is transverse to the long axis of the flake and lies at about 70° to the ventral surface. There is no evidence of use wear. It measures 157 × 73 × 27 mm.

Side Scrapers (2)

One of these specimens is an asymmetric, opposing convergent side scraper made on a thick flake of ferruginous chert (figure 4.4A). The axis of the scraper departs about 20° from the axis of the flake. One finely retouched, slightly convex scraping edge is found on the dorsal surface and runs along the entire length of the butt of the flake. A second scraping edge converges on the first from the opposite side and the face of the flake. This second edge conforms to a break in the flake that renders it highly convex. The converging edges form an unrefined tip. No use wear can be detected on either edge, and it might have been discarded before use. It measures 73 × 49 × 14 mm.

FIGURE 4.3. Transverse scrapers and quadrilateral uniface from Component I (A–C, transverse scrapers; D, quadrilateral uniface).

The second side scraper was made on a bifacial thinning flake of rhyolite (figure 4.4B) and has a straight, steeply retouched scraping or cutting edge. Retouch was also applied to one end of the flake, forming a narrow convex bifacial edge. It measures 62 × 41 × 10 mm.

End Scrapers (9)

End scrapers from Component I at Dry Creek (figure 4.5) were manufactured on flakes of dacite (3), green chert (2), gray chert (1), rhyolite (1), and chalcedony (2). Four of the scrapers were made on bladelike flakes that preserve one or two arises on the dorsal surface. On these specimens, the scraper edges are on the distal ends of the flakes, and (on four examples) the edges are transverse to the axes of the flakes. Two other specimens were made on simple flakes, and the scraper edges occupy the same distal and transverse positions. Two specimens display end scraper edges on the left side of the flakes. In each case, this edge lies at a 90° angle to the axis of the flake.

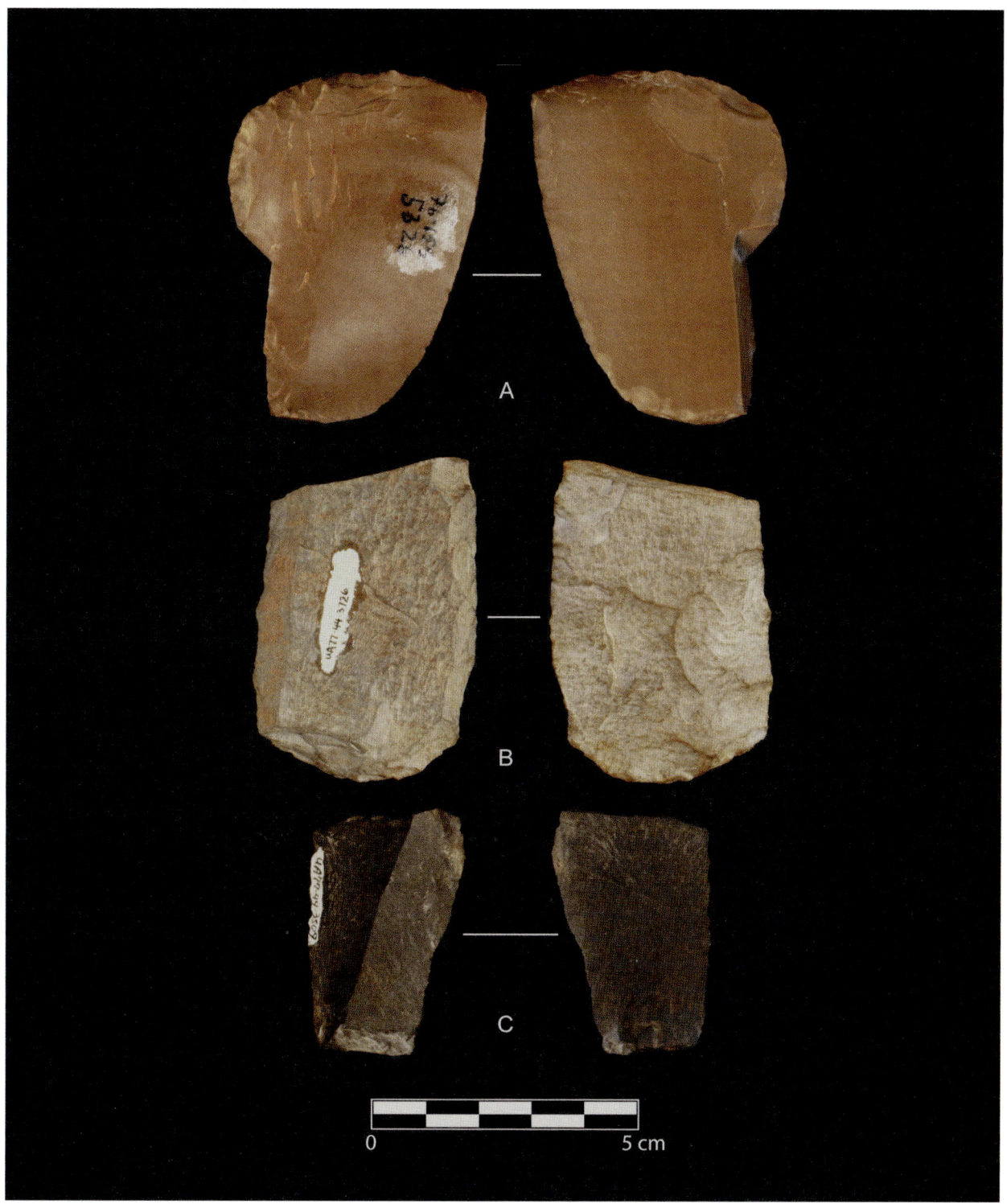

Figure 4.4. Side scrapers and unshaped flake tool from Component I (A–B, side scrapers; C, unshaped flake tool).

FIGURE 4.5. End scrapers from Component I
(A–C, E, steep end scrapers; D, flat end scraper).

Fine pressure retouch was used to form the convex scraper edges on all specimens. In each case, the length of the retouching facets depends on the variable thickness of the distal ends of the flakes.

On the basis of the scraper edge/ventral surface angle exactly at the edge of the specimen, the end scrapers can be subdivided into two categories: steep and flat. Steep end scrapers (figure 4.5A–C, E) have edges ranging from 70° to 80° to the ventral surfaces of the flakes. Flat end scrapers (figure 4.5D) have scraper edge/ventral surface angles that vary from 40° to 60°. Four of the steep end scrapers bear traces of minor, sporadic retouch along the lateral edges, and one has a well-shaped, convex scraping edge on the left margin and minor retouch at the proximal end. Two of the flat end scrapers are retouched on both lateral edges. One lacks additional working.

All specimens classified as end scrapers lack polishing on the working edges. Rather, each displays fine to microscopic crushing that could best be explained by a cutting action that was directed against a fairly resistant material such as bone. This is an important distinction, since this artifact type is consistently classified as a scraper, with that function implied, and while experimental evidence exists to support that assumption (Semenov 1964), it appears that in some instances, the end scraper can function as a knife (Aigner 1970, 1978).

Microscopic examination also indicates that all but one of the specimens were probably hafted. This is revealed by wear on the crests of flake facets beginning just behind or below the scraper edge and continuing to the proximal end of the flake.

The last specimen is the detached working edge of a flat end scraper, which was broken during either use or manufacture. It is chalcedony and measures 7 × 16 × 2 mm.

Double End Scraper (1)
This single specimen is bifacially retouched. It has two opposing scraper edges and additional retouching along both lateral edges. Enough of the ventral surface of the flake remains to distinguish proximal and distal extremities. The proximal scraping edge lies at a 90° angle to the ventral surface. Like some other end scrapers, it was probably used as a cutting implement that was applied to cut-resistant material.

The distal scraping edge might have never been used, and an attempt to prepare it miscarried. This edge lies at 75° to the ventral surface. The distal end of the scraper bears hafting wear. This scraper was made on a flake of black chert and measures 22 × 26 × 9 mm.

End Scraper/Burin (1)
This is a combination tool made on a thin flake of moss agate. It was broken during use. The short, convex scraper edge lies at the distal end of the flake at a 70° angle to the ventral surface of the flake. Three burin blows were struck at the distal end, although only one succeeded. A return blow was struck in the opposite direction. The use wear on the burin is on the ventral edge and at the origin of the return blow. Both lateral edges of the flake are formed by breaks that bear extremely fine retouch or utilization chipping. It appears that the breaks were also used as burins.

Again, the scraper edge was used as a cutter on a resistant material, but no wear from hafting could be discovered. After the piece was broken, the snapped edges were further utilized as burins. The specimen measures 26 × 18 × 3 mm.

Miscellaneous Artifacts

The miscellaneous artifacts from Dry Creek have been classified as follows: quadrilateral uniface (1), unshaped flake tools (6), split cobble tools (3), cobble cores (4), and anvil stones (2). As such, they do not constitute a typological entity but rather a convenient grouping of unrelated forms.

Quadrilateral Uniface (1)

This object (figure 4.3D) is a rectangular unifacial tool with each of the four edges processed by steep percussion flaking, resulting in a strongly plano-convex cross section. It was made on a very thick flake of degraded quartzite. The working edges lie at angles of 40°–90° on the ventral surface. The tool appears to have been hand held and used to hack, chop, or pound a resistant material, as evidenced by edge damage sporadically extending onto the ventral surface. The measurements are 105 × 87 × 43 mm.

Unshaped Flake Tools (6)

These small tools (one of which is shown in figure 4.4C) are commonly called retouched flakes. They were made on flakes of gray chert (1), ferruginous chert (1), black chert (1), devitrified volcanic glass (1), chalcedony (1), and rhyolite (1). They range in size from 32 × 19 × 4 mm to 48 × 30 × 10 mm, with a mean of 41.6 × 25.5 × 7 mm.

Two of these tools have fine retouch on one end of the flake and have been used as scrapers. The remaining four have unifacial edge retouch (one on both edges and three on a single edge) and would have served well as small flake knives.

Split Cobble Tools (3)

These heavy implements were all manufactured on pieces of split cobbles. They have a general oval to rectangular outline and a plano-convex cross section. Each is slightly different. One rhyolite implement (figure 4.6A) was flaked on the dorsal surface at one end, which formed an irregular edge. Sporadic retouch was then applied, and it carried onto the ventral surface. At the opposite end, heavy flaking formed an irregular edge. Sporadic retouch was then applied, and it carried onto the ventral surface. At the opposite end, heavy flaking formed an irregular edge that was then retouched regularly and extensively, but only on the dorsal surface. It measures 145 × 101 × 32 mm.

The second specimen, a quartzite tool (figure 4.6B), bears irregular, light retouch along the edge of the ventral surface. The dorsal surface is worked. It measures 124 × 72 × 48 mm.

The last specimen in this group is quartzite (figure 4.6C) and has flaking, which culminates in a point, concentrated at one end of the implement on the ventral surface. It measures 119 × 72 × 71 mm.

Functionally, these tools are probably identical and were used in heavy scraping or planing activity. Evidently, the desired feature being sought was the relatively long, flat ventral surface.

Cobble Cores (4)

These objects are elongate river cobbles with roughly rectangular to suboval cross sections. Three of these cores are of degraded quartzite and one is of rhyolite. They range in size from 122 × 64 × 33 mm to 178 × 121 × 96 mm.

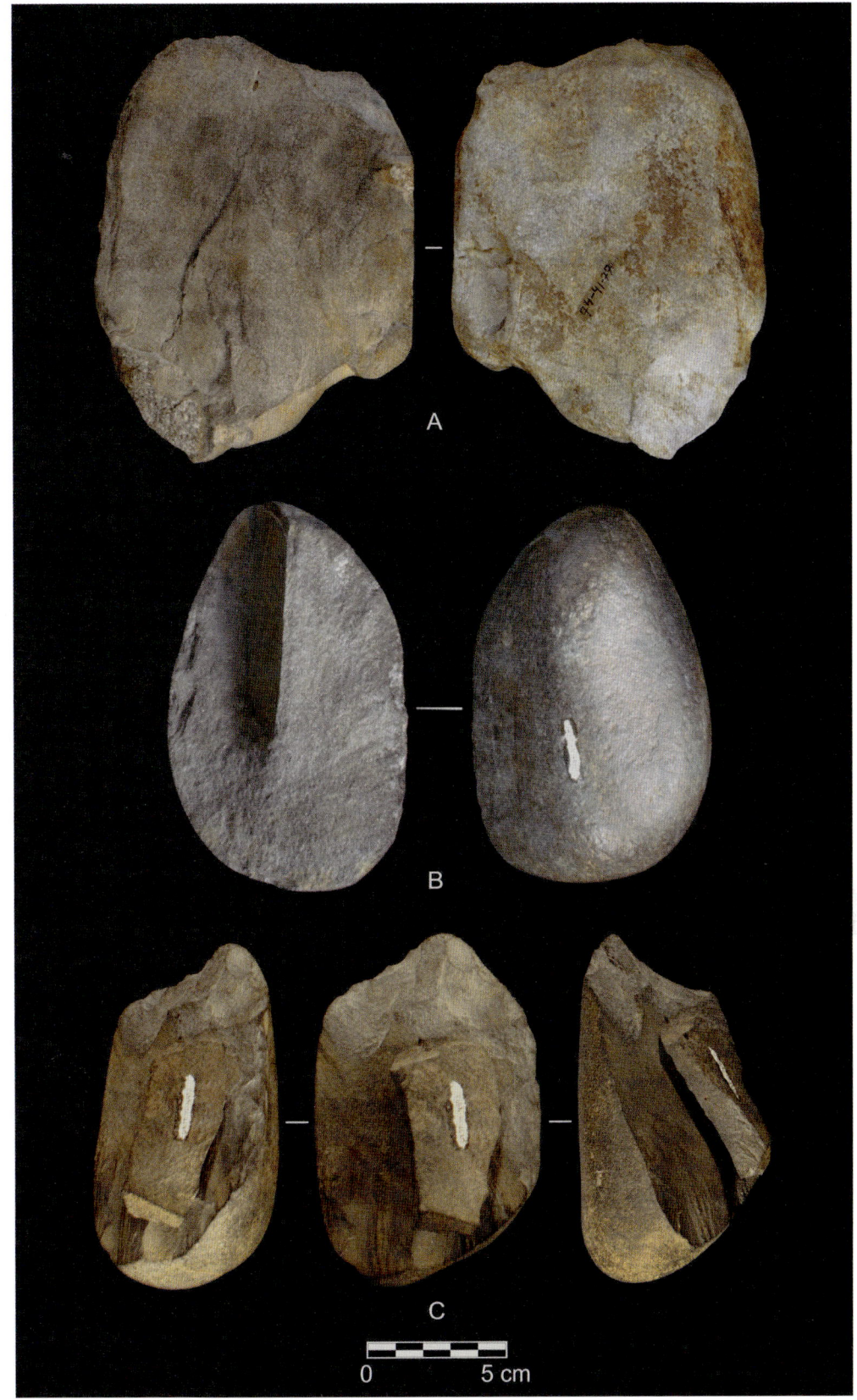

FIGURE 4.6. Split cobble tools from Component I.

On three of the pieces, one end was removed to establish a striking platform from which blows were directed down the thinner edges (parallel to the long axis) of the cobble. The fourth core (figure 4.7) illustrates the method of platform preparation. Thick sections of the cobble were struck off as if splitting a loaf of bread.

Anvil Stones (2)

Both of these artifacts are flat rhyolite boulders with evidence of heavy battering and denting on one surface. They measure 158 × 125 × 60 and 199 × 122 × 56 mm, respectively.

Split Boulder (1)

This specimen is a rounded river boulder of conglomerate measuring 205 × 155 × 150 mm. It was split in two, and one half bears unifacial flaking at one end on the dorsal surface. It is quite heavy and might have been used to weigh something down. On the other hand, it might have been brought to the site to be used as an anvil stone.

FIGURE 4.7. Cobble core from Component I.

Component II: Artifacts

Component II is composed of 28,881 lithic artifacts. Of these, 2,127 (7.4 percent) are worked, and the remaining 26,757 specimens are flakes (including 3 bladelike flakes), which accounts for 92.7 percent of the lithic assemblage. However, as discussed in chapter 5, many of the larger flakes were probably used as butchering tools.

Within the category of worked pieces, 1,966 specimens (92.4 percent) are the result of microblade production and burin utilization and are classified as follows: wedge-shaped cores (21), aberrant microblade cores (8), microcore preforms (3), miscarried microcore preforms (21), core tablets (45), miscellaneous wedge-shaped core parts (24), microblades (1,772), burins (29), core-burins (8), and burin spalls (35).

The remaining group of 161 worked pieces (7.6 percent) is composed of bifacially worked tools (44), heavy percussion flaked implements (47), scrapers (21, including 2 spokeshaves), subprismatic cores (4), flake tools (21), bladelike flake tools (18), hammerstones (3), and anvil stones (3).

Wedge-Shaped Cores, By-Products, and Microblades

The wedge-shaped microblade core has been given a great deal of attention in the northern archaeological literature (cf. Aigner 1970; Anderson 1970a, 1970b; Dikov 1977; Kobayashi 1970; Medvedev 1968; Mochanov 1977; Morlan 1965) because its distribution spans northeast Asia and northwest North America during the terminal Pleistocene and early Holocene. Since the initial discovery of this distinctive form of stone artifact more than fifty years ago in Alaska and Mongolia (Nelson 1935, 1937), it has provided a rich field for subsequent research on cultural relationships between the Old and the New Worlds (Morlan 1965, 1970). Furthermore, extensive typological and technological studies of these microcores make them a relatively well-known item with which to work (Kobayashi 1970; Morlan 1970, 1978; Yoshizaki 1961). The distinctive attributes, manufacturing sequences, and by-products of these cores are now well understood—at least technologically—but there is room for refinement, especially in our understanding of the exact methods of blade detachment and the purposes to which the microblades were put.

In the treatment of the wedge-shaped core technology presented here, these speculative matters will be discussed, and differences as well as similarities with established concepts of wedge-shaped core technology will be addressed.

At Component II of the Dry Creek site, a total of 1,894 specimens are the result of wedge-shaped microcore technology. These include the microcores themselves (21); the microblades detached from them (1,772); aberrant cores (8); preforms for microblade cores (3); miscarried preforms ("core-scrapers"; 21); core tablets (45), which are a by-product of both core manufacture and maintenance; and miscellaneous core parts (24).

In addition to these easily recognized groups of artifacts, perhaps thousands of bifacial thinning flakes and many hundreds or even a few thousand platform preparation flakes should be visualized when one attempts to appraise the amount of stonework involved in the production, use, and maintenance of wedge-shaped cores.

Wedge-Shaped Microblade Cores (21)

A summary of metric and nonmetric observations is presented in table 4.1. Of the 21 finished, utilized, and/or discarded microblade cores, 17 were made on chunks of cryptocrystalline rocks in which brown to red chalcedony (11) predominates. Other stones utilized are gray chert (4), rhyolite (5), and black chert (1). Four of the cores were made on flakes. Where chunks of stone served, bifacial reduction was employed to shape the core. Where flakes were used, unifacial beveling with minor bifacial thinning was employed, usually to reduce the thick dorsal surface of the flake. It is clear that in selecting a piece of stone, whether chunk or flake, the worker looked for at least one thick edge on which to establish the platform. If the opposite edge was thin, so much the better, and only minor adjustments were necessary. However, if this edge was thick, bifacial or unifacial working was employed. The cores are small and range from 15 to 46 mm in length, 10 to 18 mm in thickness, and 18 to 31 mm in height. The mean values for these measurements are 30 × 14 × 25 mm.

TABLE 4.1. Summary of metric and nonmetric observations on wedge-shaped cores

Catalog Number	Figure	Material	Original Metrics	Discarded Metrics	Intact	Platform Removal
76-3765	4.12E	Gray chert		29 × 12 × 23		Complete
77-2089	4.11G and 4.12G	Gray chert		26 × 18 × 18		Complete
76-1787	4.11H and 4.12H	Gray chert		20 × 13 × 22		Complete
76-5058	4.11F and 4.12F	Gray chert		17 × 15 × 22		Complete
77-637	4.13A	Rhyolite	46 × 16 × 30			Complete
76-3225	4.13A	Rhyolite	51 × 14 × ?	21 × 14 × 26		Complete and partial
76-273	4.10A and 4.11C	Rhyolite		28 × 12 × 31	X	Partial
73-24	4.10C	Rhyolite		31 × 14 × 29		Partial
77-2777	4.10E	Rhyolite		23 × 17 × 27		Complete
76-474	4.10D	Chalcedony		32 × 14 × 25		Partial
76-587	4.11D and 4.12D	Chalcedony		34 × 15 × 24		Partial
76-764	4.12A	Chalcedony		23 × 13 × 21		Partial
76-241	4.12D	Chalcedony		25 × 14 × 23	X	Partial
76-278a	4.13B	Chalcedony		35 × 18 × 23		Complete
76-278b	4.13E	Chalcedony	46 × 15 × ?	33 × 13 × 28		Complete
76-4058	4.11E and 4.10B	Chalcedony		46 × 15 × 25		Partial
76-731	4.10G	Chalcedony		23 × 10 × 26		Partial
76-757	4.10H	Chalcedony		26 × 12 × 23		Partial
76-4518	4.13C	Chalcedony	37 × 15 × 30	30 × 14 × 25		Partial
76-4097	4.10F	Black chert		32 × 10 × 25		Partial
76-588	4.11C	Chalcedony		28 × 18 × 29		Partial
Mean				30 × 14 × 25		

The number of microblade flutes can range from 3 to 8, with a mean of 5.6 flutes. The mean fluted width for individual cores ranges from 2.4 to 4.8 mm, with the mean width for all cores being 3.4 mm.

Judging from both the completed and the discarded attempts at preforms, the stoneworker established a platform on the shaped piece of stone. This was accomplished by beveling the edge and attempting to keep it as even as possible so that when finished, the prepared platform lay at about an 80°–90° angle to the sides of the core. Notwithstanding, blanks and some core tablets show that the initially prepared platform can be quite concave, resulting in a high hook or spur at the rear of the platform (see the following text and figures pertaining to microblade core tablets). The edge opposite the platform may be convex. It may also arch slightly in one direction. One side of this arch will ultimately be the terminus for microblades leaving the core base, which arches backward and upward until it intersects the prepared platform at the end opposite the area of microblade removal. This arch may be interrupted

Platform Removal Blows (N)	Striking Platform Length (mm)	Striking Platform Width (mm)	Microblade Flutes (N)	Microblade Flutes (Mean Width; mm)	Back Condition	Base Condition	Base Bevel
2	14	12	4	2.4	Cortex	Straight	Right
	26	8	5	2.5	Cortex	Arched	Left
	20	13	7	3.5	Fluted	Arched	Right
	16	10	8	3.6	Retouched	Arched	Right
1	15	10	5	3.5	Unmodified	Straight	Left
8	11	14	5	3.8	Unmodified	Straight	Both
			7	3.6	Unmodified	Arched	Right
	12	13	6	3.2	Retouched	Arched	Right
	7	13	6	5.3	Retouched	Straight	Right
	11	12	5	3.6	Unmodified	Straight	Left
	12	13	5	4.8	Retouched	Arched	Left
	6	13	6	3.0	Retouched	Arched	Right
			4	3.4	Broken	—	—
2	17	15	6	3.0	Retouched	Arched	Right
4	32	13	5	3.3	Unmodified	Straight	Left
	9	14	5	3.8	Unmodified	Arched	Left
	13	9	6	3.7	Retouched	Arched	Both
	3	9	5	3.2	Retouched	Arched	Both
2	17	13	6	3.0	Retouched	Arched	Right
1	11	7	3	3.0	Retouched	Arched	Both
1	9	13	8	3.0	Unmodified	Arched	Both
	14	11	5.6	3.4			

by an area of unworked stone or cortex that leaves the core with a distinctively shaped base and a back that is unmodified.

This situation is also found on preforms that have a rough rectangular outline when viewed from the side. In this instance, there is a distinctively retouched base that may be relatively straight and an unmodified or cortical back.

The condition of the back of the core was probably irrelevant; only half exhibit any degree of retouching. Conversely, the bases were universally treated. The purpose of shaping the base was to straighten it by bifacial retouch where feasible and by unifacial beveling—particularly on flakes, where the dorsal surface was especially bulbous. It seems that straightness rather than symmetry was the critical consideration in shaping the base for securing it into a clamp or vise. On one example, a large bifacial thinning platform remains on the base. It did not impede the use of the core.

The bases of the cores all show considerable crushing, and some exhibit wear or polish on the back. Apart from this, there is little apparent wear on the cores. This wear must be the result of the technique used to secure the core, at least initially. The variability of the core form, especially the shape of the backs and bases, leads one to conceive of a securing method that met the formal requirement of technology but likewise allowed for flexibility.

If the microblades were detached with a shoulder punch, then the critical control problem is forward rocking in the vise. The type of clamp discussed by Crabtree (1967) should leave traces on the lateral surfaces, especially if the core was not set in a support device. Since the Dry Creek cores do exhibit basal damage, they must have been set in something similar to a triangular incision in a piece of bone, which would provide stability and resistance against rocking.

This brings us to a discussion of platform modification during the microblade removal sequence, which bears directly on the question of clamping techniques.

Once the preform was ready for microblade removal (figure 4.8A), the stoneworker utilized three different options, separately or in combination, in dealing with the initially prepared platform (figures 4.8, 4.9).

The least common option at the Dry Creek site is what could be called System 1 (figures 4.8B, 4.10A, 4.11A). On two examples, blades were detached directly from the initially prepared platform. The core was discarded after an episode of microblade removal.

A second option, called System 2a, was used on eight cores to remove the entire retouched platform with a burin blow aimed directly at the front of the core (cf. core tablet). Following this, a sequence of microblades was detached (figures 4.8C, 4.11E, 4.12F–G, 4.13A).

The most common method of establishing a platform for microblade removal, System 2b, was to detach only a portion of the retouched platform (figure 4.9). Twelve Dry Creek cores were treated in this fashion. This resulted in a stepped platform, since the burin blow did not carry the full length of the core. Otherwise, the technique is the same as System 2a. In these instances, the actual working platforms are from 3 to 32 mm long and 7 to 15 mm wide. The mean length is 14 mm and the mean width is 11 mm (figures 4.10B–G, 4.11B–D, 4.12A–E).

It would be very tempting to view these variations in platform treatment as grounds for establishing morphological types of wedge-shaped cores or even manufacturing techniques. Fortunately, there are two reconstructed cores from the Dry Creek site that exhibit both forms of System 2 platform treatment. One of these (figure 4.13D) had a partially retouched

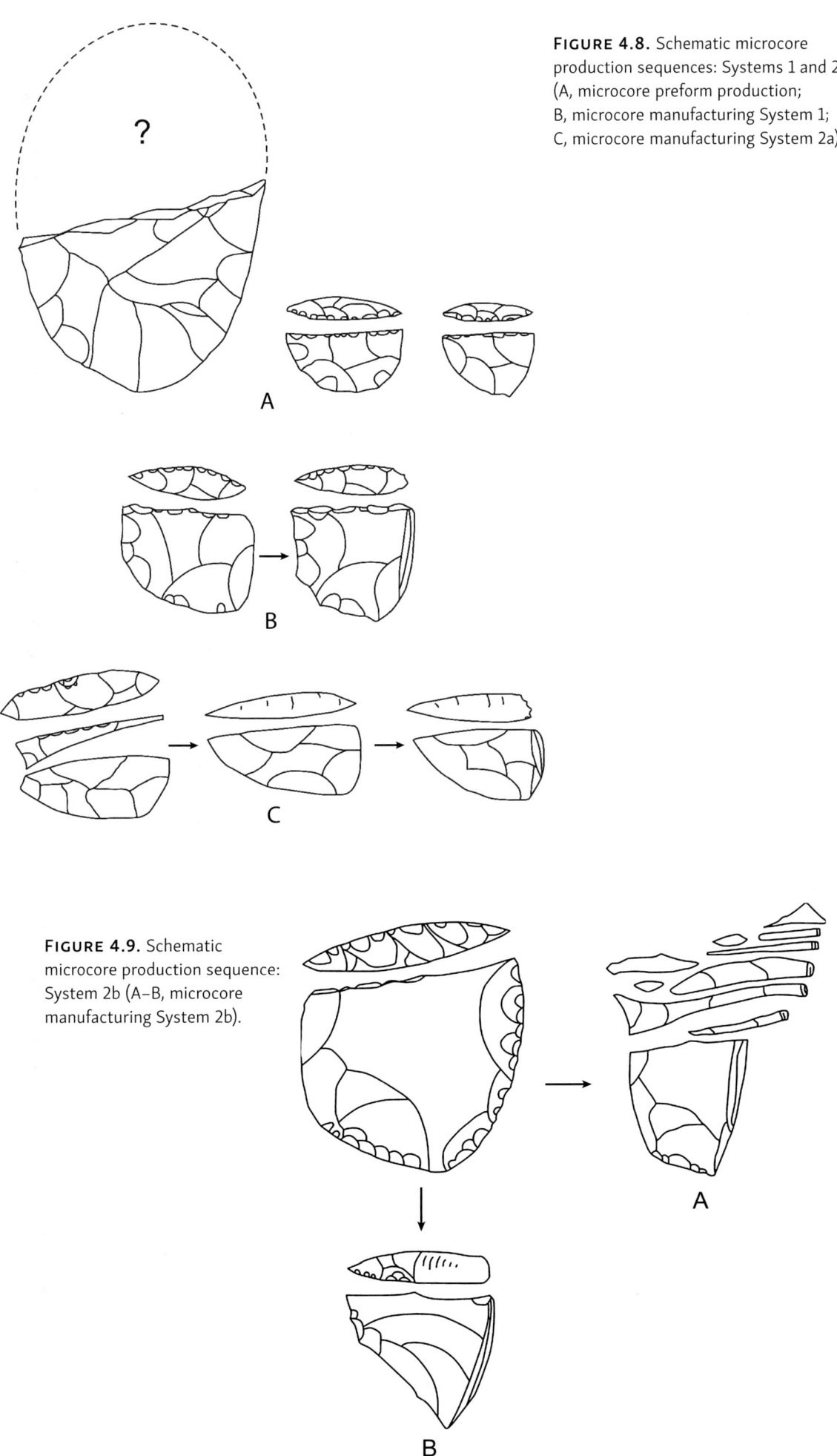

FIGURE 4.8. Schematic microcore production sequences: Systems 1 and 2a (A, microcore preform production; B, microcore manufacturing System 1; C, microcore manufacturing System 2a).

FIGURE 4.9. Schematic microcore production sequence: System 2b (A–B, microcore manufacturing System 2b).

FIGURE 4.10. Wedge-shaped microblade cores from Component II (A, System 1 microblade core; B–G, System 2b microblade cores).

Figure 4.11. Platform views of wedge-shaped microblade cores from Component II (A, System 1 microblade core; B–D, System 2b microblade cores; E, System 2a microblade core).

Figure 4.12. Wedge-shaped microblade cores from Component II (A–E, System 2b microblade cores; F–G, System 2a microblade cores).

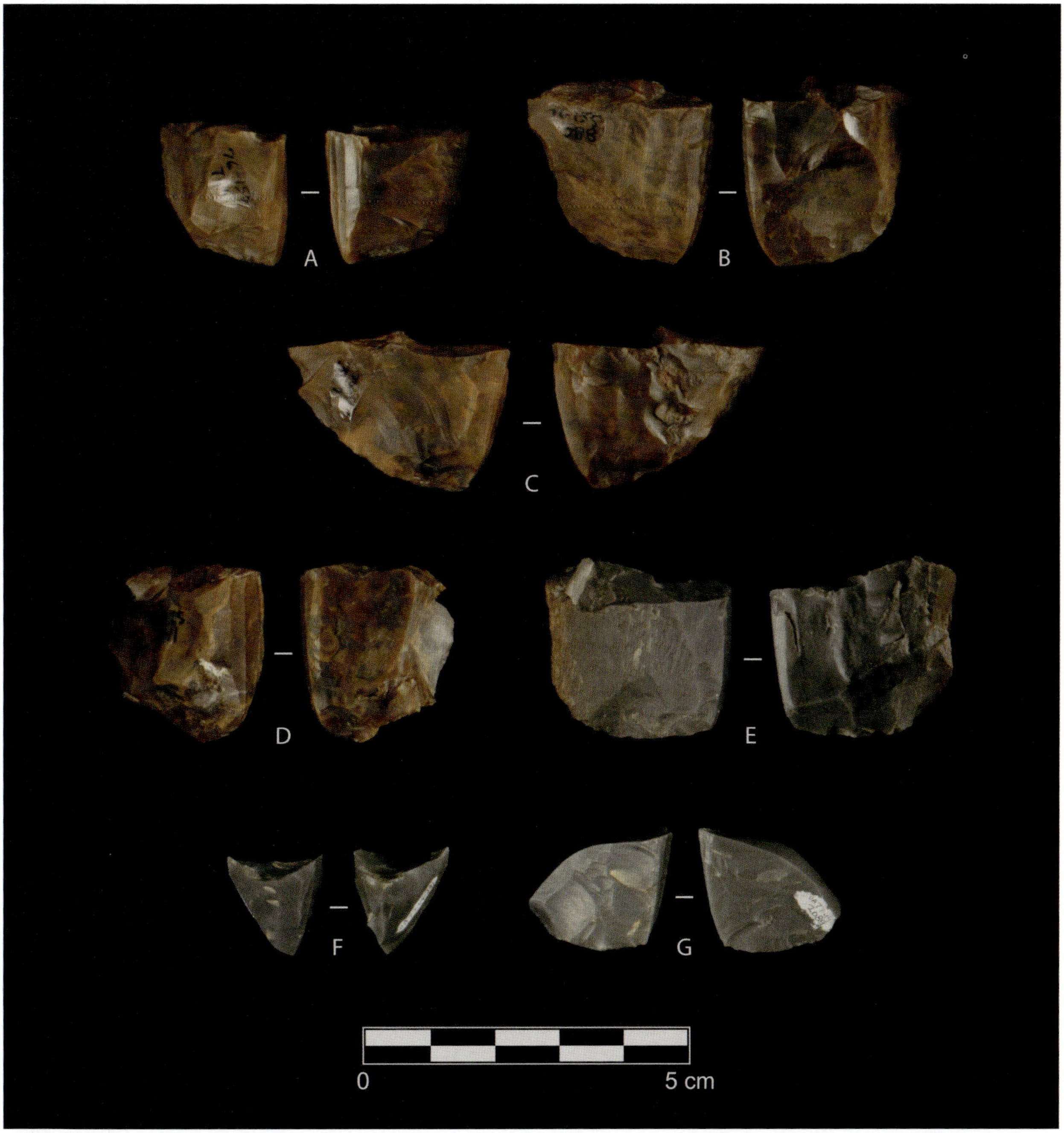

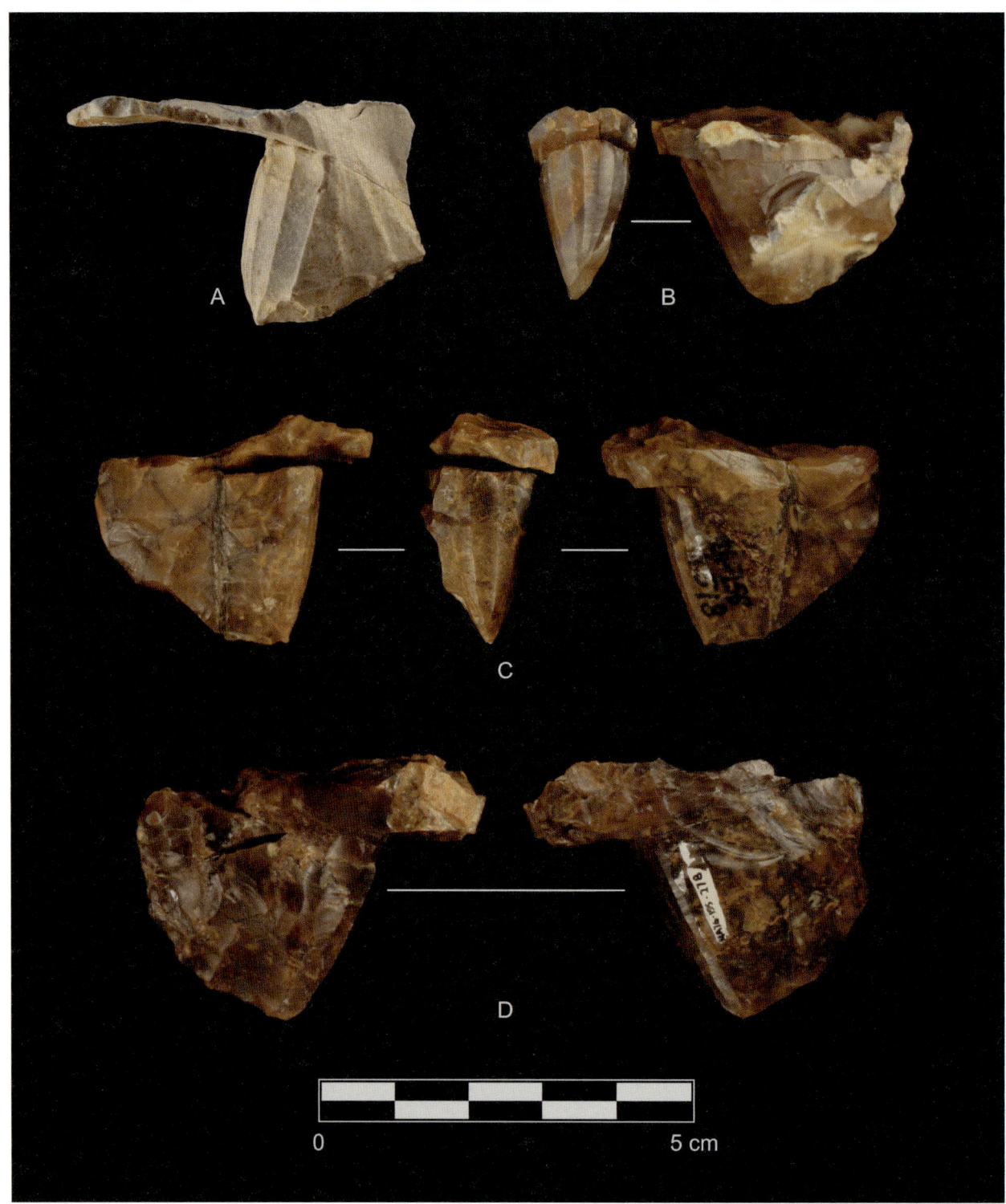

FIGURE 4.13. Microblade cores with rearticulated core tablets from Component II.

platform that was detached after a suite of microblades was removed. Following this, the platform was totally removed, and more microblades were detached. The second of these reconstructed cores underwent eight episodes of platform removal.[5] The first tablet removal created the working platform. The second through the sixth platform removals were likewise partial, but the seventh removed the entire core platform. The eighth, and last, was again partial. More microblades were detached after the last episode than from any previous attempts at platform rejuvenation.

In both examples, the stoneworker combined both System 2 methods for the removal of the retouched platform, clearly illustrating that total and partial platform removals form a technological continuum.

Another feature of core platforms (figure 4.11) is heavy crushing along one edge. This transpires as the platform is being beveled prior to partial platform removal. Once the platform is removed, a remnant of the beveling (retouching and crushing) remains at the rear of the platform area. After the core tablets have been struck off, additional fine chipping and crushing is applied to an otherwise smooth platform. This type of modification can take the form of a low concavity at the edge of the platform, and in most instances, this feature appears on the right-hand side of the core. However, it can occur sporadically on the left. As this chipping and crushing is far removed from the fluting arch, it must be related to the device used to hold the core.

It seems improbable that the core was held directly in the hand. The size alone would make this impractical, and if this was the method employed, the careful shaping of the core base would have been superfluous.

Based on the various nonmetric attributes of these cores, especially basal shaping and platform treatment (edge crushing or dulling), two methods of securing the core seem more feasible.

In one method, the core base could be fitted to a slot, and then a lateral clamp (bone and wood) could be lashed at the top. One of the lateral pieces would have to have been bent over the top of the core to prevent forward rocking. In such a clamp, microblades could be detached by either indirect percussion or applied pressure, although direct percussion might have been employed (Del Bene 1981). There are numerous examples of fairly deep negative bulbs on nearly all the cores.

A second method has been suggested by Del Bene (1981) as a result of his analysis of the cores and microblades. He argues that the core could have been grasped in a holding device, with the base of the core in a slotted anvil. The holding device was then wrapped with cordage to keep the core from slipping and rocking. Such a method appears quite feasible and makes the lateral crushing of the platform and the treatment of the base easier to understand. The common occurrence of a hinged (partially removed) platform then could be explained by the core tablet terminating at the haft. In the two examples of total platform rejuvenation, either the core was removed from the haft or the break was fortuitous.

The use of such a technique would allow the stoneworker the option of either holding the haft in the hand or weighing it under a foot.

Aberrant Microblade Cores (8)

Attempts to remove microblades from variously shaped chunks of chalcedony were fairly common at the Dry Creek site. Platform preparation and microblade removal resulted in a series of rather aberrant-looking cores (figure 4.14) so idiosyncratic that individual description would be necessary to describe the variability. In each case, a burin blow was used to establish a platform, and some microblades were detached. Some of these pieces have shaping along the bottom or rear edges, and others display some crushing along one edge of the platform area. Most display some classic wedge-shaped core attributes, but no single specimen has them all. It would be difficult to ascertain the original size of these chunks or to gauge the amount of blade detachment. These cores were probably short-lived, and the microblades removed (judging from the flutes on the cores) were quite irregular with respect to both length and width. Also, the cores seldom display a continuous fluting arch. In the majority of cases, deep irregular bulbs at the proximal ends of the flutes suggest blade detachment by direct percussion.

These specimens range from a maximum length, width, and height of 73 × 49 × 21 mm to a minimum of 29 × 17 × 13 mm. The mean values for these dimensions are 43 × 31 × 15 mm.

Core Preforms (3)

Three objects (figure 4.15) illustrate a method of preparing a wedged-shaped microblade core preform that employs bifacial reduction to form an asymmetric preform with one strongly convex edge and one straight or slightly beveled edge. The straight edge is further modified by unidirectional retouch directed across it at a right angle to the faces. This process produces a straight to slightly concave flat truncation along this side of the biface. It is this truncation that eventually is removed, either in part or entirely, by burin blows as the platform of the core is prepared for microblade detachment.

These cores were manufactured from gray chert (1), black chert (1), and chalcedony (1). Measurements range from 44 × 22 × 17 mm to 63 × 62 × 22 mm, with a mean of 54.6 × 41 × 19.6 mm.

Miscarried Microcore Preforms (21)

This group of artifacts is quite ambiguous. In a couple of cases, chunks of chalcedony were used to attempt microblade removal. The endeavor was aborted after bifacial thinning and platform development failed. Nonetheless, they all bear retouching or damage at some point along the edge of the "platform." However, it does not extend onto the "platform"—only down a lateral surface or face (figure 4.16).

These specimens were probably attempts at microblade cores (five were fire treated), but when the effort appeared hopeless, they were transferred, apparently to heavy-duty scraping or possibly to use as burins. This action was consistent with the frugal use of high-quality raw material at Dry Creek.

The maximum size range for these tools is 80 × 45 × 27 mm and the minimum is 33 × 11 × 12 mm. The mean dimensions are 49 × 34 × 19 mm.

Core Tablets (45)

Core tablets are a by-product of microblade technology. Technically, they are spalls produced by a burin blow directed at the top of the fluted end of the wedge-shaped microblade core. Such a blow truncates the plane of the core at a right angle and shears away all or part of the retouched platform. In some cases, this was done to remove impurities that impeded

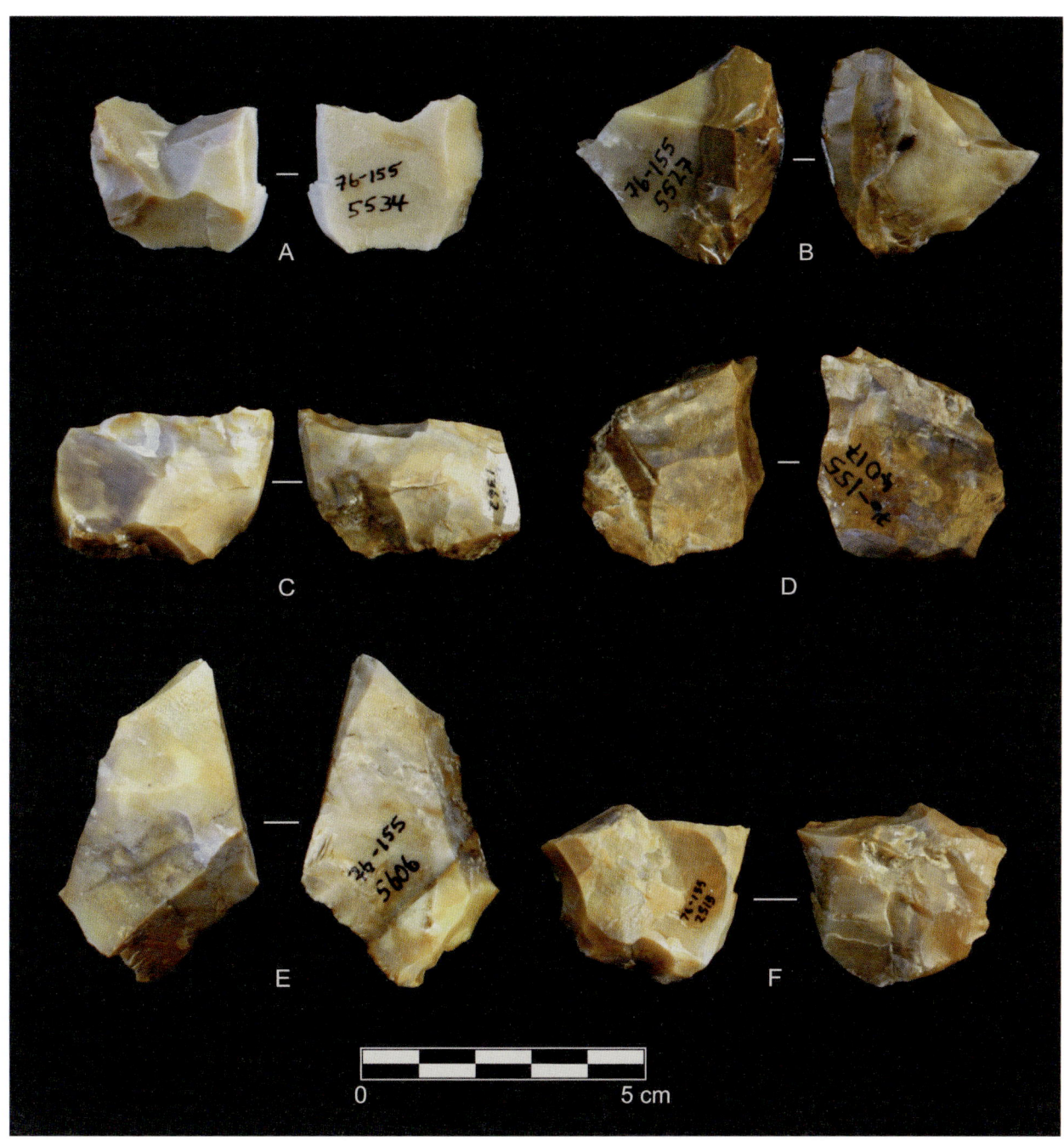

FIGURE 4.14. Aberrant microblade cores from Component II.

Figure 4.15. Microblade core preforms from Component II.

Figure 4.16. Miscarried microblade core preforms from Component II.

microblade detachment, while in others, the purpose was platform rejuvenation. All the tablets have some degree of retouch on the dorsal surface, but in many cases, the entire dorsal surface is heavily truncated by fine lateral pressure retouch. The tablets also display scrubbing and roughening across the edge where microblades were detached.

Almost all the examples represent well-executed symmetrical removals, but seven of these specimens are the result of blows that miscarried and took away only one side of the platform and a portion of the lateral surface of the core.

Of the core tablets, 45 are considered waste material, while 2 were transformed into transverse burins (cf. burins). Of the 45 waste tablets, 13 could be fitted to their respective cores (cf. wedge-shaped cores). The core tablets fall into the following raw-material groups: light rhyolite (13), gray chert (18), obsidian (4), chalcedony (1), brown chert (7), black chert (1), and a heavily fired opal-like rock (1).

Morphologically, the core tablets fall into two groups. Specimens in the first group (37; figure 4.17) have a generally rectangular outline with the exception that the proximal end is usually slightly convex and bears remnant microblade flutes. Also, the distal end can be quite irregular as a result of termination in a steep hinge fracture. Cross sections are normally

FIGURE 4.17. Microblade core tablets from Component II. Top view showing prepared platform remnant (left) and paired bottom view (right).

rectangular, since the lateral edges lie at 90° angles to the ventral and dorsal surfaces. These are the result of the partial removal of the core's platform. This type of platform removal results in a deep hinge fracture near the back of the core. The remaining portion of the platform is higher than that from which microblades are being removed. The size ranges for these pieces are 17 to 50 mm in length, 7 to 23 mm in width, and 3 to 16 mm in thickness. The mean values for these dimensions are 34 × 13 × 7 mm.

The second type of core tablet (8; figures 4.18, 4.19) resulted from the total removal of the prepared platform—in effect, the entire top of the core. In so doing, the blow sheared downward, and the upper portion of the rear of the core was cut away. The size ranges are 33 to 46 mm in length, 10 to 18 mm in width, and 7 to 28 mm in thickness. The mean values for these dimensions are 44 × 13 × 17 mm.

FIGURE 4.18. Microblade core tablets from Component II (A–D, lateral views of core tablets).

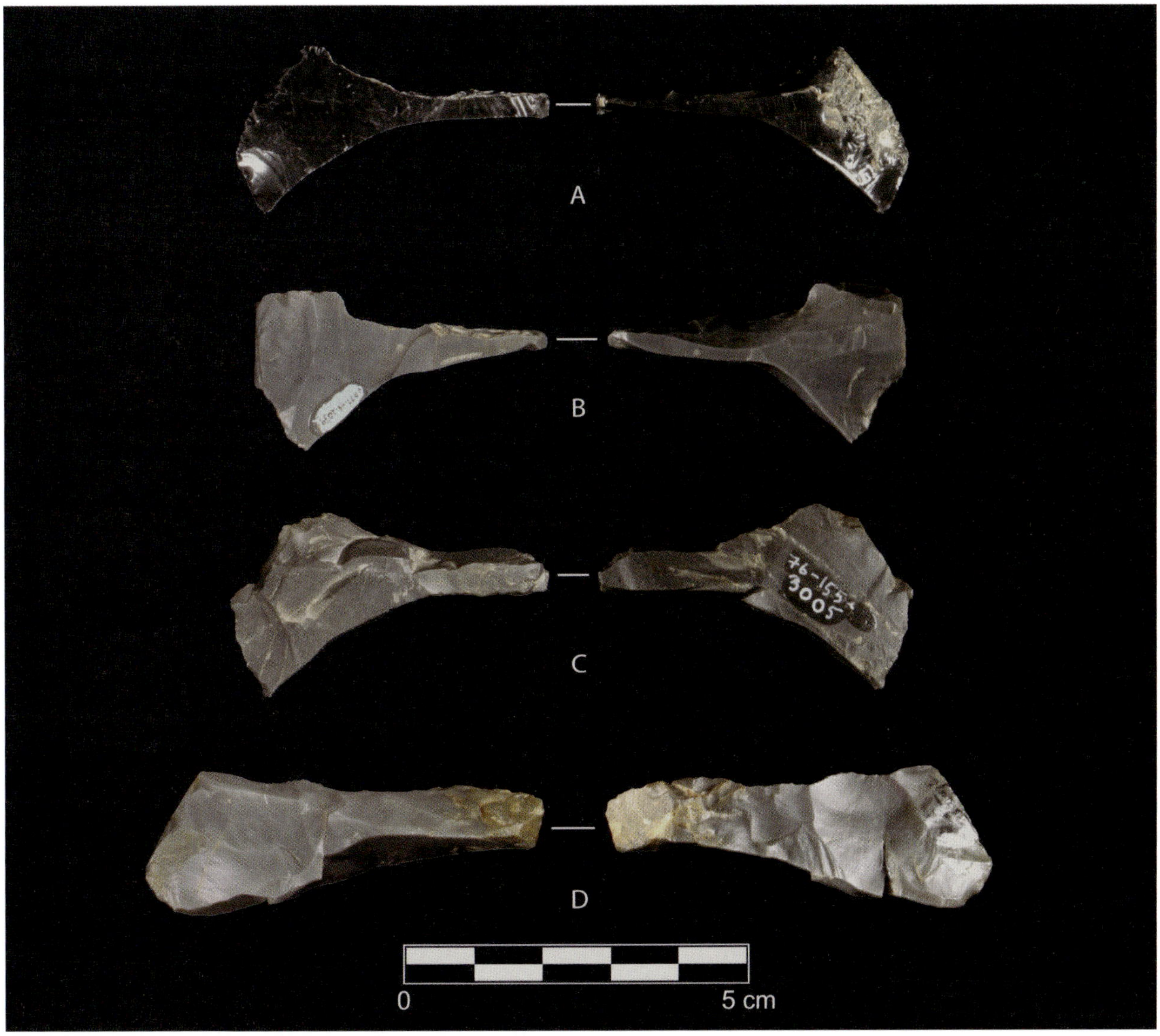

FIGURE 4.19. Microblade core tablets from Component II (top view of the tablets shown in figure 4.18).

Miscellaneous Wedge-Shaped Core Parts (24)

This group includes pieces of wedge-shaped cores resulting from "industrial accidents." They all fall well within the normal size ranges of the cores. Artifacts in this group include 4 sheared fluted surfaces (3 chalcedony, 1 gray chert), 2 sheared bases (1 chalcedony, 1 gray chert) and 18 unidentifiable pieces from the corpus of the core (7 gray chert, 11 chalcedony).

Microblades (1,772)

The entire purpose of the elaborate preparation and maintenance procedures employed in wedge-shaped microblade core technology was the production of microblades, of which 1,823 were recovered for Component II at Dry Creek. The sample for analytical purposes is 1,772, or 97.2 percent of the total sample. These were found in microblade clusters (see chapter 5), while the remainder (51, or 2.8 percent) occurred as surface finds, have poor provenience, or were isolated finds outside the defined limits of the microblade clusters.

In some instances, microblade raw material types could not be matched to a core. This information is summarized in table 4.2.

Considering the importance of chalcedony in both microblade and microcore production, very few core tablets of this material were recovered. Gray chert and rhyolite compare favorably across the categories, but there are no cores or core tablets for quartzite (64 microblades). A considerable number of brown chert microcore platform rejuvenation pieces compare curiously with low microblade recovery and an absence of microcores of this material type. The same is true of obsidian. One core tablet of opal was recovered, but no microblades or microcores of this material were discovered. The four green chert microblades might have come from a brown or gray chert microcore with greenish inclusions. Another possibility

TABLE 4.2. Raw material frequencies for microblades, core tablets, and microcores

Raw Material	Microblades		Core Tablets		Microcores	
	No.	%	No.	%	No.	%
Chalcedony	465	26.2	1	2	11	52
Gray chert	732	41.3	18	41	4	19
Rhyolite	362	20.4	13	29	5	25
Quartzite	64	3.5				
Brown chert	32	2.0	7	15		
Black chert	25	1.4	1	2	1	4
Obsidian	88	5.0	4	9		
Green chert	4	0.2				
Opal			1	2		
Totals	1,772	100	45	100	21	100

presents itself and focuses our attention on a methodological problem: these specimens might be burin spalls struck from one of the green chert burins. In terms of technology, burin spalls and microblades can be difficult to separate.

This outline indicates that in spite of the numerous examples of single core production and microblade detachment episodes recorded at the site, part of the record is still missing either as a result of sampling procedures or as loss due to bluff erosion. Also, the cores might have been taken elsewhere.

Only 184 (10.4 percent) of the microblades were complete. The remaining occurred as segments, with proximal segments accounting for 796 (45 percent), medial segments for 539 (30.3 percent), and distal segments for 253 (14.3 percent; figure 4.20).

A breakdown by raw materials is summarized in table 4.3.

It is impossible to distinguish the process of segmentation, since those segments that were purposely executed by the stoneworkers and those that were broken while detached from the cores cannot be typologically separated. The high frequency of incomplete microblade flutes on the cores is ample evidence that premature microblade termination constantly plagued the ancient stoneworker. This is an important matter, since it is assumed that the segments, especially medials and possibly proximals, were the desired end product of the entire manufacturing sequence. These, it is assumed, were employed as inset blades and set into narrow grooves or slits incised into the edge of bone or antler projectile points or knives (see the following).

The complete microblade is essentially useless for inset purposes, since it is almost always curved. Apparently, the reason for intentional segmentation was the removal of the curvature to create the straightest possible cutting edge. Such edges, in all likelihood, would be found on the medial segments, and one would expect the medial segments to have the lowest representation in a microblade count.

An examination of the frequencies of microblade segments shows that the lowest frequencies are of complete and distal segments. Low frequencies of complete microblades are expected, since complete blades were broken purposely, or accidentally, during production.

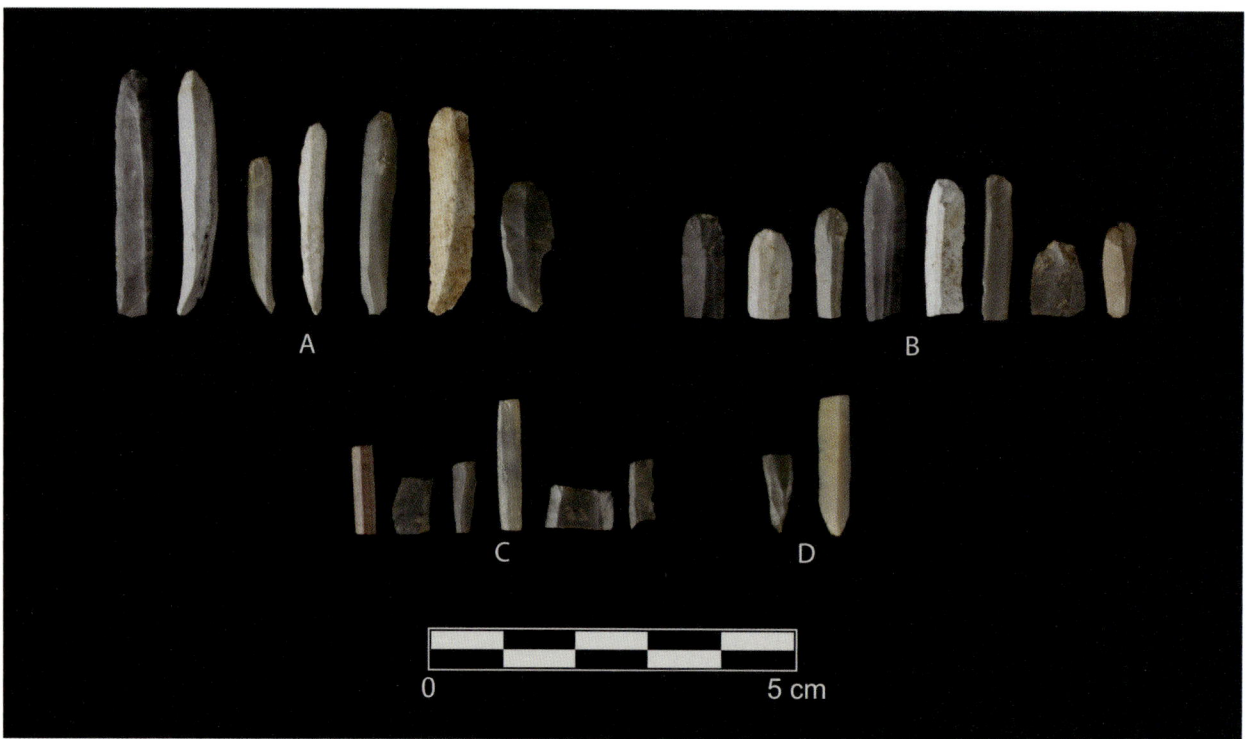

FIGURE 4.20. Microblades from Component II (A, complete; B, proximal; C, medial; D, distal).

TABLE 4.3. Raw material frequencies for microblades and microblade segments

Raw Material	Complete	Proximal	Medial	Distal	Total
Chalcedony	88	207	92	78	465
Gray chert	66	334	236	96	732
Rhyolite	19	169	134	40	362
Quartzite	2	21	29	12	64
Brown chert	4	11	12	5	32
Black chert		9	13	3	25
Obsidian	4	42	22	19	88
Green chert	1	2	1		4
Totals	184	796	539	253	1,772

Proximals and distals should be about equal, but at Dry Creek, there are three times as many proximals as distals. In fact, the lower number of distals might be the result of the microblades snapping during detachment, leaving the distal portion on the core. Obviously, this process would automatically increase the frequency of proximal segments. Most of these proximals must have been usable for segmentation, and considering that the stoneworkers could snap as many as four or more for medials from each complete microblade and at least half that many from proximal segments, the actual number of medial segments should be from two to four times the number recovered. While the actual number of medial segments is high (539, or 30.3 percent), one would expect that far more were produced, possibly 1,000–2,000. Viewed from this perspective, the number of medial segments suddenly appears low when compared to production levels, as it should if this segment was the most functional.

When considering metric variability, it should be noted that some northern archaeologists consider maximum microblade width to be the crucial variable (e.g., Cook 1968; West 1967). Following this procedure, width measurements can be briefly summarized. There is some variability in mean width among the different raw materials—for example, the widest complete microblades are rhyolite and obsidian, whereas microblades from the remaining lithic groups are narrower (table 4.4).

The mean width for all microblades from Dry Creek is 3.8 mm, which compares well with the mean width of the microblade flutes on the microcores, where the mean width is 3.6 mm. The only lithic groups that really stand apart are black and green chert, but in both instances, the samples are very small. The remaining specimens are quite uniform in width.

Only 193 (11 percent) of all microblades bear any evidence of retouching. For the most part, this occurs as irregular nicking of the fragile edges. Thirty-two (16.6 percent) complete specimens, 81 (42 percent) proximals, 56 (29 percent) medials, and 24 (12.4 percent) distals exhibit this treatment. The causes of this nicking might not be cultural, since excavation and measurement could be responsible. Some of this treatment might result from utilization, but only one medial segment of an obsidian microblade nears extensive unifacial retouching along both edges of its dorsal surface.

It is necessary, at this point, to return to the problem of microblade function. As stated earlier, the presumed use of the microblade was to obtain segments with straight edges that could be set into laterally grooved bone or antler points. This idea is also suggested by Anderson (1970b) for the Akmak microblades and by West (1967) for the microblades from the Donnelly Ridge site and the Teklanika River sites.

However, there is not universal agreement. Cook (1968) has presented data to support the argument that Campus-type cores (i.e., wedge-shaped cores with stepped platforms) were not cores but burins and that the microblades were actually waste material. He supports this idea by ascribing the crushing and flaking on the lateral margins of the platforms to the core

TABLE 4.4. Mean width measurements (mm) for complete microblades and microblade segments

Material	Complete	Proximal	Medial	Distal
Chalcedony	3.6	4.0	3.9	3.0
Gray chert	4.1	4.2	3.6	3.6
Rhyolite	4.9	4.4	3.8	4.3
Quartzite	3.8	4.4	3.4	3.5
Brown chert	3.8	4.0	3.1	5.1
Black chert		5.3	6.0	3.3
Obsidian	4.6	3.9	3.4	3.5
Green chert	2.7	3.4		

or tools being used as a burin. Furthermore, he feels that the low incidence of retouching on microblades indicates that low numbers were utilized.

It has been argued in this report that platform-edge damage results from securing the core for microblade detachment. It is agreed that some objects from Dry Creek that look like microcores were used as burins (cf. "core-burins" in this report), but not all microcores were used in that manner. Lack of retouching on microblades should not be used to infer lack of utilization.

One of the troublesome issues in dealing with this sort of controversy is the Alaskan data base: no interior Alaskan site has produced a grooved bone/antler point.[6] There is evidence indicating that the composite inset technique was employed in manufacturing weapon tips at Trail Creek Cave No. 2 (Larsen 1968), but here the age is much too young to be pertinent to our discussion.

The Siberian data are direct but distant. In the third cultural horizon (14,450 ± 150 ^{14}C yr BP [LE-628; 17,610 cal yr BP]) at the site of Kokorevo I on the middle Yenisei River, Abramova (1967: figure 4) reports and illustrates a fragment of an antler spearpoint with one lateral groove and microblade segments intact. The antler point is flattened with an oval cross section and has a sharpened tip. The base is missing. The specimen is 110 mm long, and its greatest width is 16 mm. At 15 mm from the tip, a groove begins and continues to the broken base. The groove is from 1 to 2 mm wide and up to 3 mm deep. At 33 mm from the tip, there is a tiny fragment of a microblade and two more small fragments next to it that were apparently broken during use. The actual continuous flint edge begins at 51 mm from the tip and is composed of six sections of microblades that range in length from 6 mm to 97 mm. They are 4 mm wide, have a triangular cross section, and bear no traces of secondary working.

A similar find was reported from Afontova Gora III by Sosnovskii, where a point with three intact, unretouched microblade fragments was found (Abramova 1967). An additional discovery was reported by Gromov from Afontova Gora II, where a bone point with two unretouched microblades in the groove was recovered (Abramova 1967).

In addition to these direct examples of the inset technique, both unilaterally and bilaterally grooved bone/antler points lacking microblades occur at Kokorevo I and Kokorevo II (Abramova 1967) and, in fact, are a common occurrence in the late Paleolithic of South Siberia (Chard 1974).

The bone/antler inset technique occurs within lithic complexes that feature wedge-shaped microblade cores, microblades, large side scrapers, end scrapers, and transverse burins, a tool complex that is essentially identical to the early Alaskan microblade complexes. There are differences—especially in the presence and absence of bifacial technology—that will be discussed later, but the overall similarities are sufficient to assume that a cultural-historical continuum existed in Siberia and northwestern North America at the end of the Pleistocene and that microblades were manufactured for the purpose of obtaining insets for composite tools and possibly for other functions as well.

Burins

Burins from Component II at Dry Creek are represented by 28 specimens. While many burins possess the same nonmetric, technological attributes (platform preparation sequence) as a microcore, particularly those made on flakes, they can be distinguished functionally.

Those specimens classified as burins bear both macroscopic and microscopic wear patterns. Normally, the edges of the burin facets show fine chipping, crushing, or microscopic damage. It is interesting that little use wear can be isolated at the junction of platform and facets, although burin blow damage might well obscure such evidence.

Along with microcores on flakes, the burins form one end of a technological continuum, with "core/burins" lying at midpoint on an imaginary curve and bifacial microcores occupying the opposite extreme. While the extremes (burins and microcores) are obvious, microcores on flakes are problematic and can only be separated from burins by the absence of edge damage.

The Dry Creek burins can be divided into four groups: burins on snaps ("single blow burins"; 10), dihedral burins (ordinary burins; 3), angle burins (2), and transverse burins (13). Each group is discussed in the following sections.

Burins on Snaps (10)

These pieces were formed by a burin blow or blows struck on a snap or hinge fracture of a flake or other suitable piece (figure 4.21). In fact, three were made on core tablets, utilizing the distal hinge fracture as a striking platform (figure 4.21A, C).

The remaining examples (figure 4.21B) were made on flakes with the burin facets confined to one edge and running parallel to the axis of the flake. One of these has a well-retouched scraper edge lying opposite the burin facets. It also displays use wear on the base of the flake. Gray chert, brown chert, and black chert served as raw materials for these specimens, and measurements range from 13 × 10 × 3 mm to 40 × 39 × 7 mm, with a mean value of 27.3 × 18.5 × 5.2 mm.

Dihedral or Ordinary Burins (3)

Of these three burins (figure 4.22A–C), two were made on flakes of gray chert. One of these (figure 4.22B) appears to have been made on a bifacial thinning flake. Brown chert served for the remaining specimen. Dihedral burins are distinguished by a burin facet, which functions as a platform for small removals. The platform and the spall facets form an intersection at the corner of the flake.

On one specimen (figure 4.22A), the same technique of platform preparation was utilized as one would find on a microcore: initial lateral preparation, removal of preparation by burin blow (tablet removal), and detachment of burin spall from the platform. At the opposite end

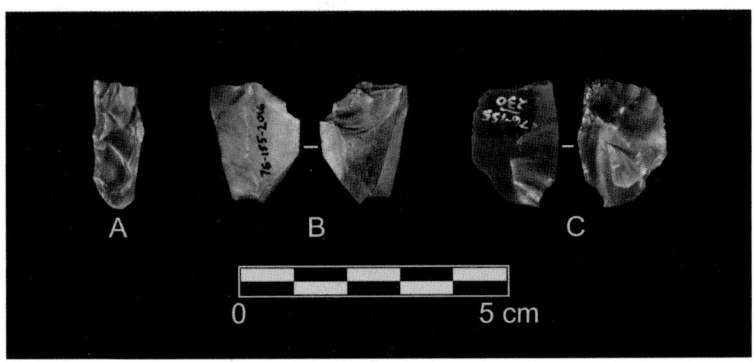

FIGURE 4.21. Burins on snaps from Component II.

FIGURE 4.22. Burins from Component II (A–C, dihedral burins; D, angle burin).

of the fluted surface, there is a notch that was possibly intended to control the length of the spalls. This specimen might be a core, although damage along the edge of the fluted surface indicates use as a burin—or possibly a scraper—on hard, resistant material.

A second flake (bifacial thinning flake) has a dihedral burin at both ends (figure 4.22B), and damage is visible along all edges.

The measurements range from 22 × 12 × 4 mm to 42 × 28 × 9 mm, with a mean of 35 × 19.6 × 7 mm.

Angle Burins (2)

The first specimen in this group is a multiple-angle burin with spall facets running down both lateral edges from a transverse truncation at one end of the piece. The opposite end bears a small notch at the lower right corner, and spalls were also struck along this edge, utilizing the notch as a platform. One spall struck down the left edge traversed the lower left corner of the flake and terminated at the notch. All dorsal edges display damage. The second angle burin (figure 4.22D) was made on a core tablet. The burin blow was struck on a notch formed at the distal end of the tablet. It bears use wear at one locality. Both were manufactured on gray chert flakes and measure 27 × 21 × 4 mm and 33 × 10 × 5 mm, respectively.

Transverse Burins (15)

Transverse burins were made on flakes (13) and core tablets (2). Their distinguishing characteristic is a facet(s) resulting from a blow struck from the side and running transverse to the axis of the piece. In all our examples (figure 4.23), the facet(s) traverses the distal end of the flake or spall. The striking platform for the burin blow can be a straight truncation (9), snapped edge (2), convex truncation (1), or an actual notch (3). Five have single facets, and the remainder have two or more.

FIGURE 4.23. Transverse burins from Component II.

The raw materials used for manufacture are gray chert (5), brown chert (6), jasper (1), brown chalcedony (1), rhyolite (1), and moss agate (1). Measurements range from 18 × 14 × 4 mm to 32 × 31 × 11 mm, with a mean of 23.4 × 23.8 × 5.8 mm.

Core-Burins (8)
These specimens are very difficult to classify (figure 4.24). Technologically, they were prepared in the same manufacturing sequence as a microblade core. A platform was prepared by either retouching one or two edges of a flake (3 specimens) or striking a burin blow on the edge (4 specimens). From these platforms, burin blows were struck. The resulting artifact resembles a small, chunky, wedge-shaped core. A distinguishing characteristic of these specimens is minor crushing and use wear along one edge of the burin facet(s), possibly indicating that these specimens were used as end scrapers. They were made on thick flakes of chalcedony (3), green chert (1), gray chert (1), brown chert (1), and rhyolite (2). Measurements range from 20 × 17 × 5 mm to 32 × 29 × 14 mm, with a mean of 24.5 × 24 × 7 mm.

Burin Spalls (35)
These distinctive spalls are distinguished from microblades by their generally smaller size and triangular to rectangular cross sections. However, burin spalls lacking evidence of use as a burin edge can technologically grade directly into microblades. There are probably many more true spalls in the collection, but only those that could be definitely identified have been classified. Information on the spalls is summarized in table 4.5.

Of the 35 burin spalls, 22 have utilized edges and are broken down by raw material in table 4.6.

FIGURE 4.24. Core-burins from Component II.

TABLE 4.5. Raw material frequencies for burin spalls

Material	Complete	Proximal	Medial	Distal	Total
Rhyolite	1			1	2
Obsidian	1				1
Brown chert	7	2			9
Chalcedony	10	1		1	12
Gray chert	5			6	11
Totals	24	3	0	8	35

TABLE 4.6. Raw material frequencies for burin spalls with utilized edges

Material	Complete	Proximal	Medial	Distal	Total
Obsidian	1				1
Chalcedony	10	1			11
Gray chert	5			5	10
Totals	16	1	0	5	22

The spalls are very uniform in size, regardless of raw materials (figure 4.25). The range for complete spalls is 15 × 2 × 1 mm to 34 × 8 × 5 mm. The mean size is 24 × 3 × 2 mm.

The proximal sections of the spalls range from 14 × 2 × 1 mm to 16 × 4 × 3 mm. The mean size is 15 × 3 × 2 mm.

The distal spall fragments range from 6 × 2 × 1 mm to 20 × 6 × 3 mm, with the mean value for the dimensions being 14 × 4 × 2 mm.

FIGURE 4.25. Burin spalls from Component II.

Projectile Points

There is one complete specimen of a projectile point from Component II and six projectile point bases. The six bases represent a uniform group, although the finished shape of these specimens is unknown. The basal portion of the complete specimen is unlike the six basal fragments.

Projectile Point (1)

In an earlier draft of this report, this specimen was mistakenly assigned to Component I. This very unfortunate situation was discovered quite late in final manuscript preparation. This mistake was due either to incorrect bagging in the field or to a cataloging error.

This single complete specimen of a point (figure 4.26) has a lanceolate outline. The point was produced by the bifacial reduction of a gray chert flake. The edges are excurvate, slightly serrated, and ground on the lower one-quarter of the point. The unifacially retouched base is asymmetrically concave, and the spurs at both corners of the base are highly polished. The tip is sharp but well pointed. It measures 34 × 13 × 3 mm. During analysis, Del Bene noted that this point might never have been used, unless the blunted tip represents an impact fracture. Unfortunately, the evidence is equivocal.

FIGURE 4.26. Projectile point from Component II.

This point appears to be fluted on one face, although careful examination of the facet outlines demonstrates that it was actually formed by the termination of lateral thinning flakes that have truncated a previous basal thinning facet. While the outline of the point, shape at the base, and apparent fluting would fit into a continuum of fluted points in Alaska and farther south, we would caution against implying a historical link between this specimen and Clovis points. At Dry Creek, it is an isolated and technologically marginal artifact.

Projectile Point Bases (6)
Six basal portions of projectile points have been isolated at Dry Creek. Raw materials for these bases are dark rhyolite (1), chalcedony (1), degraded quartzite (2), pumice (1), and light rhyolite (1). It is unfortunate that complete specimens could not be located, since it is impossible to determine exactly to which type of points they should be assigned. There are only a few attributes available, but these should serve to both narrow the range of possibility and provide a tentative suggestion as to which early projectile point types the Dry Creek specimens are most typologically similar. The specimens range in size from 19+ × 17+ × 6+ mm to 45+ × 27+ × 8+ mm.

These basal fragments are from some type of stemmed or lanceolate point. They have been prepared with fine bifacial retouch and have symmetrical lenticular cross sections. The edges are straight and expanding, and the bases are straight on four examples and slightly convex on two bases. Stem angles range from 13° to 43° with a mean angle of 27°. The edges are ground on five of the specimens. Only one lacks grinding, and there is evidence that resharpening was attempted. Hafting wear is exhibited on all specimens except the one lacking grinding.

All the point bases appear to have been broken in the haft. The length of the specimen is probably a good indicator of the depth of hafting (figure 4.27B–E). Two of the bases display simple hinge fractures. One of these shows use wear on one edge of the fracture, indicating that this piece saw future service as a scraper or burin. The other four bases display more complex fracturing; on two of these, it can be attributed to impact damage. One of these pieces (figure 4.27B) had an edge sheared away in addition to snapping. Another (figure 4.27C) was sheared twice and displays several burin-like fractures at the broken end. The uppermost sheared fragment was not recovered. The remaining bases display complex fracturing that presumably resulted from impact.

This category of artifacts displays a uniform, albeit incomplete, set of attributes, which provide the only basis for attempting comparisons with point types that might be related historically.

The basal fragments possess no attributes that would make comparison with fluted point complexes fruitful. While the age of Component II at Dry Creek is broadly synchronous with fluted point complexes on the Great Plains of temperate North America, the point bases find their closest similarities to certain point styles that fall under the term *Plano*. More specifically, they appear to be very similar to the basal extremities of Hell Gap points. In particular, the morphology of the base, the thickness of the stem, and the minimum width of the stem compare favorably with Hell Gap points from the Casper site (Frison 1974: table 1.4). Even the longer examples of the basal segments of the points are very similar in length (Frison 1974: figures 1.35–1.43). In comparing stem angles, it is obvious that three of the Dry Creek

FIGURE 4.27. Projectile point bases from Component II.

specimens fall within the range of Colby-site Hell Gap points. However, three Dry Creek specimens exceed the range (30°–40°).

The Dry Creek specimens also show similarity to Haskett points from the upper Snake River Plain of Idaho (Butler 1978). Although the stem angles of the Dry Creek specimens appear to overlap slightly at the upper range of Haskett points, the minimum basal width of Haskett points is narrower than that of the Dry Creek specimens.

One further observation should be noted: the flaking technique displayed on the Dry Creek specimens differs markedly from both Haskett and Hell Gap points. This is probably a stylistic rather than a functional attribute.

In summary, the Dry Creek specimens are probably bases of projectile points with expanding stems. They are broadly similar to Plano points, and of these, they appear closest to Hell Gap points. However, without the tips, certainty is impossible.

Projectile Point Tips (2)
Two small, bifacial, triangular segments with lenticular cross sections are probably tips of projectile points. They are flatter and have tip angles that are narrower than those observed on knife categories. One is chalcedony and one is chert. They measure 9 × 8 × 2 mm and 30 × 23 × 3 mm.

Knives

Twenty-six specimens from Component II at Dry Creek are defined as knives on the basis of wear patterns on the tips and morphological asymmetry. Originally, some of these specimens might have been projectile points that became broken or damaged. Many of the tools in this category have been reworked and/or repaired, possibly from pieces of damaged points.

They can be subdivided on the basis of morphology into seven groups: spatulate or weakly stemmed, elliptical, oblong, asymmetric triangular, oval, ovate, and lanceolate. In addition to these, there are five bases, three tips, and three midsections that can be assigned to the knife category. Also, there are three bifaces (one discoid and two deltoid) for which a function is not apparent and five miscellaneous bifaces.

Oblong Knife (1)

This single specimen of a black chert biface is at the pressure retouch stage. It has an oblong or crudely rectangular outline with slightly excurvate edges and a rounded tip. The base retains the hinge fractures of the flake on which the tool was made. It has a symmetrical, lenticular cross section. The ventral surface displays a longitudinal flake facet that looks like a flute. However, it is simply a previous flake facet that was truncated by later trimming. The edges are ground from the base to about the midpoint, but there is no obvious sign of hafting. Isolated portions of the edges above the grinding indicate use on a cut-yielding material. This piece measures 65 × 25 × 9 mm.

Asymmetric Triangular Knife (1)

This little biface was made on a flake of translucent chalcedony and displays an asymmetric triangular outline with a lenticular cross section (figure 4.28). There is a slight twist running from the base to the tip. One edge and the base are straight, while the opposite edge is slightly excurvate. No use or hafting wear is evident. A series of hinge fractures near the tip and impurities in the stone along one edge made further thinning impractical. It seems that this piece was discarded. Its dimensions are 33 × 21 × 5 mm.

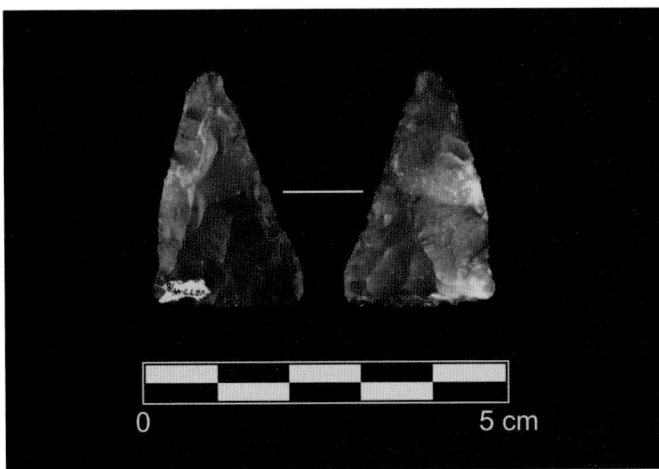

FIGURE 4.28. Asymmetric triangular knife from Component II.

Small Spatulate or Slightly Stemmed Knives (8)

This category is represented by one complete piece, three that are essentially complete, two bases, one midsection, and one tip. All are bifacial and appear to have been made on flakes. Dark rhyolite (1), light rhyolite (5), gray chert (1), and obsidian (1) served as the raw materials. They have a spatulate outline and a narrow base that expands to a point of maximum width that is closer to the tip than the base. This expanding base has the appearance of a weak stem on one specimen (figure 4.29A). The tips have been thinned, while the basal areas retain a thicker cross section. The cross sections are symmetrical and lenticular, although thickness measurements indicate different degrees of thinning. Complete or nearly complete specimens have a size range of 46 × 24 × 8 mm to 54 × 26 × 9 mm, with a mean of 50 × 25 × 9 mm. Minimum width at the base is 7 mm. Base angles range from 28° to 47°, with a mean of 39°.

The one complete specimen (figure 4.29C) displays fairly fresh edges. It was probably abandoned after the unsuccessful attempt at removing a knob on the dorsal surface near the base. This impediment would surely have been a problem if hafting had been attempted.

Two of these knives are missing the basal extremity and display extremely worn and ground edges extending from the base to the point of maximum width. One of these (figure 4.29A) has use wear on both edges of the tip, whereas the other has wear only on one edge. Its opposite edge is very fresh due to resharpening. Both of these knives bear heavy hafting wear. The missing basal extremities might have resulted from snapping in the haft.

The base of the remaining early complete specimen was probably snapped off in the haft (figure 4.29B) and appears to have been in a phase of resharpening. Both edges of the base are ground, but the edges of the tip are very fresh. However, during resharpening, nearly the whole ventral surface of the tip was sheared away, and while there was an attempt to recover the edge, this accident probably rendered the piece useless.

It appears that these knives were applied to substances of varying hardness. Two of the knives (figure 4.29A) were used on a cut-yielding material, while the third knife (figure 4.29B), judging from crushing on the edges, was used on a cut-resistant substance.

Three bases have been assigned to this group on the basis of morphology. Two are complete and indicate that the bases of these knives were narrow, straight, sharp, and unground. The third specimen is missing the basal extremity. All three have both edge grinding and hafting wear. Two of the basal sections (figure 4.29D) saw further use as burins or scrapers on some cut-resistant material, an action that formed heavily crushed shallow concavities along one edge of the snapped end. These pieces were made from obsidian, light rhyolite, and gray chert and have a dimensional range of 16+ × 17+ × 5+ mm to 37+ × 25+ × 7+ mm. Minimum width at the base is 6–7 mm. Base angles range from 35° to 43°, with a mean of 39°.

The dark rhyolite midsection (figure 4.29E) is heavily ground on both edges and shows hafting wear. Its base was broken by a clean snap, but a more complex hinge fracture removed the tip. It measures 18+ × 23+ × 7+ mm.

The light rhyolite tip shows heavy dulling and use polish on both edges and appears to have been resharpened at least once. It measures 16+ × 16+ × 6+ mm.

Elliptical Knives (7)

Five complete specimens and two tips comprise this group (figure 4.30). Of the five complete specimens, one was made from a piece of light rhyolite, two from black chert, one from

FIGURE 4.29. Small spatulate knives from Component II (A–C, small spatulate knives; D, small spatulate knife base; E, small spatulate knife midsection).

quartzite, and one from pumice. Measurements range from 55 × 21 × 8 mm to 119 × 32 × 13 mm, with a mean of 83 × 27 × 11 mm.

These knives have elliptical or bipointed outlines and are fairly symmetrical, with slightly rounded bases. The cross sections are symmetrically lenticular except for one specimen that is plano-convex. The tip areas have been thinned, leaving the greatest thickness either at midpoint or close to the base.

One of these knives has edge grinding, hafting wear, and heavy edge wear near the tip. One specimen was probably discarded after flaking destroyed one edge. Another knife (figure 4.30A)

FIGURE 4.30. Elliptical knives from Component II (A–C, elliptical knives; D–E, elliptical knife tips).

bears no trace of edge grinding or hafting wear but does have minor edge modification from application to cut-resistant material. It might have been hand held. This is also the case with the last two complete specimens (figure 4.30B–C). Some edge damage is evident, but the coarseness of the raw material makes this difficult to determine. One of these (figure 4.30C) was broken, although the two pieces were found together.

The knife tips made of quartzite were both applied to cut-resistant materials, and one (figure 4.30E) that retains edge grinding up to the snap appears to have snapped in the haft. They measure 43+ × 27+ × 8+ mm and 76+ × 29+ × 19+ mm, respectively.

Oval Knives (5)
This category contains large bifaces that have been manufactured on thick flakes of brown chert (1), argillite (1), light rhyolite (1), and dark rhyolite (2).

They measure from 95 × 39 × 12 mm to 128 × 69 × 37 mm, with a mean of 113 × 57 × 22 mm. They have elongate oval outlines and lenticular cross sections. All the specimens are unfinished, as they retain patches of cortex. Three have thick, flat butts, and another has an irregular edge. One of these bifaces bears evidence of use along some portions of its edges, while other portions were resharpened. Two oval bifaces that look like preforms of some sort also display heavily used edges. One (figure 4.31B) bears widespread use wear on both the edges and faces that might have resulted from rotating the piece in a haft. It was probably used on a cut-yielding substance. The other (figure 4.31D) displays no evidence of hafting but shows edge damage or dulling from use on a cut-resistant material.

One interesting piece was broken, and its two halves were recovered from different parts of the site. Prior to breaking, it appears to have been hand held and used to cut both yielding and resistant materials. After the break occurred, an attempt was made to further reduce one half of the piece, but the effort was fruitless.

The remaining specimen is in an early bifacial reduction stage (figure 4.31C). Even this piece shows isolated instances of use wear that probably resulted from working a relatively yielding substance.

The evidence indicates that these larger bifaces were used as cutting implements at the stage of production in which they were found. This observation is not intended to deny the possibility that further bifacial reduction might have been intended in order to produce more refined tools.

Ovate Knives (3)
Knives assigned to this category were manufactured on bifacially reduced flakes of gray chert (1), dark rhyolite (1), and diabase (1). Their dimensions range from 79 × 43 × 16 mm to 99+ × 54 × 19 mm, with a mean of 89 × 52 × 18 mm (figure 4.32).

These knives have an ovate outline. The bases are strongly convex, and the edges are excurvate and converge to form a tip. Their cross sections vary from lenticular to plano-convex.

Two of the specimens display signs of use. On both, the edge of the convex base is even and finished. Grinding is apparent on this edge, and wear—which probably indicates hafting—is widespread on both of the faces over the lower one-third of the tool. The converging edges of the tip are sinuous and, while showing use-wear polish, also display clear signs of resharpening. Damage to the edges is quite minimal, and the use wear present resulted from contact with a resistant material. One of these knives (figure 4.32A) is missing the tip, and the other (figure 4.32B) still retains a striking platform on one edge of the tip.

The third specimen in this category does not appear to have been hafted. A striking platform lies on one edge of the tip, and a thick platform can be seen on one side of the base. One

FIGURE 4.31. Oval knives from Component II.

edge was badly damaged during flaking and then abandoned. The opposite edge is unfinished but does have isolated areas of use wear resulting from working a resistant material.

Lanceolate Bifaces (2)
These specimens are large, relatively crude bifaces that have evidence of use on their edges (figure 4.33). They have broad convex bases and excurvate edges that taper to a tip. The cross sections are variable on both specimens but are generally lenticular.

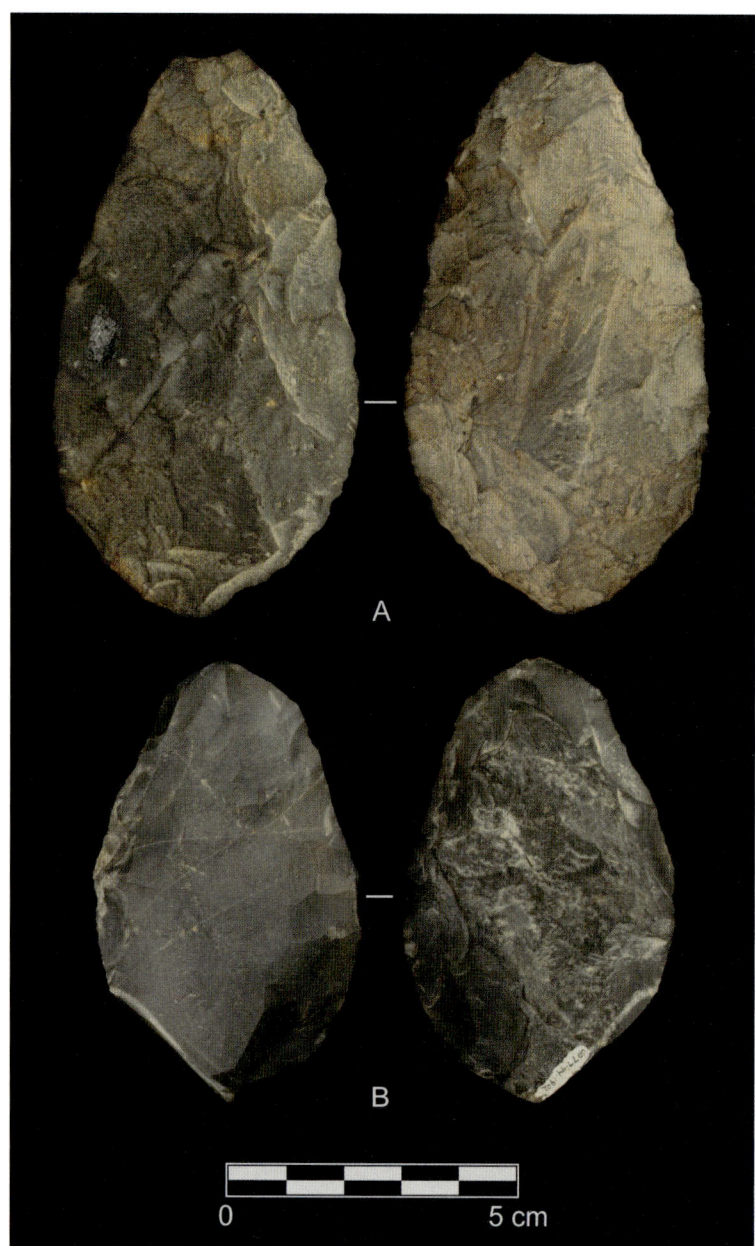

Figure 4.32. Ovate knives from Component II.

The first is an elongate biface that displays a convex base (figure 4.33B). The edges converge to form this tool. The implement still retains part of the cortex on the dorsal face and a portion of the fracture on the ventral face near the tip. An attempt to flatten a thick bulb of percussion at the base was partially successful.

There are no signs of hafting, but possible use wear can be seen on both edges, although the coarseness of the material makes this determination very difficult. The measurements of this biface are 176 × 66 × 28 mm.

The second biface in this category, which was found in two pieces, is a larger version of the previous specimen (figure 4.33A). It was manufactured on an enormous flake of degraded

FIGURE 4.33. Lanceolate bifaces from Component II.

quartzite by bifacial percussion flaking. One edge is fairly straight, but the other edge and the base are irregular and sinuous. The tip was only partially formed and is fairly thick. Isolated use wear is evident on the tip, the lower part of the edges, and the base. The tool could have been used for working through the joints of an animal, or it could have functioned as a digging implement. It appears to have been held in the hand and was broken in two during early stages of use. It measures 257 × 119 × 41 mm.

Deltoid Biface (1)
This piece was an attempt at bifacially reducing a flake of argillite, and it was not very successful. The flaking process produced a small biface that remained incomplete and was probably discarded. No traces of use wear could be detected. It measures 58 × 53 × 25 mm.

Discoidal Biface (1)
In this case, an attempt was made to bifacially reduce a flake of gray chert. Impurities in the stone prevented further refinement. However, isolated areas of the edge show evidence of utilization on a hard, resistant material. No evidence of hafting can be detected; presumably, it was hand held. It measures 51 × 42 × 16 mm (figure 4.34A).

FIGURE 4.34. Bifaces from Component II (A, discoidal biface; B–D, base fragments of bifaces).

Base Fragments of Bifaces (3)

These basal fragments might have been part of ovate or oval bifaces (figure 4.34B–D). Only one appears to have been finished (figure 4.34D). They were made of pumice, dark rhyolite, and argillite. The finished piece was made on a thick flake of dark rhyolite and retains a patch of cortex on the dorsal surface at the base. The edges are straight and well trimmed. There is some edge abrasion, and the crests of facial flake facets show polishing from the haft. It measures 61 × 52 × 20 mm.

Another of these specimens (figure 4.34B) is a fragment of either an ovate or an oval biface that was abandoned in a bifacial reduction stage. It was made on a flake of argillite and does show edge wear and possible evidence of utilization of hafting. The third, on pumice, was probably the base of a lanceolate knife (figure 4.34C). It measures 68 × 32 × 9 mm.

Biface Tips (5)

These tips are of degraded quartzite (figure 4.35A–D). Three are biface fragments that were probably finished and subsequently broken in the haft (figure 4.35B-D). They have straight edges with varying degrees of use wear, and two of the tips have hafting wear on the faces. Some minor resharpening can be seen on all the specimens.

The fourth tip (figure 4.35A) has a sinuous edge formed by bifacial flaking, and no final trimming is evident. But it was used and shows cut-yielding wear along both edges. It also displays hafting wear on both faces. It was also probably broken in the haft. The last tip snapped along a vein of impurity in the stone. It displays some minor use wear and was probably broken early in its career. The size ranges are from 55 × 46 × 10 mm to 107 × 57 × 25 mm, with a mean of 70 × 51 × 14 mm.

Biface Midsection (1)

This light rhyolite biface midsection was snapped at both ends. One edge is heavily worn through use on a cut-yielding material, while the opposite edge is fresh and appears to have been in the process of resharpening. Hafting wear is present on both faces. Its measurements are 49 × 50 × 16 mm.

Miscellaneous Bifaces (5)

This category includes objects that display some degree of bifacial flaking but do not fall into established formal categories (figure 4.36). Three of these (figure 4.36B-C; dark rhyolite [1], banded rhyolite [1], and degraded quartzite [1]) have edges formed by bifacial percussion flaking. While these edges are crude and sinuous, they do display abundant evidence of nicking and abrasion and have undergone substantial utilization as cutting implements. The material being processed was fairly resistant and durable. Heavy wear on the crests of facial flake facets can be found over most of the surfaces of these tools, and this is interpreted as having resulted from rotation in a haft. In these examples, the haft probably was designed to cover one edge of the tool, leaving the opposite edge free for use.

Another piece resulted from an attempt at bifacially reducing a degraded quartzite flake. Most of the flaking was directed at the dorsal surface and resulted in so many deep hinge fractures that the effort was abandoned.

The last item in this group (figure 4.36A) is a thick piece of light rhyolite that still has a patch of cortex along one edge. One end has been flaked to produce a thick, short, broad-angled (slightly pointed) cutting edge. The edge is fresh and bears no serious damage or dulling. The tool recalls a modern splitting wedge. Measurements for this category range from 90 × 50 × 15 mm to 113 × 68 × 36 mm, with a mean 102 × 58 × 24 mm.

FIGURE 4.35. Biface tips from Component II.

Heavy Percussion Flaked Implements

These large percussion flaked tools are very crude with respect to both appearance and techniques of manufacture. Of these, 37 were made on water-rounded river cobbles and 10 were made on flakes struck from cobbles. The cobbles used for tool manufacture are of degraded quartzite (16), rhyolite (16), sandstone (12), and a poor-quality brown chert (3), all of which are available in the debris apron of the site and in the bed load of Dry Creek.

FIGURE 4.36. Miscellaneous biface fragments from Component II.

These tools could have been made on the spot and discarded quickly, since the working edges undoubtedly dulled rapidly during use. Availability of raw material and ease of manufacture resulted in a relatively large number of these implements in the site, most of which were found in a tight concentration together with the debris resulting from their manufacture (see chapter 5). This suggests a very specific activity conducted in a limited area.

These implements are divided into the following categories: cobbles with lateral working edges, cobbles with working edges on the end and side, miscellaneous cobble artifacts, and miscellaneous large flake artifacts.

Cobbles with a Lateral Working Edge (7)

These tools have a crescentic outline and a cuneate cross section. One thick edge retains the cobble cortex, and the opposite side has been shaped by bifacial percussion step-flaking to form a thick, convex working edge. Four of these tools were made on cobbles and two on flakes. The flake tools maintain the same attribute as the cobble tools. Their efficacy as cutting implements is very doubtful. The edges are heavily crushed, indicating use as either cleavers or wedges on a resistant material (e.g., bone or joints).

These implements range in size from 126 × 52 × 30 mm to 164 × 121 × 77 mm. Mean values are 149 × 95 × 61 × mm.

Cobbles with Working Edges on the End and Side (29)

These tools have two working edges: one on the side and another on the end. Of these, 25 were made directly on cobbles and 4 were made on large flakes actually representing longitudinally split cobbles. The outlines are generally ovate but sometimes tend toward discoidal. The cross sections range from plano-convex to cuneate. All were made by heavy percussion flaking and retain a substantial amount of cortex on both the dorsal and the ventral surfaces. Also, on 22 of the tools, the side opposite the lateral edge has been left intact, retaining cortex and the original curvature of the cobble. This might have been done intentionally to provide a broad area to absorb shock in the palm of the hand while the tool was being used. The working edges have been worked bifacially but to differing degrees. In most cases, substantial bifacial reduction on the ventral surfaces was necessary to flatten the bottom of the tool. This feature was definitely desired and surely had special functional significance. The lateral edges are quite uniform and display very steep, almost vertical, step-flaking. Some are simply thinner examples of the aforementioned tools, and their edges are extremely dull from use, but others were damaged from flaking that miscarried.

The ends of the tools have been treated in two separate ways: 18 have convex edges (figure 4.37B), while 11 (figure 4.37A) have a pointed end resulting from the convergence of two lateral edges.

Those tools with convex edges also vary somewhat. Five (figure 4.37B) have thick, broad, convex noses that give the tool the appearance of an enormous end scraper. The remaining 13 also have broad, convex noses, but substantial biface reduction has resulted in a much flatter, spatula-like working edge. Considerable dorsal keels have resulted from the formation of the working edges on these examples.

These implements probably were used in heavy butchering work—for example, dismemberment—and our field experimentation has shown that tools such as these, manufactured from the same raw materials, accomplished the task of separating joints better than any other stone tool. Since the raw materials are so coarse, we were unable to detect use wear on these tools. Also, several tools exhibit percussion damage on the end opposite the convex edge, which suggests that they were used as gigantic wedges. These also would make an ideal tool for working through joints.

These tools range in size from 122 × 82 × 41 mm to 175 × 140 × 75 mm. Mean values for these dimensions are 150 × 102 × 59 mm.

The 11 remaining tools in this category, as mentioned previously, have pointed rather than convex ends (figure 4.37A). Except for this feature, they are morphologically identical.

FIGURE 4.37. Heavy percussion flaked implements from Component II (cobbles with working edges on the end and side).

They also differ in the degree of flattening of the end of the implement. On these tools, the converging edges form a fairly thick point. Percussion flaking is heaviest on the dorsal surface, but some flaking is present on the ventral surface, which served to flatten this area of the tool. Like the preceding category, coarseness of the raw material creates problems in interpreting both use wear and function. Functionally, they are probably similar to the tools with convex ends, and the development of a point on the end of the tool is most likely the result of situational requirements. Like the preceding category, they would be well suited for separating joints.

The size ranges for length, width, and thickness are 135 × 71 × 44 mm to 210 × 120 × 90 mm. Mean values for the same dimensions are 162 × 98 × 71 mm.

Miscellaneous Large Bifacial Tools

This group of implements is represented by miscellaneous cobble tools (5) and large flake tools (7).

Miscellaneous Cobble Tools (5)
These artifacts were manufactured on cobbles of sandstone (3), quartzite (1), and rhyolite (1). The degree of heavy crushing on the flaked end and battering on the opposite end indicate application of the working edge to a very resistant material. The size range of these objects is 125 × 146 × 76 mm to 146 × 98 × 80 mm.

The third item in this category is unique in the site (figure 4.38). A flat, water-rounded river cobble of sandstone, which probably had an original crescentic outline, was percussion flaked bifacially to form a broad, deep concavity, giving the tool the appearance of an enormous spokeshave. The working edge is dull, worn, and even crushed from what was probably a pounding action. It measures 191 × 86 × 27 mm.

Miscellaneous Large Flake Tools (7)
Large flakes of rhyolite (4), pumice (1), quartzite (1), and low-grade brown chert (1) illustrate differing degrees of alteration and edge damage.

Two display some retouch along the most acute edge, while the opposite edge is much thicker and suitable for holding in the hand. A single specimen displays a thin, sharp edge with some nicking and an opposing edge that was dulled or backed by bifacial retouch. These tools would have made excellent knives. These three tools range in size from 97 × 61 × 15 mm to 122 × 74 × 40 mm.

Two other specimens are large, thick, bifacial thinning flakes that have fairly sinuous edges, which are crushed from pounding and bashing a hard substance. These measure from 120 × 82 × 32 mm to 123 × 100 × 34 mm.

The last piece in this group has a broad, unifacially prepared concave working edge at one end of the flake. This edge is dull and partially crushed and was probably used in heavy butchering activity. It measures 16 × 83 × 42 mm.

Scrapers

This class of implements displays unifacial working edges and some degree of formal shaping. While unifaciality is a definite criterion, some degree of bifaciality does occur for obvious

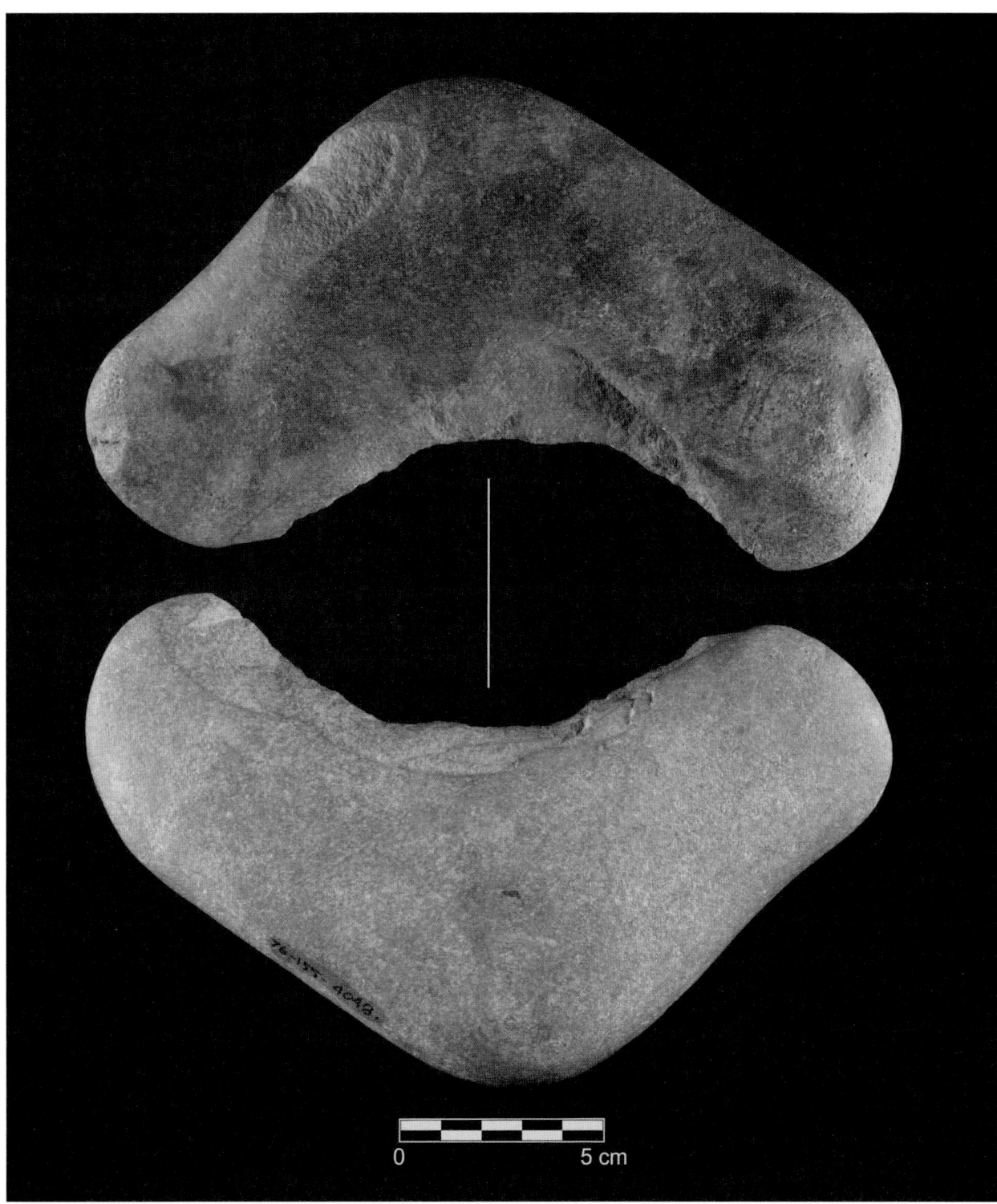

FIGURE 4.38. Miscellaneous cobble tool from Component II.

technological reasons—that is, straightening the curvature of a flake or reducing bulbs of percussion or other impediments to the thinning and shaping process.

All the scrapers were made on flakes of various dimensions and thicknesses and can be separated into the following categories: transverse scrapers (3), spokeshaves (2), side scrapers (10), and convergent side scrapers (6).

Transverse Scrapers (3)

These scrapers are characterized by a straight or convex working edge that is transverse to the axis of the flake. The working edge lies at the opposite end of the flake from the bulb of percussion (figure 4.39A–B).

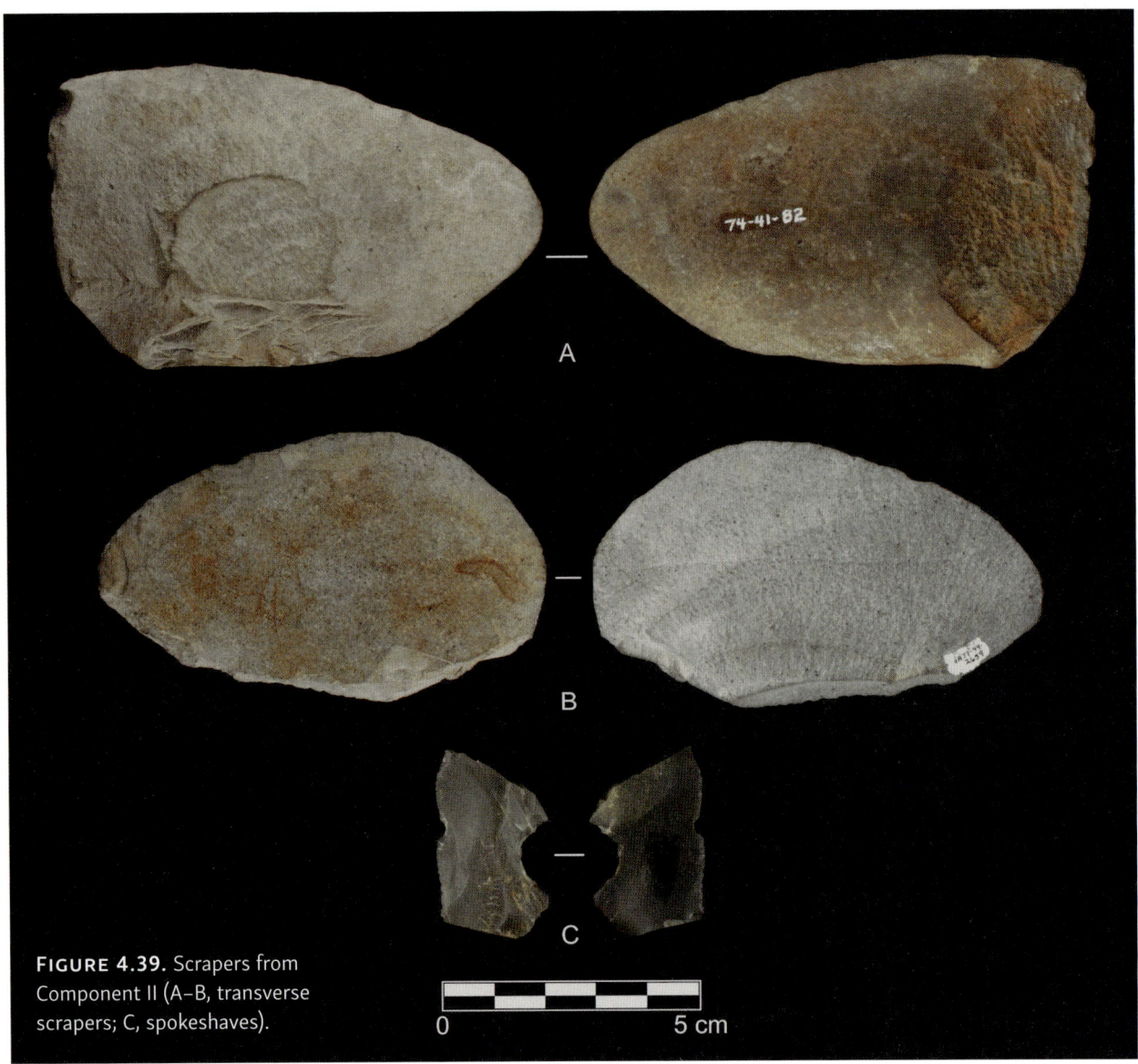

FIGURE 4.39. Scrapers from Component II (A–B, transverse scrapers; C, spokeshaves).

Raw materials for this category are light rhyolite (2) and dark rhyolite (1). The size ranges from 52 × 87 × 9 mm to 75 × 124 × 19 mm. Mean dimensions are 62 × 101 × 16 mm. Scraper edge angles vary from 12° to 20°, with a mean of 14°.

Of the scraper edges, one bears no obvious evidence of use or hafting, although it might have been resharpened. A second (figure 4.39A) shows use wear polishing and micronicking, probably indicating use on substances of differing degrees of hardness. This piece also was used in a haft that held the side opposite the working edge and covered about two-thirds of the tool. The third scraper of this group likewise shares differential use on both scrape-yielding and scrape-resistant substances but shows no evidence of hafting (figure 4.39B).

Spokeshaves (2)

Both of these spokeshaves were made on flakes. One has a deep unifacial concavity in one edge and minor degrees of fine chipping along the opposite edge (figure 4.39C). It measures 39 × 22 × 7 mm and is of gray chert.

The second spokeshave is chalcedony and bears a unifacial concavity. Adjacent to the concavity is a finely retouched nose. At the opposite end of the flake, there is another such nose with a vertical, crushed edge along one side. It measures 53 × 26 × 14 mm.

On both specimens, the deepest part of the concavity is nearly vertical as a result of heavy crushing in this area. It is assumed that they were employed to shape shafts for tools.

Side Scrapers (10)

In this category, the scraper edge lies parallel to the axis of the flake. All but one of these has the scraper edge on the dorsal surface of the flake. Six of these (figure 4.40) are single side scrapers, and they all have convex working edges. Raw materials for these scrapers are diabase (2), black chert (2), gray chert (1), and quartzite (1). Measurements range from 56 × 41 × 8 mm to 118 × 102 × 31 mm, with a mean of 88 × 62 × 181 mm. Edge angles vary from 10° to 20°, with a mean of 21°. Four are double side scrapers—that is, they have two opposing scraper edges situated on alternate faces (figure 4.41). In this subgroup, three have convex working edges, one has one convex and one straight edge, and one has two straight edges. Raw material for these are black chert (1), light rhyolite (1), banded rhyolite (1), and quartzite (1). Measurements are 50 × 32 × 7 mm to 140 × 100 × 42 mm, with a mean of 99 × 63 × 19 mm. Scraper edge angles vary from 5° to 70°, with a mean of 26°.

Among the single side scrapers, three have edges dulled from use on a fairly yielding material and might have been held in the hand. The remaining three might also have been hand held, since no further hafting wear can be detected. Two of these exhibit edge damage that must have resulted from working a scrape-resistant substance, whereas the remaining specimens appear to have been used partially for cutting and scraping. Most of the nick marks occur on the dorsal surfaces, but some isolated marks can be found on the ventral face.

The double side scrapers show similar variability. One (figure 4.41B) has isolated wear interrupted by resharpening, two (figure 4.41A, C) appear to have both cut and scraped, and one shows heavy dulling of both edges from working scrape-yielding material. None of the scrapers display any evidence of hafting.

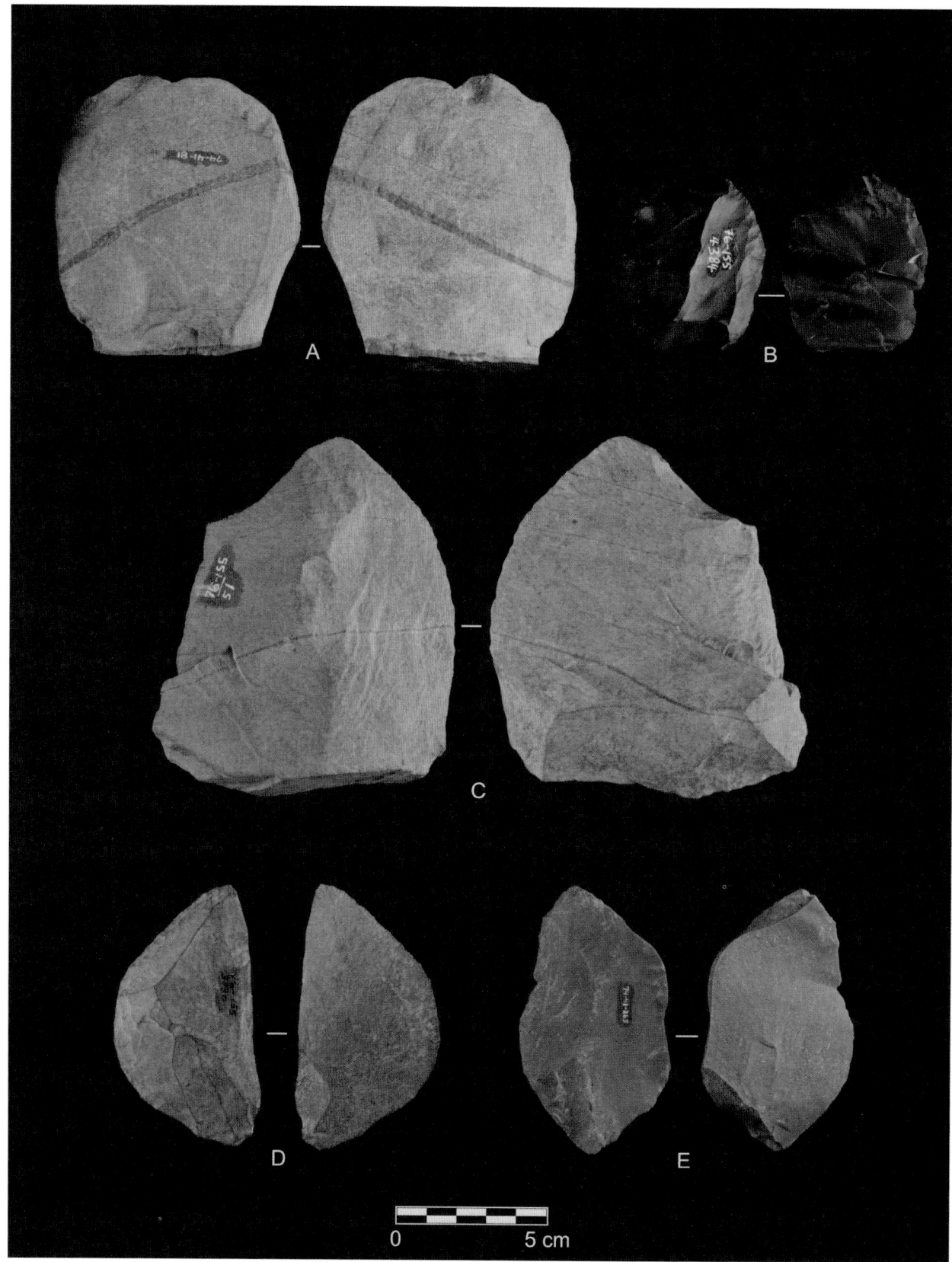

FIGURE 4.40. Single side scrapers from Component II.

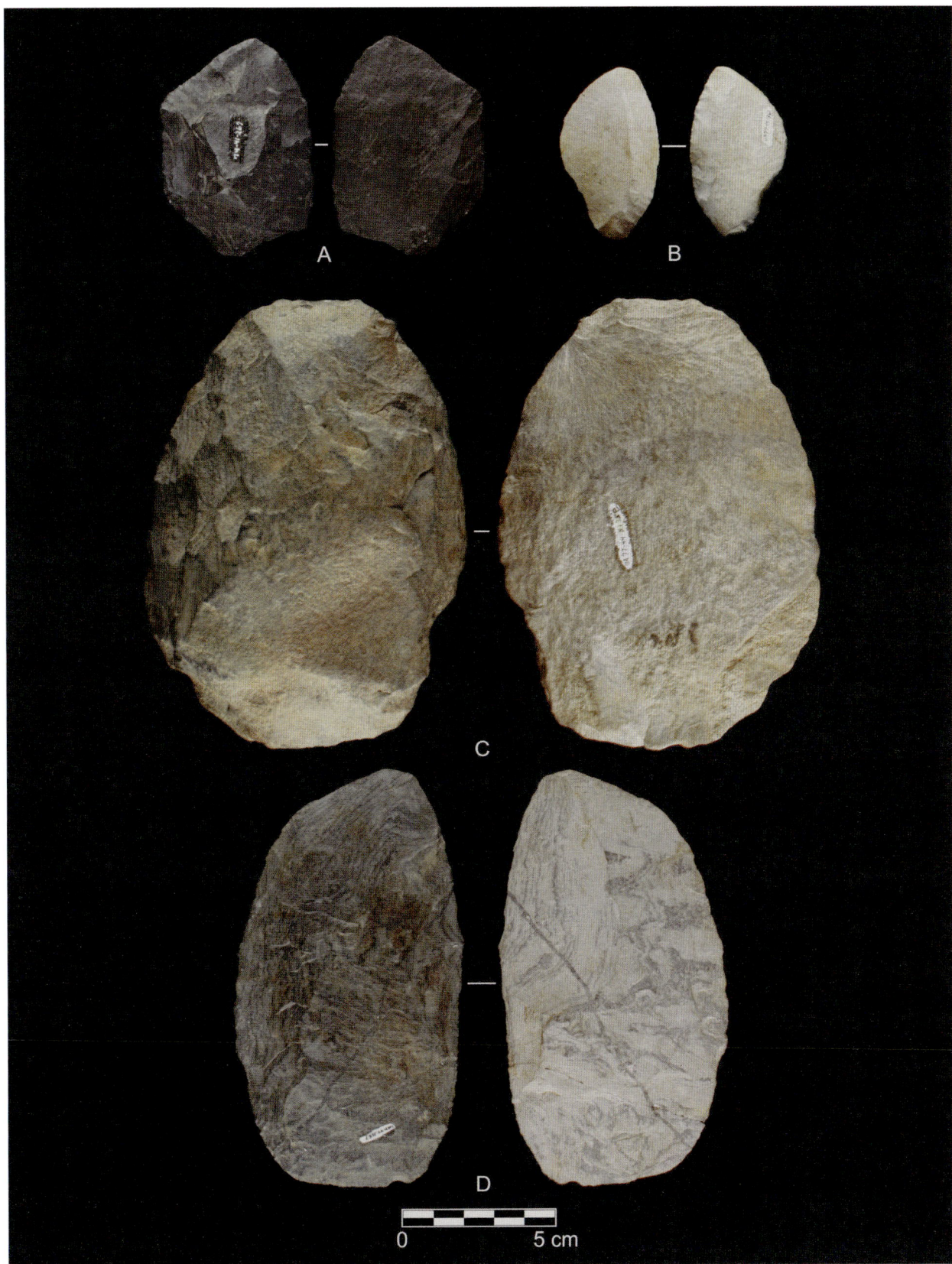

FIGURE 4.41. Double side scrapers from Component II.

Convergent Side Scrapers (6)
These scrapers have two scraper edges that converge to form a tip (figure 4.42). On two examples, the steep flaking used to form the edges created a keel on the dorsal surface (figure 4.42E–F). The other four specimens are broad and lack keels. Each scraper has both a relatively straight and a convex edge. Large flakes of chalcedony (2), quartzite (2), light rhyolite (1), and dark rhyolite (1) were used to make these scrapers. Size ranges run from 46 × 24 × 8 mm to 74 × 48 × 27 mm, with a mean of 65 × 36 × 12 mm. Edge angles range from 30° to 60°, with a mean of 48°.

In five instances, these scrapers were manufactured on the edges of the dorsal surface of flakes. The remaining specimen has convergent edges, but they lie on alternative faces. Four of the flakes display edges that are parallel to the long axis of the flake, with the bulb of percussion at the base of the tool. The other two scrapers were simply made on pieces of flakes lacking the bulbar end.

One of these scrapers (figure 4.42C) bears no evidence of use wear; it was probably discarded after flaking miscarried and destroyed one edge.

The remaining five pieces display varying degrees of use wear, and one shows evidence of resharpening. None of the scrapers show any clear hafting wear, so they might have been held in the hand.

One specimen (figure 4.42A) has clear hafting wear on both faces. It was secured so that both edges were exposed, and it appears to have been broken in the haft.

The evidence of utilization is variable on these items and would appear to indicate that both cutting and scraping affected the edges. While classified as convergent side scrapers on morphological and technological grounds, these tools were probably used, at least in part, as knives.

Other Core Technology
Subprismatic Cores (4)
Subprismatic cores were manufactured on water-rounded cobbles of gray chert (2), diabase (1), and degraded quartzite (1) that were probably taken from local gravel deposits.

Their measurements range from 87 × 93 × 65 mm to 137 × 106 × 140 mm, with a mean of 121 × 100 × 87 mm. They are crudely prismatic and are characterized by the removal of flakes from one side of the cores. In one example (figure 4.43B), about three-quarters of the perimeter of the core was used for flake removal. The opposite side, or back of the core, preserves areas of cortex. A few flakes were struck from this side and from the end opposite the platform. On two of the cores (figure 4.43), one frontal blow removed the entire top, and massive flakes were then removed without further modification of the striking platform. The remaining two specimens have platforms with substantial preparation, and one of these (figure 4.44) displays edge preparation along one side of the face from which the flakes were removed. This core also displays an exceptionally well-developed platform. A flake struck from this core (figure 4.44) has been reattached. The proximal end of the flake displays heavy faceting. This is a remnant of the preparation on the striking platform of the core. Technologically and morphologically, this specimen is a Levallois flake; it was detached from a core type that Soviet investigators working in Siberia classify as Epi-Levallois, and this type of core-flake technology is a common occurrence in sites dating to the late Paleolithic in Siberia (Powers 1973).

Figure 4.42. Convergent side scrapers from Component II.

FIGURE 4.43. Chert subprismatic cores from Component II.

FIGURE 4.44. Diabase subprismatic core from Component II. The rearticulated flake has been removed.

Bladelike Flakes (3)

Three unworked bladelike flakes were found, of which three are quartzite and one is pumice (figure 4.45C–D). They range in size from 63 × 10 × 4 mm to 85 × 31 × 8 mm. The mean measurements are 75 × 23 × 6 mm. No blade cores from which such blades/flakes could have been struck were found at Dry Creek. The existence of a separate microblade technology is possible, but accidentally produced bladelike flakes are likewise a reasonable expectation. Larger or more complete samples are required to settle this particular conundrum.

Bladelike Flake Tools (18)

Eighteen bladelike flakes have retouched or utilized edges and are summarized in table 4.7.

One complete rhyolite specimen is quite large and measures 120 × 54 × 18 mm. The remaining nine complete pieces are more uniform in size and range from 42 × 13 × 5 mm to 60 × 33 × 7 mm. The mean size for these specimens is 47 × 22 × 6 mm (figures 4.45A–B, 4.46B–G).

The proximal portions vary from 12 × 7 × 2 mm to 71 × 29 × 10 mm. Their mean size is 40 × 19 × 5 mm. The medial portions range from 21 × 7 × 4 mm to 68 × 27 × 6 mm. The mean is 41 × 19 × 5 mm. The single distal fragment measures 40 × 16 × 4 mm.

LITHIC TECHNOLOGY OF THE DRY CREEK SITE

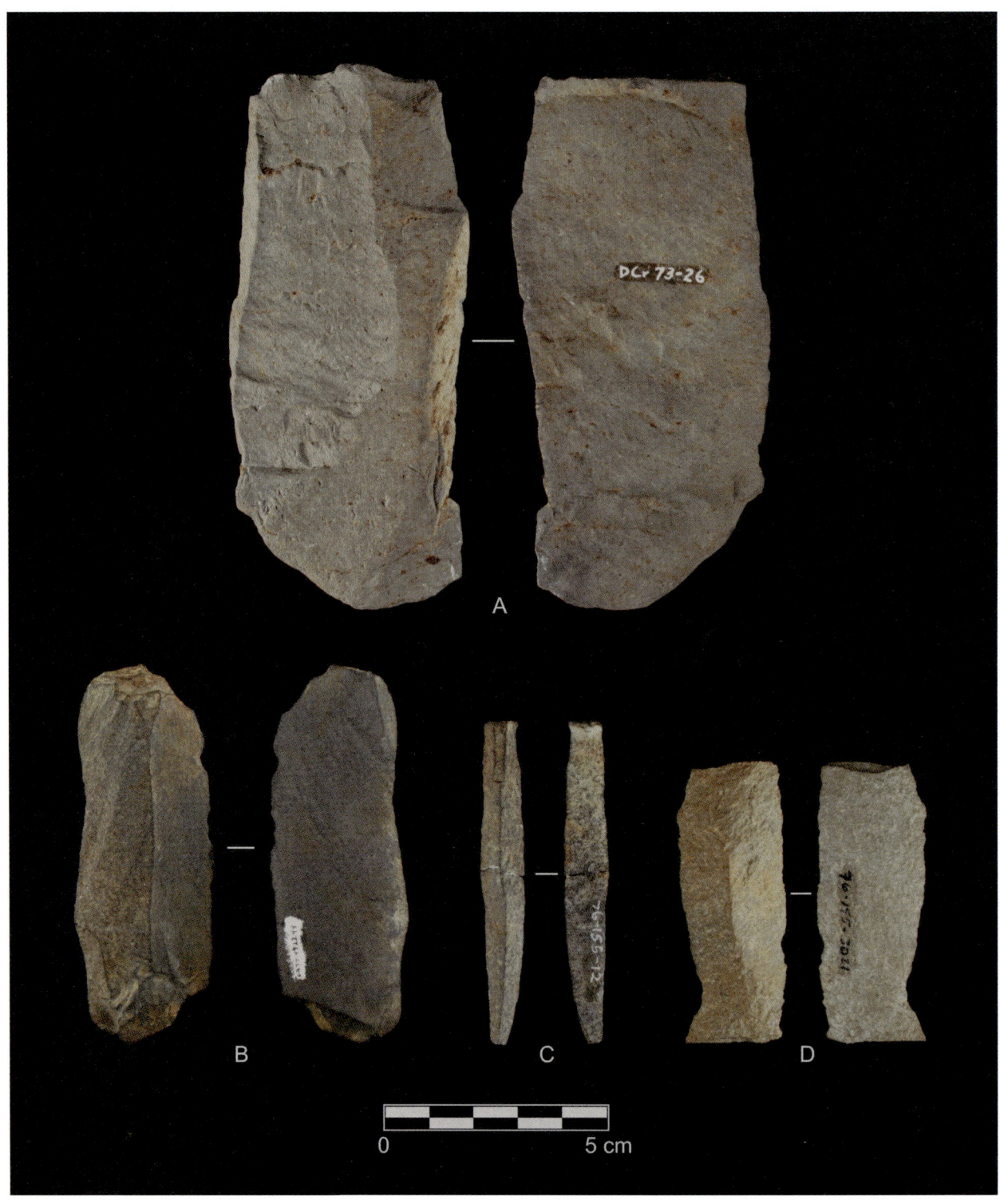

FIGURE 4.45. Bladelike flakes from Component II (A–B, bladelike flake tools; C–D, unmodified bladelike flakes).

TABLE 4.7. Raw material frequencies for bladelike flake tools

Material	Complete	Proximal	Medial	Distal	Total
Gray chert	4	1			5
Rhyolite	1			1	2
Pumice	1				1
Black chert	1	1			2
Obsidian	3				3
Brown chert		2	3		5
Totals	10	4	3	1	18

All but one, as indicated previously, have fine retouch and edge damage. They probably functioned as small knives and appear to have been very resistant to edge damage.

The last item (figure 4.45A), the largest, is a complete rhyolite piece with no evidence of flaking. However, one edge and the base are heavily polished, and there are weakly developed striations running at a 90° angle to the working edge. This tool appears to have been unhafted and used as both a scraper (the end) and a knife (the side) on a fairly soft, yielding material, possibly hide.

Unshaped Flake Tools (21)

These small flake implements have been given no formal shaping other than fine retouching along the edges. Twenty of these pieces have unifacial working, and one has bifacial edge retouch. Among the raw materials employed are rhyolite (4), quartzite (4), brown chert (2), gray chert (3), black chert (2), obsidian (2), and chalcedony (1). The sizes range from a maximum of 63 × 57 × 13 mm to a minimum of 14 × 7 × 2 mm.

These small, fairly fragile tools were probably employed as light-duty scrapers or knives (figure 4.46A).

Miscellany

Hammerstones (3)

These tools are retouched river cobbles with heavy battering and crushing on both ends as a result of direct percussion (figure 4.47). Two are coarse-grained quartzite and measure 101 × 68 × 60 mm and 94 × 78 × 51 mm. The third is rhyolite. It is long and narrow and might have been used for either pressure retouch or light percussion flaking, as a hard blow would surely have broken it. It measures 202 × 53 × 30 mm.

Anvil Stones (3)

These stones are generally oval and have flattened cross sections. They display heavy battering on one face, which probably resulted from stones being smashed against them. This battering could have also resulted from stone being placed on an anvil and struck with a hammerstone. Two are quartzite and measure 240 × 170 × 126 mm and 440 × 249 × 175 mm. The last is rhyolite and measures 167 × 123 × 77 mm.

FIGURE 4.46. Flake and bladelike flake tools from Component II (A, unshaped flake tool; B–G, bladelike flake tools).

Figure 4.47. Hammerstones from Component II.

Summary

Excavations at Dry Creek have demonstrated the existence of human groups in the Nenana Valley 10,000–11,000 ^{14}C yr BP (11,000–13,000 cal yr BP) and have shown that these groups were hunting, in part, remnants of the steppe-adapted grazing fauna that had existed in Alaska in the late Pleistocene. These hunters left a residue of their activity at Dry Creek in the form of two temporally separate sets of occupations. Component I dates to the late twelfth millennium ^{14}C yr BP (fourteenth millennium cal yr BP) and is confined to Loess 2. Component II, which dates to the mid-eleventh millennium ^{14}C yr BP (thirteenth millennium cal yr BP), lies in Loess 3 (Paleosol 2) and is separated from the underlying component by Sand 1.

It is difficult to tell how many times the site was visited during Component I time because there are far fewer cultural remains than in Component II, but during this latter occupation, the site was heavily used—probably on many different occasions.

These occupations are typified by tool kits that were probably directed at two major activities, regardless of component or cluster: (1) the production or maintenance of hunting equipment and (2) the procurement and processing of game. These activities were carried out with artifacts that fall into two general categories: (1) large, crudely fabricated, opportunistic implements made from locally available, low-grade raw materials and (2) small, lightweight tools produced from medium- to high-quality materials that were either brought to the site from the hinterlands or derived from trade networks.

The heavy, crude implements are shared by both components and were probably used in butchering activities—more specifically, in the dismemberment of carcasses.

The small, lightweight tools from Component I are difficult to characterize because there are so few of them and many appear fresh or unused. However, this same category of tools from Component II represents something quite different, as evidenced by extensive maintenance activity (reworking and resharpening).

Many of the small bifaces are highly curated and, we think, represent weapon tips as well as cutting tools. Possibly, many of the cutting tools first served as weapon tips, but as they sustained damage and were reworked and reduced to the point that they were ineffective, they were transformed into knives. All the remaining categories of small tools (burins, scrapers, etc.) likewise show evidence of multiple purposes.

During Component II time, microblade production was a major activity at the site. We assume that the purpose of this activity was the production, or at least the maintenance, of composite points. One of the attractions of composite points is the portability of the points themselves and the tools necessary for their production. The microblade core can be viewed as a perfect adaptation to mobility. It is small, lightweight, and produces the maximum amount of sharp linear edge per unit of stone. A technology such as this, plus the other small tools from the site, would have made excellent lightweight tool kits for small groups of people pursuing a highly mobile hunting strategy (see also Del Bene 1981).

While there are general similarities in the activities conducted at the site, Components I and II differ markedly in technological emphasis.

The technological emphasis of Component I was in the production of bifaces (both projectile points and knives), side scrapers, transverse scrapers, burins, cobble tools, and unshaped flake tools. This set of artifacts, or what we can see of it, constitutes the tool kit. The absence of microblade cores, core tablets, and microblades is notable. This situation might be the result of a sampling error at the site or regional level, and differences in site-specific tasks or seasonal activities beyond the site could account for the absence of a microblade technology at this time in the Dry Creek site. However, for the time being, verification of these hypothetical sampling problems is beyond reach, as there are no adequately sampled sites of this age in the area with which to compare this particular stratigraphic situation.

Microblade technology is reported from the Chindadn Complex at Healy Lake, which is probably penecontemporaneous with Component I at Dry Creek. However, this site should be assessed cautiously, as the entire sequence lies in a compressed loess section with ample possibility for internal mixing and attendant sampling problems. Besides Healy Lake and Dry Creek, the Moose Creek site in the Nenana Valley (Hoffecker 1982) is the only other locality that has produced data that relate to this early lithic horizon. It occupied a similar topographic position as Dry Creek. The artifacts are also contained in an eolian section. Component I at this site is associated with a set of paleosol stringers (Unit 6) and an underlying silt (Unit 7) that, in turn, rests on a till or outwash deposit of a pre-Wisconsinan glaciation of undetermined age. There is a set of radiocarbon dates run on soil organics from the paleosol stringers (Unit 6) that range from $8,160 \pm 260$ to $11,730 \pm 250$ ^{14}C yr BP. These are viewed as upper limiting dates for Component I. The artifacts are vertically distributed through about 30 cm of silt down to the surface of the glacial deposits. To date, fragments of six bifaces have been recovered from this component, two of which appear to be bases of lanceolate points. One possible microblade fragment is also reported (Hoffecker 1982). An expansion of the excavations at this site should provide badly needed information on the possible existence of a premicroblade lithic horizon in Alaska.[7]

Component I at Dry Creek is a small but distinctive assemblage; as such, it represents one aspect of the earliest lithic horizon in Alaska. While specific typological similarities are lacking for the projectile points, the general appearance of this component is close to established Paleoindian tool kits much farther to the south on the Great Plains of interior North America. It is entirely possible, albeit speculative, that Component I at Dry Creek might be a northern variant of the Paleoindian Plains adaptation of the twelfth millennium ^{14}C yr BP (fourteenth millennium cal yr BP).

The artifacts recovered from Component II occurred in tight, relatively well-defined horizontal concentrations, or clusters (cf. chapter 5). Here, however, it is necessary only to note that there are two basic types of clusters: (1) those with microblades and (2) those lacking microblades.

The microblade clusters contain, of course, the microblades plus the by-products of the entire production sequence (cores, preforms, core tablets, and broken parts of cores), numerous bifacial knives, and flake tools. Also occurring in these concentrations are burins and burin spalls.

This complex of artifacts can be assigned to the Denali Complex (West 1967), which is a widely distributed early complex in parts of interior Alaska. Beyond the interior, the Denali Complex becomes part of a broader lithic continuum that includes the Ugashik Narrows Phase in the Alaska Peninsula (Dumond 1977) and the Akmak Complex from Onion Portage on the Kobuk River in northwestern Alaska (Anderson 1970a). These early dated complexes appear to be related to the Diuktai Tradition of central Siberia (Mochanov 1977). As a result, this spatially discontinuous series of lithic complexes, distributed in the circum-Beringian region during the late Pleistocene, has been called the Siberian-American Paleoarctic Tradition by Dumond (1977) and the Beringian Tradition by West (1981).

The nonmicroblade clusters are quite a different matter. Not only do they lack microblades, but the by-products of the microblade production sequence are absent. Instead, the nonmicroblade clusters contain crude bifacial implements, flake tools including shaped scrapers, and the bases of projectile points of general Plano appearance. These points are not found in microblade clusters.

This situation creates an interpretive dilemma presently impossible to resolve but with options that are fairly straightforward: (1) all clusters are part of the same culture and can be explained by differences that are activity specific (composite point manufacture/repair and butchering), (2) the differences are due to seasonal technological variability (different weapon systems for different game species and butchering), or (3) two separate cultures are present in the area at the same time conducting the same activities.

These options, or hypotheses, are testable given that certain conditions are met—namely, that more sites from this time period are excavated with these problems in mind and that the sites have good faunal preservation so that potential seasonal shifts in subsistence activity can be better documented.

Notes

1. More recent studies suggest that most of this dark gray rock is basalt (chapter 8).
2. Reuther et al. (2011) provide an update on the sources of the obsidian; some may have come from a closer source than Batza Tena.
3. The biface tip was later conjoined with its basal portion; figure 4.1E shows both fragments rearticulated to form a complete lanceolate-shaped biface preform.
4. This bifacial knife was stolen from Powers's lab sometime in the late 1970s or early 1980s and could not be photographed.
5. Unfortunately, this rearticulated artifact was not available for photography but is illustrated in Hoffecker et al. (1996a: figure 7-9:b).
6. Since preparation of this report, slotted points have been recovered from at least two interior Alaskan/Yukon sites, Lime Hills Cave (Ackerman 2011) and the Gladstone ice patch (Hare et al. 2004); however, these were not intact with microblades still inset.
7. Since preparation of the original Dry Creek report, renewed excavations at Moose Creek identified two stratigraphically separate components as at Dry Creek, Component I dating to about 13,000 cal yr BP and Component II, to about 12,000 cal yr BP (Pearson 1999). Lithic assemblages from the two components correspond to those from Dry Creek. The lanceolate points and microblade mentioned here could correspond to Component II.

CHAPTER FIVE

The Occupation Floors at the Dry Creek Site[1]

JOHN F. HOFFECKER

Introduction

While the eolian sedimentary context of the Dry Creek site provides important information on the temporal and stratigraphic (i.e., vertical) relationships among the remains, the broad-scale excavations provided much information on the horizontal spatial relationships. These horizontal spatial relationships may be said to comprise "occupation floors," one of which has been defined for each component. It is the objective of this chapter to provide a preliminary description and analysis of each of the late Pleistocene occupation floors (the uppermost occupation level is not included), which potentially contain much information on the social and economic significance of the site.

The occupation floors at Dry Creek consist primarily of lithic artifactual debris. Remains of former hearths exist in the form of scattered charcoal fragments, while faunal remains, which originally might have been abundant, are represented by a few severely weathered but identifiable teeth and a collection of unidentifiable bone fragments. There is no evidence of former structures (e.g., tent rings, postmold arrangements, large depressions). Nevertheless, horizontal distribution maps of the remains of each occupation level reveal a clear pattern of faunal remains; both floors are characterized by a series of dense concentrations of debris. These concentrations are the prime focus of the description and analysis that follow.

Each of the concentrations is described in terms of area, lithic raw material composition, quantity and size of debitage, quantity and size of worked implements, macroscopic and microscopic edge wear on implements, and associated faunal remains and other material. These descriptions provide a basis for comparisons both within and between components and for overall characterization of the occupation floors. It should be noted that the most appropriate units of analysis in this context are the lithic reduction sequences; the core and tool sequences reflect real functional interrelationships, not merely spatial associations that might be fortuitous (see, e.g., Cahen and Keeley 1980). However, the time-consuming process of reduction sequence reassembly has been completed only for Component I (Smith 1985); few sequences have been reconstructed among the much larger quantity of remains in Component II.

Geologic Context of Occupation Floors

Geoarchaeological studies of sites in eolian sedimentary contexts (and studies of open-air sites in cold climates generally) have been comparatively rare (Hahn 1977; Schweger 1985). This is unfortunate, because a significant body of archaeological data has been retrieved from these contexts, especially from the Paleolithic of northern Eurasia. It is important to understand how remains buried in loess and eolian sand (or their colluvial derivatives) have been affected by the action of depositional and disturbance processes (and the interaction among these processes and human occupation activities) under cold-climate conditions in order to interpret sites effectively. Site formation and modification processes can clearly have a profound influence on archaeological remains in these contexts by altering spatial provenience (vertical and horizontal), damaging artifacts (e.g., abrasion of stone surfaces), distorting site-fauna samples (e.g., destruction of small mammal remains), and so on (Butzer 1982:98–122; Keeley 1980:28–35; Wood and Johnson 1978).

The geologic context of the lower two levels at Dry Creek is chiefly the product of eolian deposition of fine-grained sediment (Loesses 2 and 3) weathered by intermittent soil formation (Paleosol 1). The sediments have undoubtedly been affected by the introduction of large quantities of lithic and (probably) organic debris by the site's occupants. Overall, primary loess provides an unusually sympathetic depositional environment for open-air archaeological sites, causing minimal disturbance to remains compared with burial in fluvial or colluvial sediments, where high- or low-energy water flow often removes, displaces, or damages materials (Butzer 1982:52–53).

However, there are a variety of site modification processes in this type of depositional environment that cause potentially significant disturbance and damage to remains and that must be controlled for as much as possible. At Dry Creek, these include frost action, rodent burrowing, and microfaulting.

Preburial frost sorting might disperse debris concentrations (Bowers et al. 1983) and even create rings (on a level surface) of larger items (Butzer 1982:102–4). It is likely that the artifact concentrations at Dry Creek have been enlarged to some degree by preburial frost dispersal of material. Cryoturbation may also be assumed to have affected the remains, causing displacement and possibly edge damage to artifacts (Keeley 1980:30–35) and fragmentation of faunal debris; upfreezing (Washburn 1979:80–91) has probably generated upward migration of some artifacts (Schweger 1985; Wood and Johnson 1978:333–46). There is no evidence of ice-wedge formation at the site.

Rodent burrowing is evident from the presence of krotovinas, and its effects on the position of archaeological remains have been observed at many sites (Butzer 1982:111–13; Wood and Johnson 1978:318–20). For example, Voevodskij (1952:104) reports downward displacement of artifacts and bones up to 15–20 cm at Chulatovo II on the middle Desna River in the European USSR. Krotovinas at Dry Creek (10–20 cm in diameter) are concentrated in

Loess 2 and present in Loess 3 and are chiefly concentrated near the bluff edge (Thorson and Hamilton 1977:162–63). They probably represent fossil burrows of ground squirrels (*Citellus* sp.), which are often deep and extensive (Ognev 1963). It appears likely that some artifacts from Component II have been redeposited in Component I through burrows in the easternmost portion of the site.

Unlike frost action and rodent burrowing, microfaulting is not a common factor in open-air site modification (Wood and Johnson 1978:366–69), and it reflects the active tectonism of the region. Microfaulting has been observed not only at Dry Creek but also at the Moose Creek site located on the east side of the Nenana Valley (Hoffecker 1985). At Dry Creek, Thorson and Hamilton (1977:162) observed dip-slip displacements of up to 50 cm. Near the bluff edge, the dips exhibit angles of as much as 30°, but angles decrease with distance from the edge, becoming progressively more vertical. Examination of the sediment profiles indicates that disturbance of archaeological remains was most pronounced near the bluff edge, where both horizontal and vertical displacement of up to 10–15 cm appears to have occurred.

Postdepositional geobiochemical weathering of the sediments must be accounted for in interpretation of the site (Butzer 1982:114–17). The Dry Creek loess beds are generally low in calcium carbonate (Thorson and Hamilton 1977:158), which often creates a favorable medium for bone preservation in such deposits. Furthermore, percolating soil acids from overlying middle and late Holocene forest soils have likely added to the corrosion of the faunal remains (Shipman 1981:42). When these factors are combined with probable pre-burial fragmentation of material due to frost action, trampling, and carnivore damage, it is not surprising that only the most resistant body parts of large mammals appear to have been preserved (see chapter 6; Brain 1976, 1981; Shipman 1981). It is conceivable that a much broader spectrum of animal remains was originally present at the site, including smaller mammals and other vertebrates (avian gastroliths have been recovered).

The human occupants of the site have also probably contributed to the modification of archaeological residues. The most significant effects might have been generated by trampling, which can displace objects both vertically and horizontally. The results of experimental work by Gifford-Gonzalez et al. (1985) indicate that horizontal movement (including some dispersal of debris concentration) is more likely to have occurred at Dry Creek given the fine texture (sandy silt with minor clay) of Loesses 2 and 3 (Thorson and Hamilton 1977:156). If the site was occupied during colder months, freezing of sediment would have further inhibited downward vertical migration. The other effects of human occupation were probably geobiochemical (e.g., increased presence of phosphate in the loess; Butzer 1982:116).

Overall, I would suggest that the geologic context of the Dry Creek site indicates that the fundamental spatial integrity of the occupation floors remains intact, although some displacement of materials has occurred. Vertical movement of artifacts appears likely to have been the most significant effect. This is essentially a function of scale: 10–20 cm displacement of items in a 2-m vertical section is of more consequence than comparable horizontal displacement of items across a 10-m-wide occupation area. Although the subdivision of Components I and II (separated by approximately 10–20 cm of largely sterile sediment) is well founded, microstratigraphic distinctions with the components (e.g., among individual artifact clusters) are likely to have been compromised. In a few cases, artifacts from Component II might have been reworked down into Component I through rodent burrows; conversely, some items

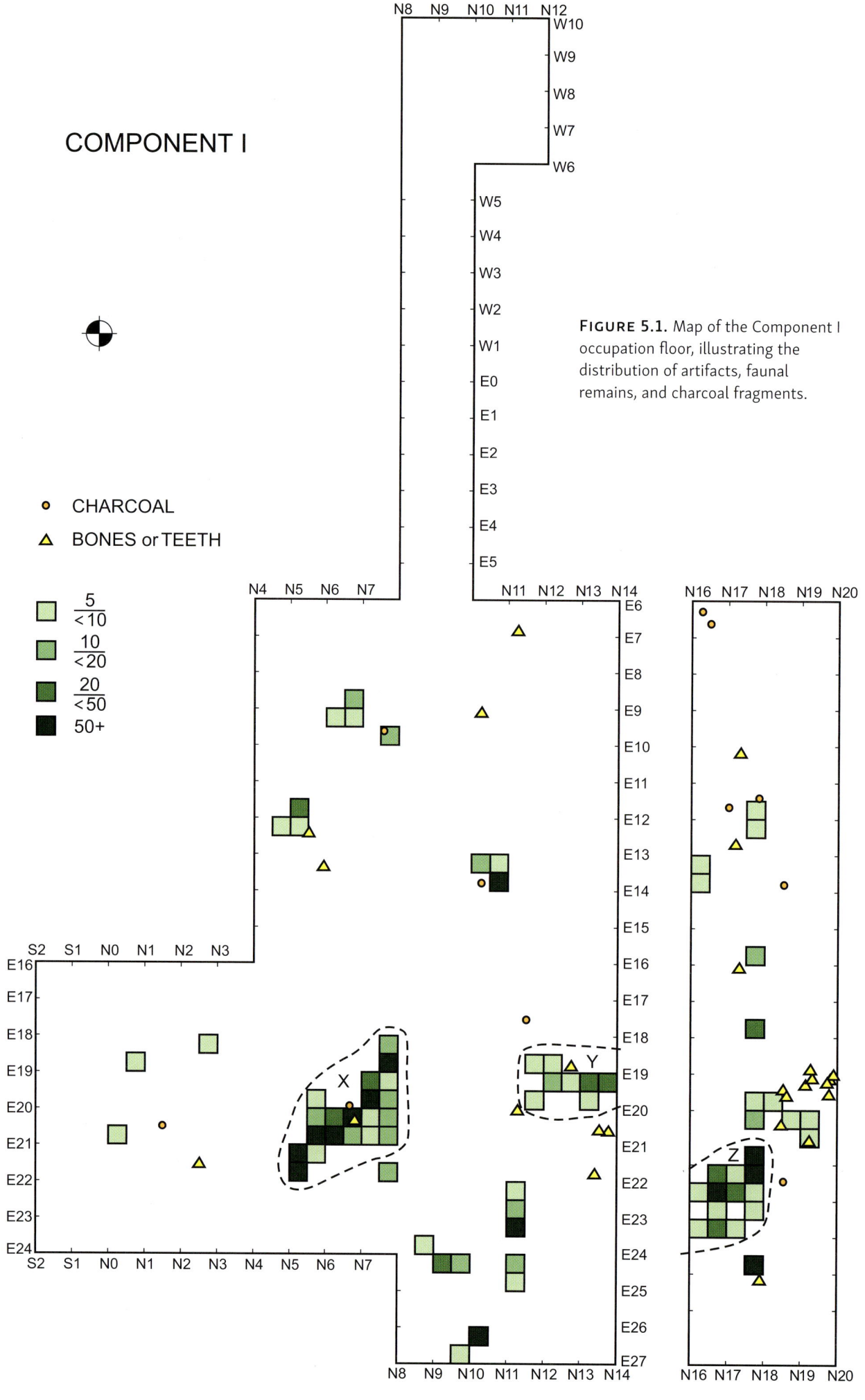

FIGURE 5.1. Map of the Component I occupation floor, illustrating the distribution of artifacts, faunal remains, and charcoal fragments.

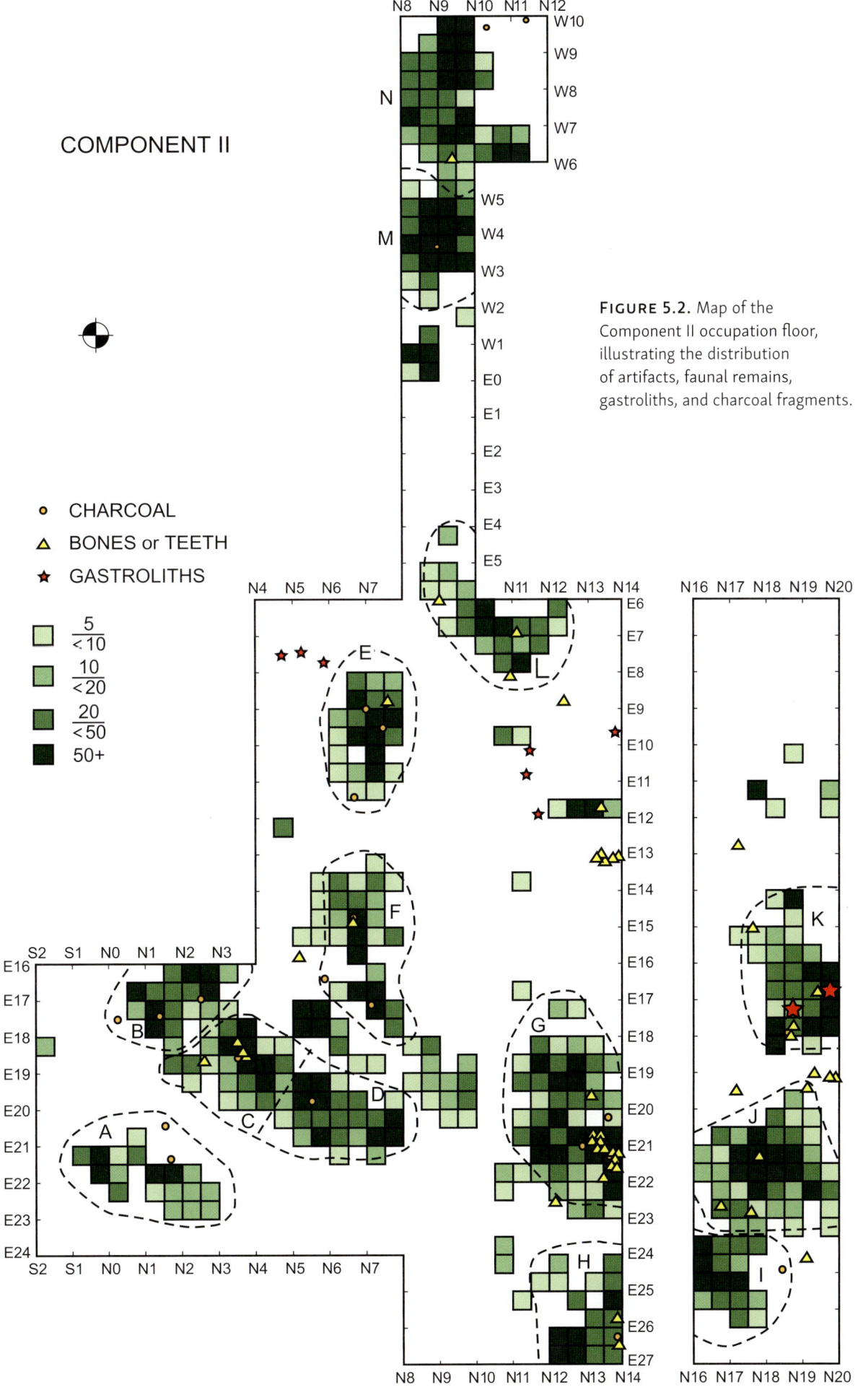

FIGURE 5.2. Map of the Component II occupation floor, illustrating the distribution of artifacts, faunal remains, gastroliths, and charcoal fragments.

might have been frost driven from Component I up into Component II. The disturbance has probably been most severe along the bluff edge (easternmost 8 m of the occupation area).

The most important conclusion to be drawn here is that the debris concentrations cannot be accounted for through processes of site formation or modification, including postdepositional human disturbance. It is probable, in fact, that the concentrations have been dispersed to some extent prior to burial through frost-sorting and trampling. Nevertheless, it is difficult to systematically measure the effects of the various processes known or suspected to have been operative at the site. The need for application of more sensitive field techniques (e.g., recording dip and strike of artifacts) to monitor these processes, although time-consuming, has been noted by Butzer (1982:114), Schweger (1985:139), and others.

Methods of Study

In order to elucidate spatial patterning, horizontal artifact distribution maps for Components I and II were constructed (figures 5.1, 5.2). The provenience of cores, flakes, blades, and tools was plotted by quarter-meter square; more precise locations did not seem necessary for the purposes of my analysis. All items were color-coded for raw material type. Rocks, bone and teeth fragments, and gastroliths were also plotted.

The picture that emerged was clear and dramatic. The debitage (and, to a large extent, the tools) resolved into a number of dense concentrations. These concentrations or clusters varied in size (from approximately 100 to more than 2,000 items) and raw material composition. In some cases, two or more clusters were adjacent (e.g., C and D), but differences in raw materials (not reproduced in figures 5.1, 5.2) made for easy separation.

Horizontal clustering of occupation debris is a common phenomenon in archaeological sites, but procedures for its recognition vary and deserve comment. At some sites, significant deviations from random spatial patterning cannot be determined reliably by a visual inspection of the distribution maps, and investigators have employed a variety of quantitative techniques to accomplish this (Hesse 1973; Whallon 1973a). Where the data are in the form of grid counts, dimensional analysis of variance (Whallon 1973b), variance mean ratio (Dacey 1973), and simple phi coefficient and chi-square tests (Freeman 1978:66–67) are applicable. Where point provenience data are available, nearest neighbor analysis may be used (Whallon 1974). At other sites, spatial clumping is readily observable (e.g., Leroi-Gourhan and Brezillon 1966; Wheat 1972), and in these situations, quantitative methods are unnecessary (e.g., Morlan 1974:92).

Once isolated on the basis of visual inspection, and determined to be unaccountable for the terms of disturbance processes (see previous section), the artifact clusters at Dry Creek became the unit of analysis within each component. A complete list of all items in each concentration (rocks, debitage, tools, and tool fragments) was compiled. Raw material composition was determined and, although subject to some error during cataloging, should be essentially accurate.

While all cores, tools, and core/tool fragments (and the larger rocks and rock fragments) were examined, it was necessary to employ sampling procedures in the study of flakes. All flakes equal to or larger than 6 cm were arbitrarily classified as "large flakes" and included in the analysis of the tools and cores. From the remainder, a systematic sample—varying according to cluster size but generally averaging about 150—was drawn from each list. Only whole flakes, roughly half of each sample, were measured. Individual sample sizes are given by cluster in figures 5.3-5.19. Length was measured as the largest axis perpendicular to the striking platform, while width was measured as the greatest distance perpendicular to the length axis. Thickness was obtained by measuring the greatest distance orthogonal to the plane created by the length-width axes (following Keeley 1980:17, figure 2).

Some observations were made on flake morphology (see Frison 1968) and patterns of primary and secondary retouch, and these are noted in the discussions of individual clusters.

All tools within or in close proximity to a cluster (< 0.5 m) have been listed in conjunction with that cluster (see the following). Besides raw material,[2] three types of data were used in the description of the tools:

1. Typological classification (following Powers [see chapter 4])
2. Macroscopic edge wear (my observations)
3. Microscopic edge wear (observations made by Terry A. Del Bene [unpublished notes; personal communication])

Component I Occupation Floor

Only three large artifact concentrations were uncovered on the Component I occupation floor (figure 5.1); these account for slightly more than 50 percent of the total number of artifacts in this component. The major clusters (labeled X–Z), which are all located within 10 m of the bluff edge, range in size from approximately 2 m^2 to 5 m^2 and comprise roughly 100 to 1,000 items. Smaller accumulations of debris, including a concentration of faunal remains, are scattered across the occupation area (although material is lacking in the westernmost excavated portion of the site). Isolated faunal remains are associated with all three of the larger clusters (identifiable in two cases as Dall sheep [*Ovis dalli*] and wapiti [*Cervus* sp.]), and charcoal fragments were recovered within at least a meter of their peripheries.

Smith's (1985) reconstruction of lithic reduction sequences confirms the assumption that the clusters are not simply spatial aggregations of debris; they comprise a set of functionally related remains. Each cluster, in fact, contains several such sets, representing specific acts of flake production and tool manufacture and/or modification. According to Smith (1985:7–8), the Component I clusters reflect the manufacture of bifaces and possibly scrapers, along with the reduction of several large cobbles of poor-quality stone.

The major concentrations are discussed individually in the following sections.

Cluster X

Cluster X consists of more than 1,160 flakes (table 5.1; figure 5.3). The predominant raw material is a moderately good-quality brown chert; some degraded quartzite and sandstone are also present. Worked implements include a brown chert bifacial knife, the tip of a wide unifacial brown chert knife, a gray chert rectangular point base, and a crude biface fragment of poor-quality chert. A brown chert retouched flake was recovered within two meters of the periphery of this cluster.

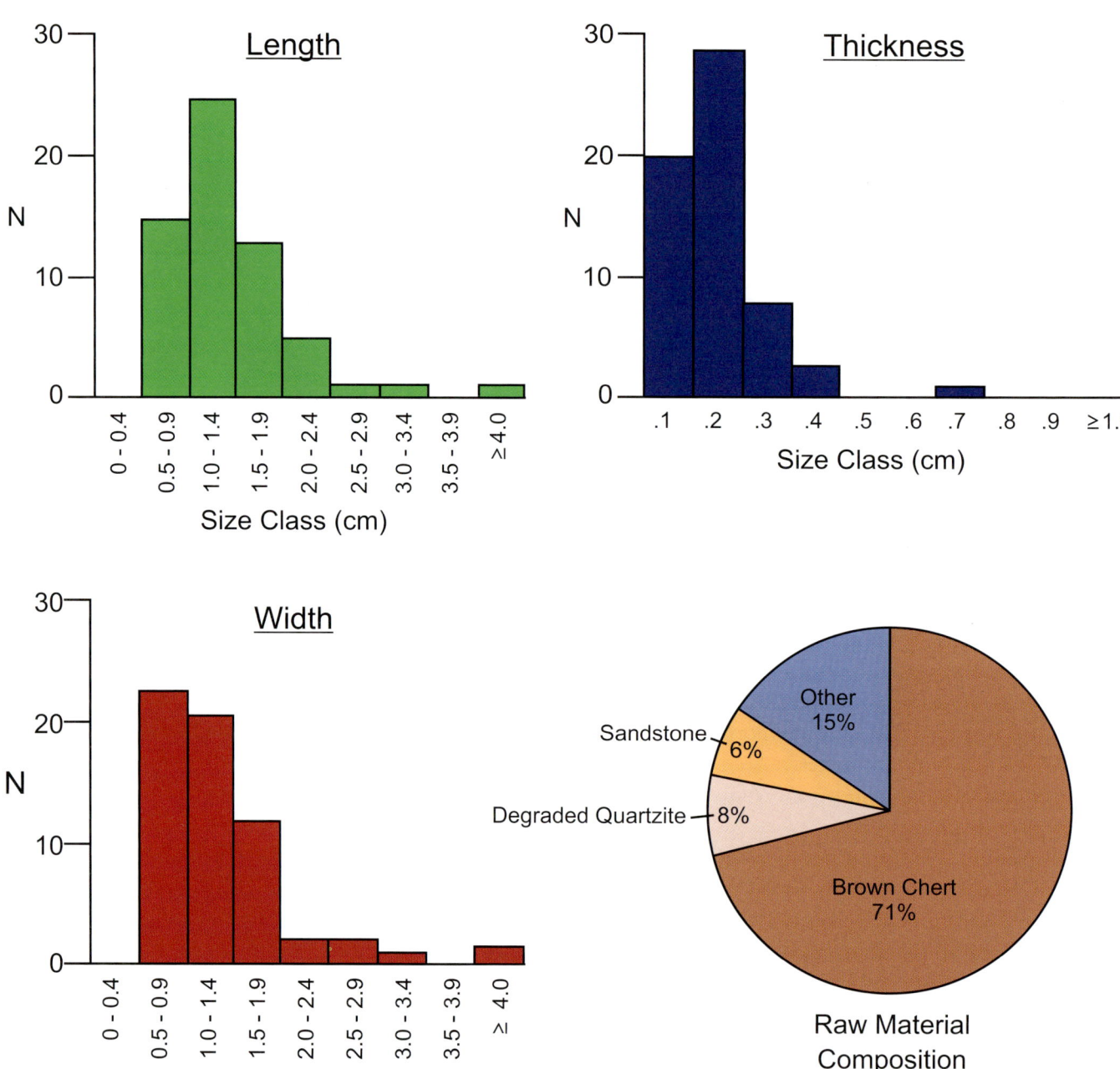

FIGURE 5.3. Raw material composition and flake sample dimensions for Cluster X, Component I.

TABLE 5.1. Cluster X: Tools and large flakes

Catalog No.	Description	Material	Dimensions (cm)
76-4382	Knife fragment	Brown chert	3.6 × 3.0 × 0.5
76-5311	Point base	Gray chert	0.8 × 1.8 × 0.3
76-1674	Biface fragment	Gray chert	3.1 × 6.0 × 2.0
76-5320	Large flake	Degraded quartzite	8.1 × 3.4 × 0.6
76-5327	Biface	Brown chert	6.2 × 3.3 × 0.8

None of the tools reflects clear indications of wear, and according to Del Bene (unpublished notes), the point base exhibits no edge-grinding and thus no suggestion of hafting. Flake size is small, being influenced by the large number of small brown chert waste flakes.

Smith (1985:7, unpublished notes) reconstructed a reduction sequence for the brown chert bifacial knife. This tool was apparently broken prior to completion (Smith 1985:7); both halves were recovered from the concentration.

Faunal remains are present but unidentifiable.

Cluster Y

This cluster is small, being composed of little more than 110 flakes (table 5.2; figure 5.4). The raw materials used here appear to have been of local origin and include gray and brown chert of poor quality and degraded quartzite. Tools comprise a small end scraper, a large chopping tool manufactured on a flat cobble of dark rhyolite, a split cobble scraping tool, and a utilized flake. A rectangular point base of good-quality chert is also present.

The end scraper possesses a steep edge angle (75°), and according to Del Bene (unpublished notes), exhibits the type of wear patterning characteristic of scraping hard surfaces. The narrow edge of the utilized flake (35° angle) bears mild damage. The chopping tool possesses a steeper edge (75° angle), which displays heavy step-flaking retouch wear (see Keeley 1980:24, figure 10). Del Bene observes that the edges of the point base show no sign of grinding. Flake size is in the medium/large range for the site. Many of the waste flakes retain portions of cortex, and some manifest step-flaking retouch on their proximal ends.

Smith (1985:7–8, unpublished notes) records the presence of a small concentration of "honey colored chert" (21 flakes) in this cluster, which he suggests might have been produced by scraper resharpening; no finished implements of this material have been recovered from Component I.

TABLE 5.2. Cluster Y: Tools and large flakes

Catalog No.	Description	Material	Dimensions (cm)
76-4563	Point base	Gray chert	0.8 × 2.4 × 0.3
76-4270	End scraper	Tan chert	2.1 × 1.7 × 0.6
76-4632	Utilized flake	Black chert	4.2 × 2.3 × 0.8
76-4516	Chopping tool	Dark rhyolite	17.5 × 10.9 × 3.1
76-4273	Scraping tool	Degraded quartzite	10.7 × 6.3 × 1.9
76-4210	Large flake	Degraded quartzite	7.4 × 5.3 × 1.7

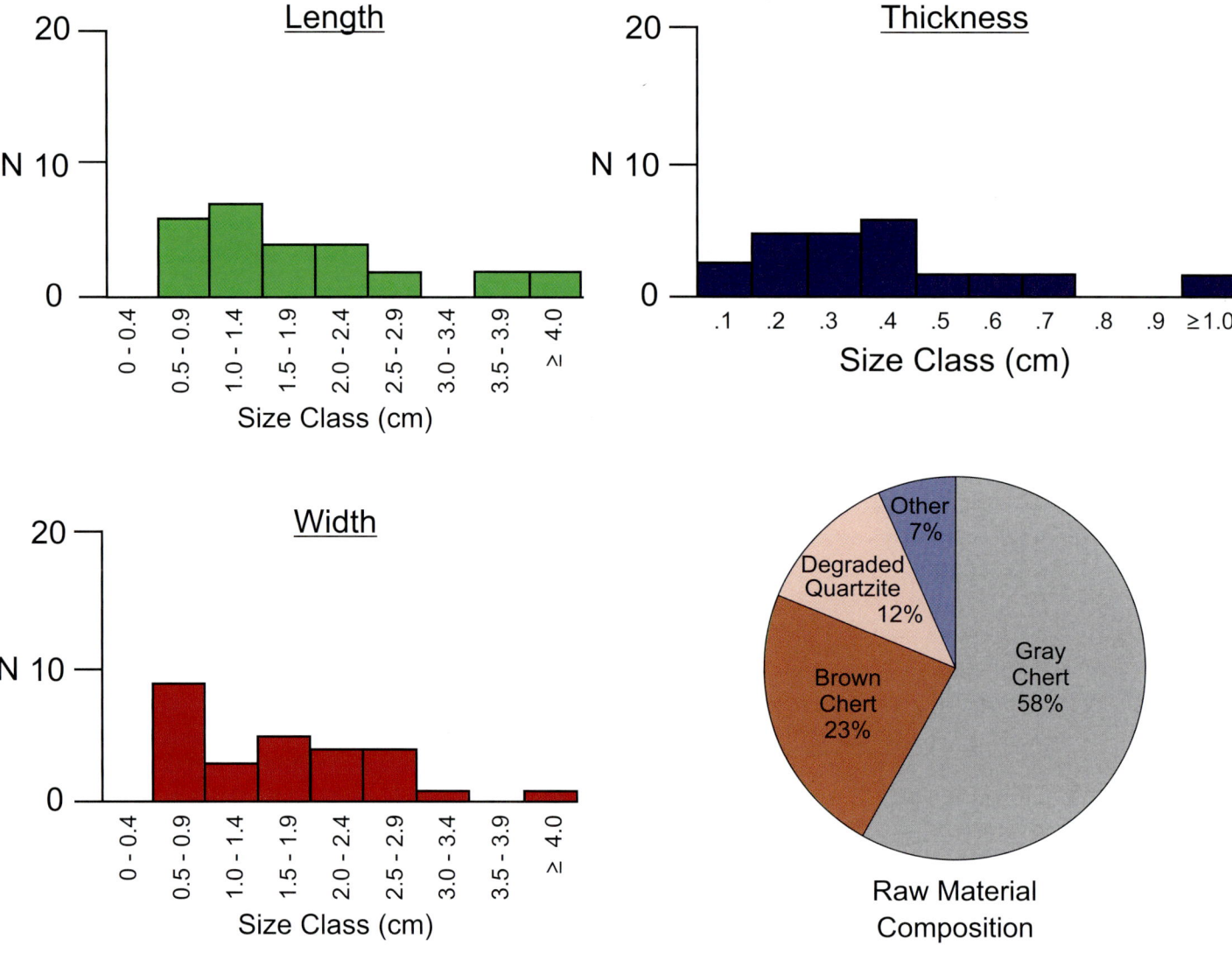

FIGURE 5.4. Raw material composition and flake sample dimensions for Cluster Y, Component I.

Faunal remains are present, and teeth fragments have been identified as belonging to *Cervus*.

Cluster Z

Cluster Z is composed of more than 500 flakes (table 5.3; figure 5.5). Most of these are of degraded quartzite; some light rhyolite is present. The tools are limited to a medium-sized scraper of crude manufacture and two broken cobbles, one of which has been worked into a chopping tool. Another broken cobble has been classified as a core.

The worked cobble possesses a steep edge (80° angle) and bears heavy step-flaking retouch. The other cobble does not manifest significant wear. One of the flakes also possesses a steep edge angle (approaching 90°) and exhibits step-flaking. The side scraper displays a medium edge angle (45°–60°) and no macroscopically visible wear. The flakes are generally large, and most do not exhibit the characteristics of resharpening flakes.

Smith (1985:7–8, unpublished notes) was able to rejoin many of the degraded quartzite flakes from this cluster and concludes that they reflect the reduction of two large cobble cores. Smith (1985:7–8, unpublished notes) also identifies a small dense concentration of "gray-speckled chert" flakes, tentatively attributed to biface production, and a group of light rhyolite flakes, attributed to unspecified tool sharpening.

The largest concentration of faunal remains on the Component I occupation floor was discovered within 1–3 m of the northwestern periphery of this cluster. These teeth fragments were identified as belonging to *Ovis dalli*.

Other Areas

Several small debris concentrations are located in the easternmost portion of the excavated area (N8-12 E22-27). These include a mass of coarse-grained, degraded quartzite flakes, many of them conjoinable, which, according to Smith (1985:7, unpublished notes), were produced by reduction of a large cobble core. Near the modern bluff edge (N9-10 E26-27), a dense cluster of gray chalcedony flakes (> 600) represents manufacture of a biface from a subangular cobble core. Fragments of the biface, which apparently broke prior to completion, were recovered at the same location. No faunal remains or charcoal fragments were found in this part of the site.

Several small concentrations of isolated fragments of faunal debris and charcoal occur in the northwestern area of the excavations (N16-20 E10-18). A small cluster of light-gray rhyolite flakes (> 40) in N17-18 E12-13 has been related by Smith (1985:8, unpublished

TABLE 5.3. Cluster Z: Cores, tools, and large flakes

Catalog No.	Description	Material	Dimensions (cm)
77-2732	Core	Degraded quartzite	12.4 × 11.8 × 6.4
77-3726	Scraper	Degraded quartzite	6.2 × 4.1 × 1.0
77-2728	Percussion tool	Degraded quartzite	15.4 × 7.0 × 4.1
77-2701	Percussion tool	Degraded quartzite	18.0 × 8.2 × 6.4
77-3743	Large flakes (n = 17)	Degraded quartzite	7.3 × 5.6 × 1.6[a]

[a]Mean dimension for 17 large flakes.

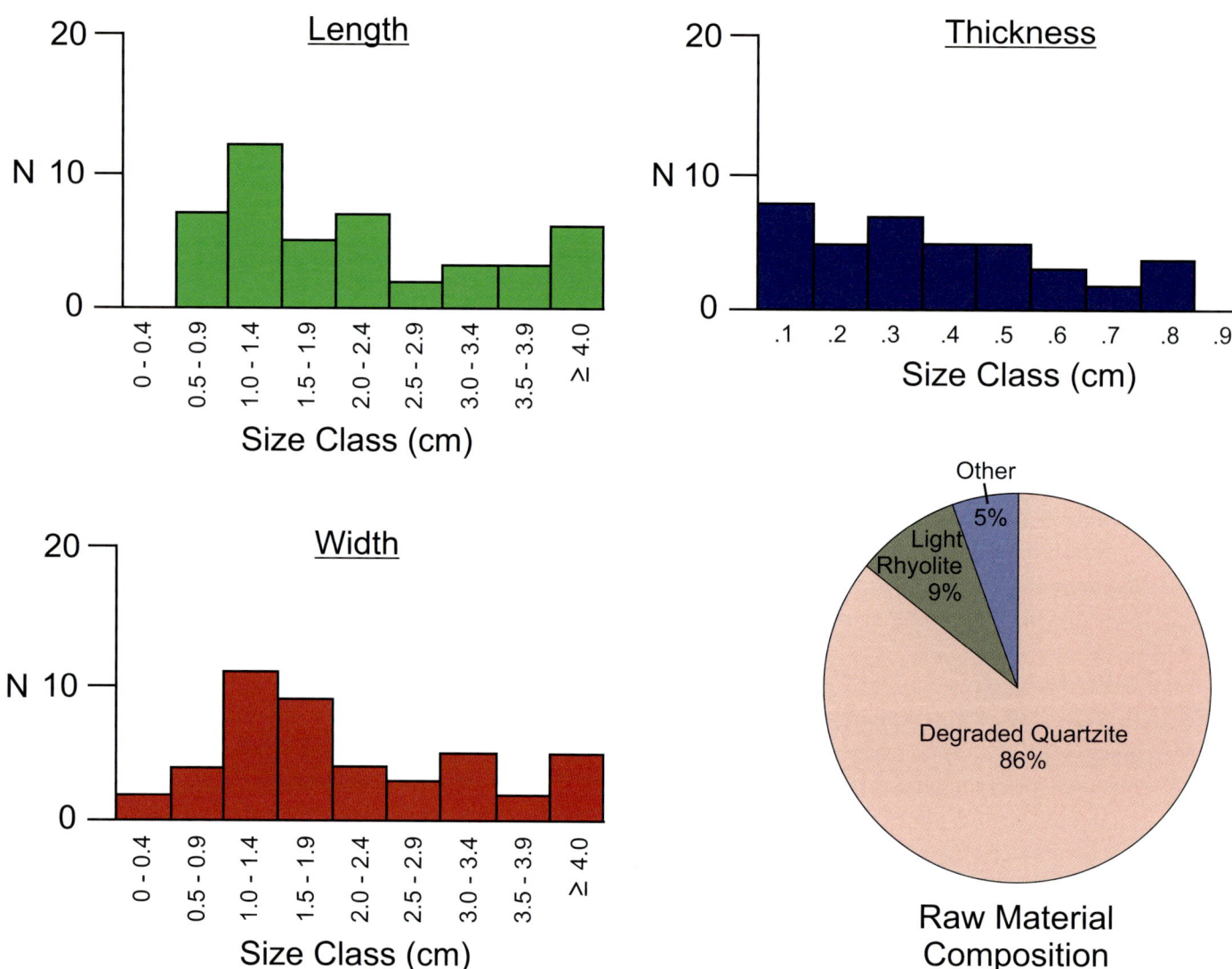

FIGURE 5.5. Raw material composition and flake sample dimensions for Cluster Z, Component I.

notes) to the production of a scraper (located in N8-9 E6-7). Eight more flakes of the same material in N17-18 E18-19 were apparently struck off this tool during resharpening.

Small concentrations and isolated fragments of charcoal and faunal debris are also present in the central (N4-14 E6-16) and southern (S2-N4 E16-24) areas of the site. No major lithic reduction sequences were reassembled from these areas.

Component II Occupation Floor

Fourteen large concentrations have been defined for the Component II occupation floor (figure 5.2), accounting for roughly 70 percent of the total number of artifacts in this component. A few small accumulations of material (including faunal remains) are found in other portions of the occupation area. The major concentrations (labeled A through N) range from approximately 4 m^2 to more than 13 m^2 and contain from less than 350 to more than 3,500 items. Ten of the clusters are associated with faunal remains (identifiable in two cases as steppe bison [*Bison priscus*]), and charcoal fragments were found within a meter of all but one (partially unexcavated) cluster. Avian gastroliths were found on this occupation floor as well, but in association with only two of the debris concentrations.

With the exception of several microblade core reduction sequences, refitting of cores and tools has not been undertaken with the Component II assemblage, and while the assumption that the contents of the debris concentrations are largely interrelated is reasonable, it remains undemonstrated. Nevertheless, there are a number of significant observations that can be made on the basis of available data.

The clusters may be divided into at least two and possibly three distinct types. The most striking difference emerges between artifact concentrations containing microblade cores, microblade core parts, and microblades (clusters A, B, C, G, and N), and concentrations lacking such remains (clusters D, E, F, H, I, J, K, L, and M). With the exception of a single microblade fragment in Cluster H and an isolated microblade core in Cluster M, the separation is complete. In several cases, microblade cores have been reassembled from the debris of particular concentrations (see chapter 4). The microblade clusters are also characterized by the presence of burin technology; other than a simple burin on-a-snap in Cluster H and a possible burin spall in Cluster F, burins and burin spalls are confined to the microblade concentration.

A further subdivision might be possible among the nonmicroblade clusters. These include two clusters (E and K) in which raw materials of comparatively high quality (e.g., chert, chalcedony, rhyolite) predominate, flake size is small to medium (relative to other clusters), and bifacial points and point fragments occur. However, bifacial tool reduction sequences have yet to be reassembled. The nonmicroblade clusters also include two examples (L and M) in which poor-quality raw materials (e.g., degraded quartzite, pumice, sandstone) predominate, and flake size is large. Relatively crude tools are associated with these concentrations. The remaining nonmicroblade clusters (D, F, H, I, and J) do not manifest clear patterns with respect to these characteristics. No patterns were observable with regard to the distribution of associated faunal remains (absent, present, or abundant) and the contents of the concentrations.

The major clusters are described individually in the following sections.

Cluster A

This cluster is composed of less than 350 artifacts (table 5.4; figure 5.6). Light rhyolite and chert predominate among the raw materials, although some degraded quartzite and sandstone are present. A single light rhyolite microblade core was reconstructed, and a total of 5 complete microblades and 105 microblade fragments of the same material were recovered. Also present were 2 microblade fragments of obsidian and gray chert. The cluster contains only 1 tool, a gray chert burin, and there are few, if any, utilized flakes.

The flake dimension profiles are skewed toward the smaller end of the spectrum, the model length and width categories being 1.0–1.4 cm and 0.5–0.9 cm, respectively. Among these are a number of light rhyolite flakes exhibiting heavy wear in the form of step-flaking along the edge created by the intersection of the platform and dorsal surface. The angle of this edge typically approaches 90° on these flakes. No faunal remains were found in this cluster.

Cluster B

Cluster B is composed of more than 800 flakes and 199 microblades and microblade fragments (table 5.5; figure 5.7). Gray chert and obsidian account for most of the raw material. There are 2 gray chert microblade cores and numerous core tablets (platform rejuvenation flakes) present, and although obsidian cores are absent, there are core tablets and microblades

TABLE 5.4. Cluster A: Cores, tools, and large flakes

Catalog No.	Description	Material	Dimensions (cm)
76-3225	Microblade core	Light rhyolite	2.2 × 1.3 × 2.5
76-3269	Burin	Gray chert	2.7 × 2.1 × 0.4
76-3219	Large blade	Brown chert	6.9 × 6.0 × 0.8

TABLE 5.5. Cluster B: Cores and tools

Catalog No.	Description	Material	Dimensions (cm)
76-3765	Microblade core	Gray chert	2.9 × 1.2 × 2.3
77-2089	Microblade core	Gray chert	2.6 × 1.8 × 1.8
77-1433	Burin	Jasper	2.1 × 2.2 × 0.8
77-1435	Burin	Green chert	3.1 × 2.4 × 1.0
77-2042	Burin	Gray chert	2.0 × 1.7 × 0.8
77-2045	Burin	Green chert	2.2 × 0.9 × 1.9
77-3429	Core-scraper	Green chert	4.7 × 4.7 × 1.9
77-2040	Core-scraper fragment	Gray chert	4.3 × 4.0 × 1.1
76-3475	Core-scraper fragment	Gray chert	3.2 × 2.0 × 0.9
77-2057	Core-scraper fragment	Gray chert	4.3 × 2.5 × 1.2
76-3447	Core-scraper fragment	Gray chert	3.6 × 2.2 × 1.1
77-1773	Utilized flake	Black chert	2.9 × 2.3 × 0.5

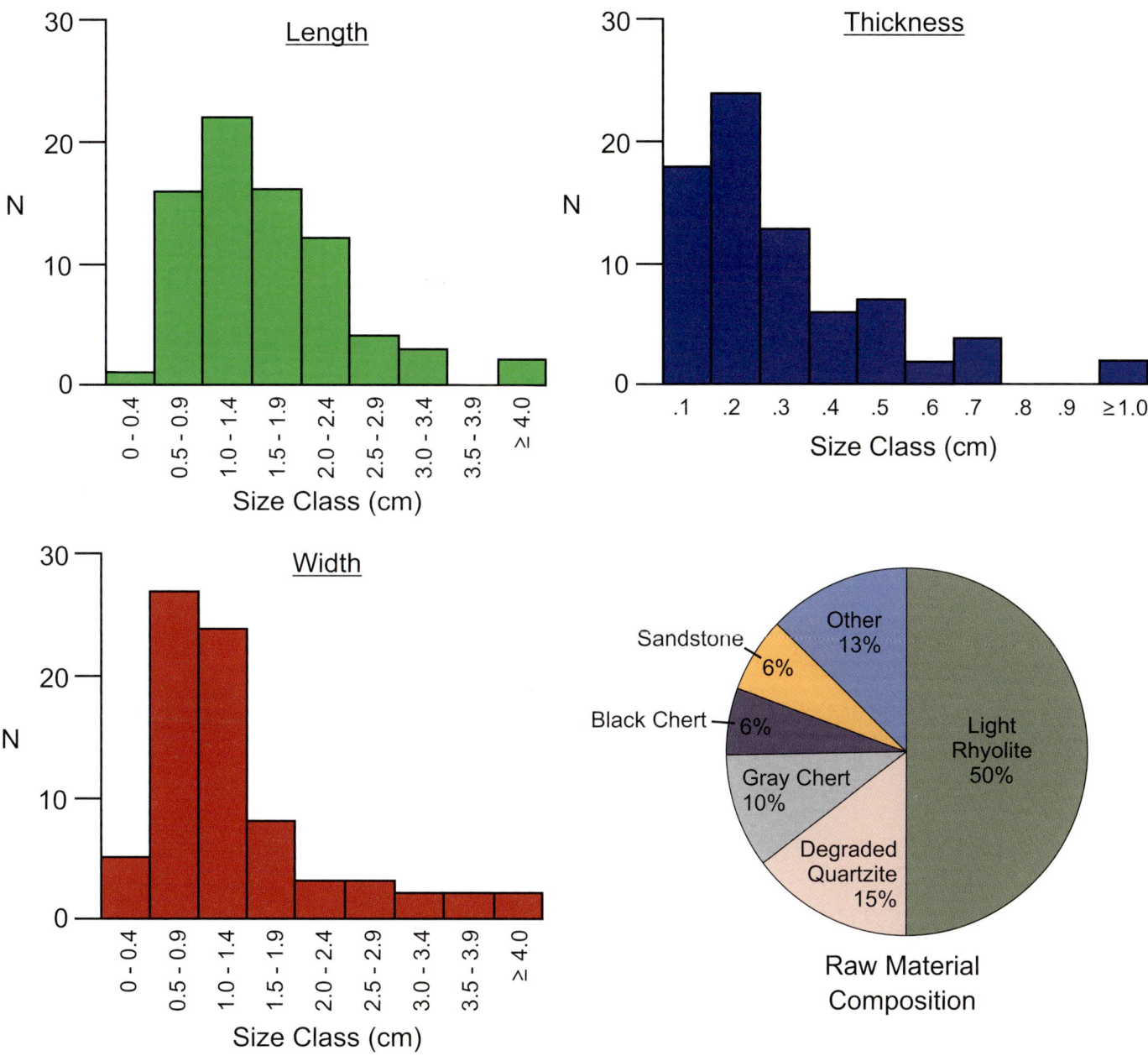

FIGURE 5.6. Raw material composition and flake sample dimensions for Cluster A, Component II.

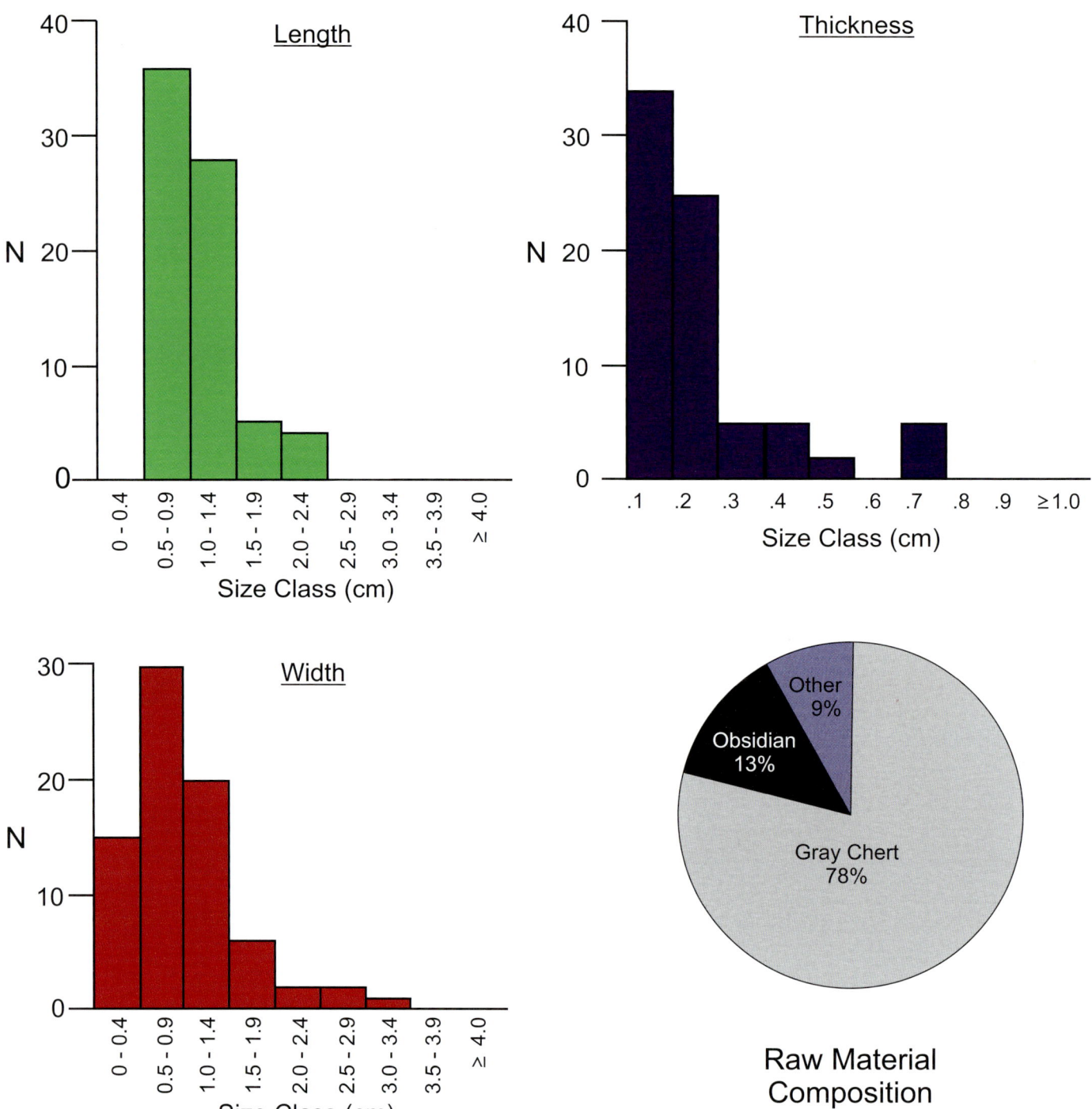

FIGURE 5.7. Raw material composition and flake sample dimensions for Cluster B, Component II.

of this material. Several chalcedony microblade fragments were also recovered. Tools include 4 burins, 1 core-scraper, and 4 core-scraper fragments.

The core-scrapers all exhibit heavy wear in the form of step-flaking retouch, as do some of the waste flakes. Edge angles on these tools range from 45° to 90°. Flake size in this cluster is unusually small, the modal size class for length and width being 0.5–0.9 cm. Faunal remains are not present.

After excavations were completed in 1977, it became clear that a substantial portion of this cluster remained unexcavated (see figure 5.2).

Cluster C

Cluster C contains more than 700 flakes, as well as 146 microblades and microblade fragments (table 5.6; figure 5.8). The three most important raw materials are chalcedony, degraded quartzite, and light rhyolite. There are three light rhyolite microblade cores and four aberrant

TABLE 5.6. Cluster C: Cores and tools

Catalog No.	Description	Material	Dimensions (cm)
77-637	Microblade core	Light rhyolite	1.5 × 1.1 × 2.4
77-364	Microblade core	Light rhyolite	3.6 × 1.7 × 2.6
77-669	Microblade core	Light rhyolite	3.2 × 1.6 × 2.8
77-638	Transverse burin	Light rhyolite	2.5 × 2.4 × 0.6
77-667	Burin	Light rhyolite	2.6 × 1.8 × 1.3
76-5480	Burin	Chalcedony	2.2 × 2.6 × 0.7
77-574	Burin	Light rhyolite	3.2 × 2.4 × 0.6
76-5496	Burin	Light rhyolite	2.7 × 2.0 × 0.8
77-308	Burin	Light rhyolite	1.8 × 2.4 × 1.1
77-370	Burin	Light rhyolite	2.2 × 3.6 × 0.5
77-604	Burin	Light rhyolite	2.1 × 2.0 × 0.7
76-1362	Aberrant core	Chalcedony	3.7 × 1.7 × 3.0
76-2513	Aberrant core	Chalcedony	3.7 × 1.4 × 3.3
76-5527	Aberrant core	Chalcedony	3.9 × 2.1 × 3.2
76-5534	Aberrant core	Chalcedony	3.0 × 1.3 × 2.7
77-362	Core-scraper fragment	Chalcedony	3.2 × 3.0 × 1.5
76-5533	Core-scraper	Chalcedony	5.6 × 3.3 × 1.5
76-5098	Core-scraper fragment	Chalcedony	4.0 × 2.2 × 1.0
77-411	Core-scraper fragment	Chalcedony	3.2 × 2.6 × 1.4
76-4364	Biface fragment	Degraded quartzite	6.8 × 3.9 × 1.0
77-365	Denticulate	Chalcedony	3.8 × 2.6 × 1.2
77-369	Utilized flake	Gray chert	2.2 × 3.3 × 0.4
77-800	Utilized flake	Degraded quartzite	4.7 × 1.9 × 0.6
76-5125	Bladelike flake fragment	Degraded quartzite	6.2 × 3.1 × 0.8

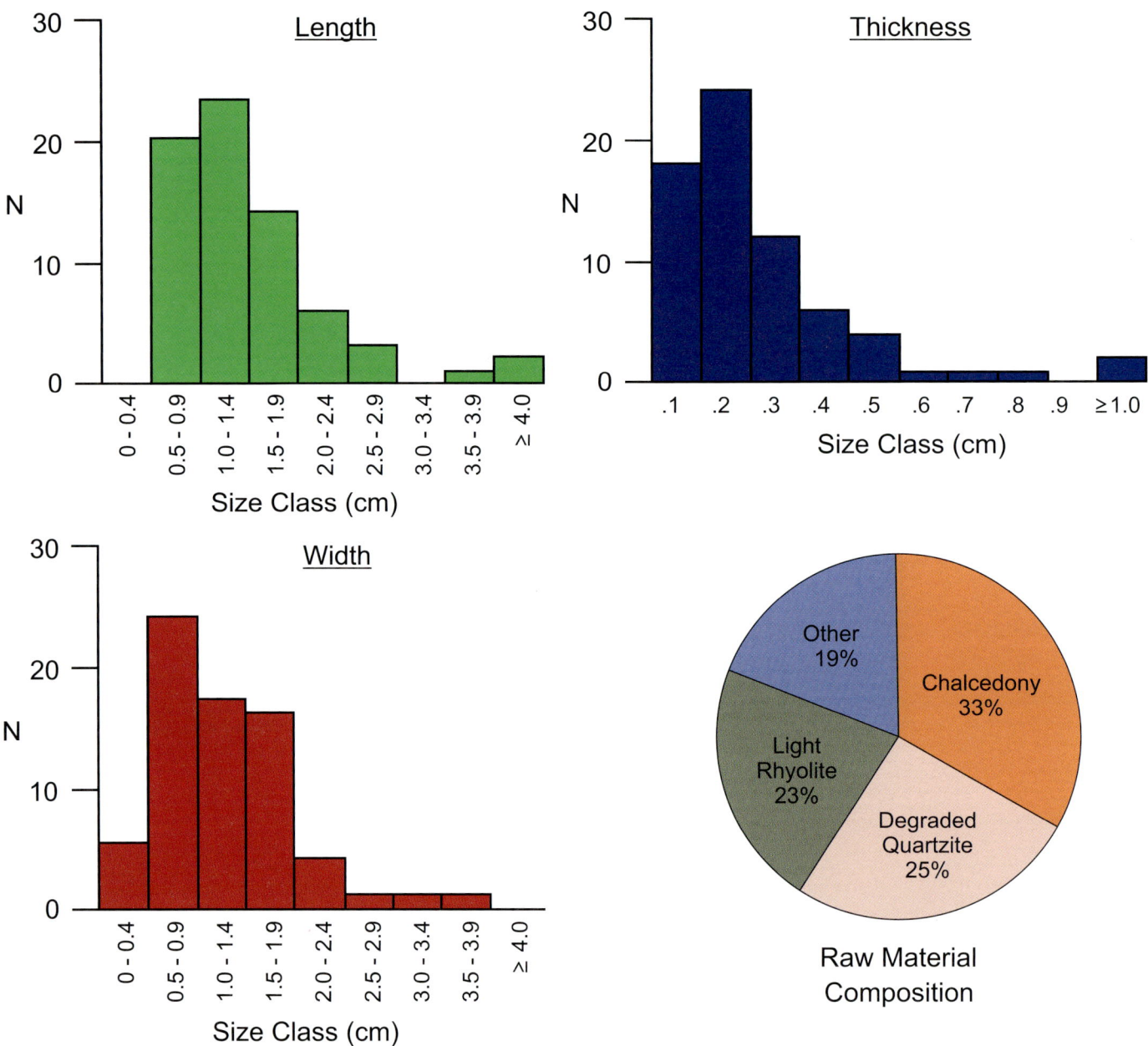

FIGURE 5.8. Raw material composition and flake sample dimensions for Cluster C, Component II.

chalcedony cores from which one or two blades have been struck. Most of the microblades are rhyolite; there are only 18 chalcedony blades and blade fragments. Eight burins were found, as well as one denticulate tool and a degraded quartzite biface fragment.

Heavy wear in the form of step-flaking retouch is visible on two of the rhyolite and all of the chalcedony cores. The burins also bear signs of substantial use, and the denticulate shows signs of moderate use. The flake dimension profiles reflect a general tendency toward small size. Step-flaking is present on some of the chalcedony and rhyolite waste flakes, which manifest edge angles between 45° and 90°. There are more than one hundred small degraded quartzite flakes, some of which reflect the characteristics of bifacial edge trimming flakes (see Bordes 1961:6–7). Two flakes, one of degraded quartzite and the other of gray chert, exhibit traces of heavy utilization (crushed edges). Faunal remains were found in association with the artifacts, and teeth fragments were identified as belonging to *Bison priscus*.

Cluster D

Cluster D is large, composed of more than 1,900 flakes (table 5.7; figure 5.9). Raw materials of poor quality predominate, chiefly degraded quartzite and diabase, although some obsidian, brown chert, and rhyolite were also used. Microblades, microblade cores, and core parts are entirely absent from this cluster. A single obsidian bifacial knife fragment was found on the southern periphery. A light rhyolite knife of similar size and shape was recovered approximately 1 m beyond the northeastern periphery. On the northwestern periphery, there is a large sandstone cobble from which several flakes have been struck on one corner.

The obsidian knife fragment exhibits heavy wear in the form of crushed edges. The edges of the light rhyolite knife, according to Del Bene (unpublished notes), bear evidence of hafting and use on soft material. The edge angles on these tools are low; the sandstone cobble–worked edge is approximately 45°. In terms of size, the flakes lie in the medium range for the site as a whole, as illustrated by the dimension profiles. Some of the larger flakes possess signs of moderate damage along their sharp edges, while other flakes (of degraded quartzite and diabase) bear the faceted platforms and dorsal scars characteristic of bifacial sharpening flakes. Some of these are large and manifest step-flaking retouch. Faunal remains are not present.

Cluster E

This cluster is composed of more than 1,000 flakes (table 5.8; figure 5.10). Although there is a substantial amount of degraded quartzite present, much of the raw material is of moderately good quality (rhyolite and medium-gray chert), and there is a sizable proportion of high-quality chalcedony. Worked implements include a delicately flaked chalcedony projectile point tip, a small triangular chalcedony biface, and a large dark rhyolite bifacial tool. Microblades, microblade cores, and burins are completely absent.

TABLE 5.7. Cluster D: Tools

Catalog No.	Description	Material	Dimensions (cm)
76-1361	Bifacial knife fragment	Obsidian	3.7 × 2.5 × 0.7
74-199	Bifacial knife	Light rhyolite	5.1 × 2.6 × 0.9
76-5298	Chopping tool	Sandstone	17.0 × 14.3 × 4.6

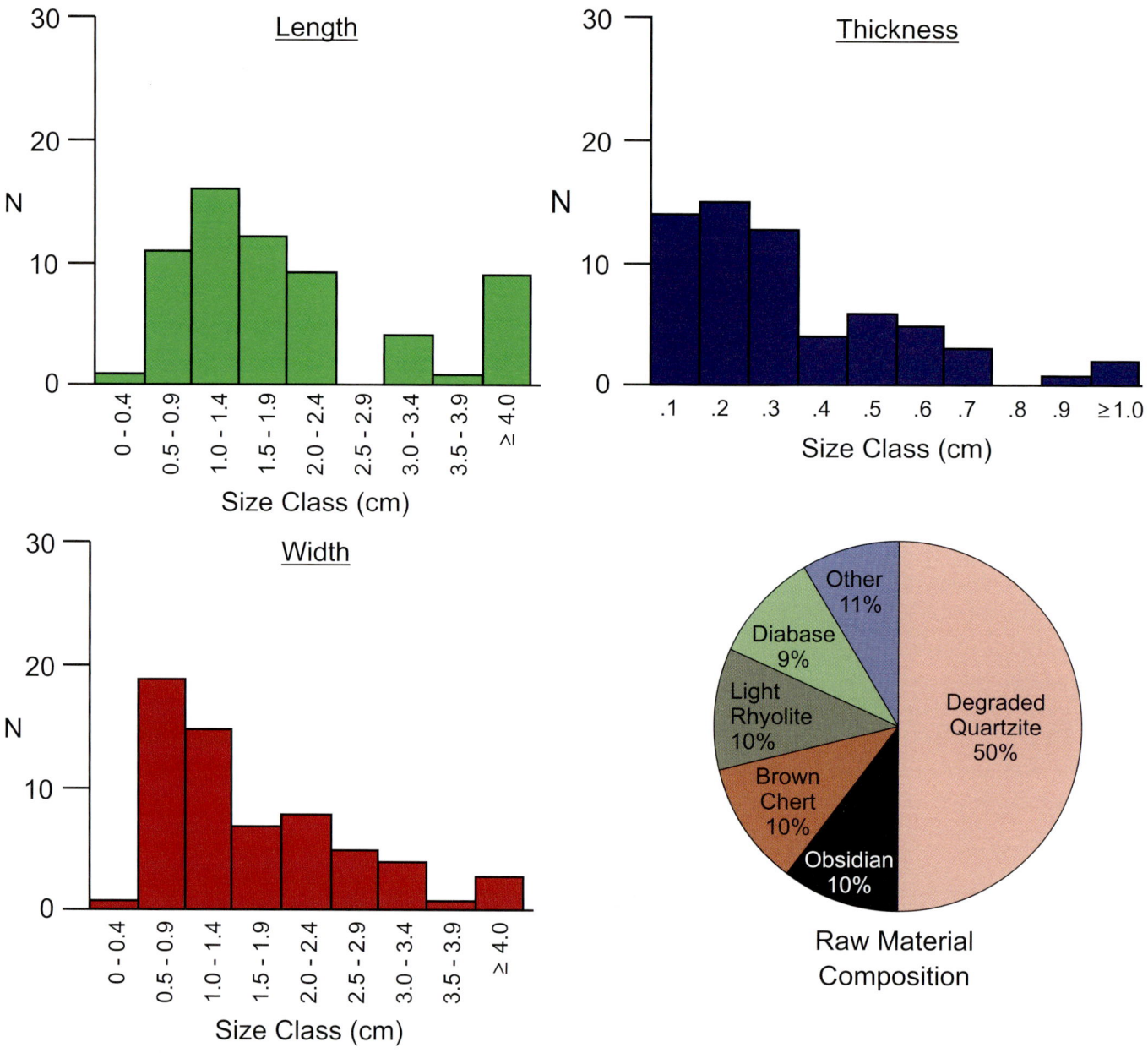

FIGURE 5.9. Raw material composition and flake sample dimensions for Cluster D, Component II.

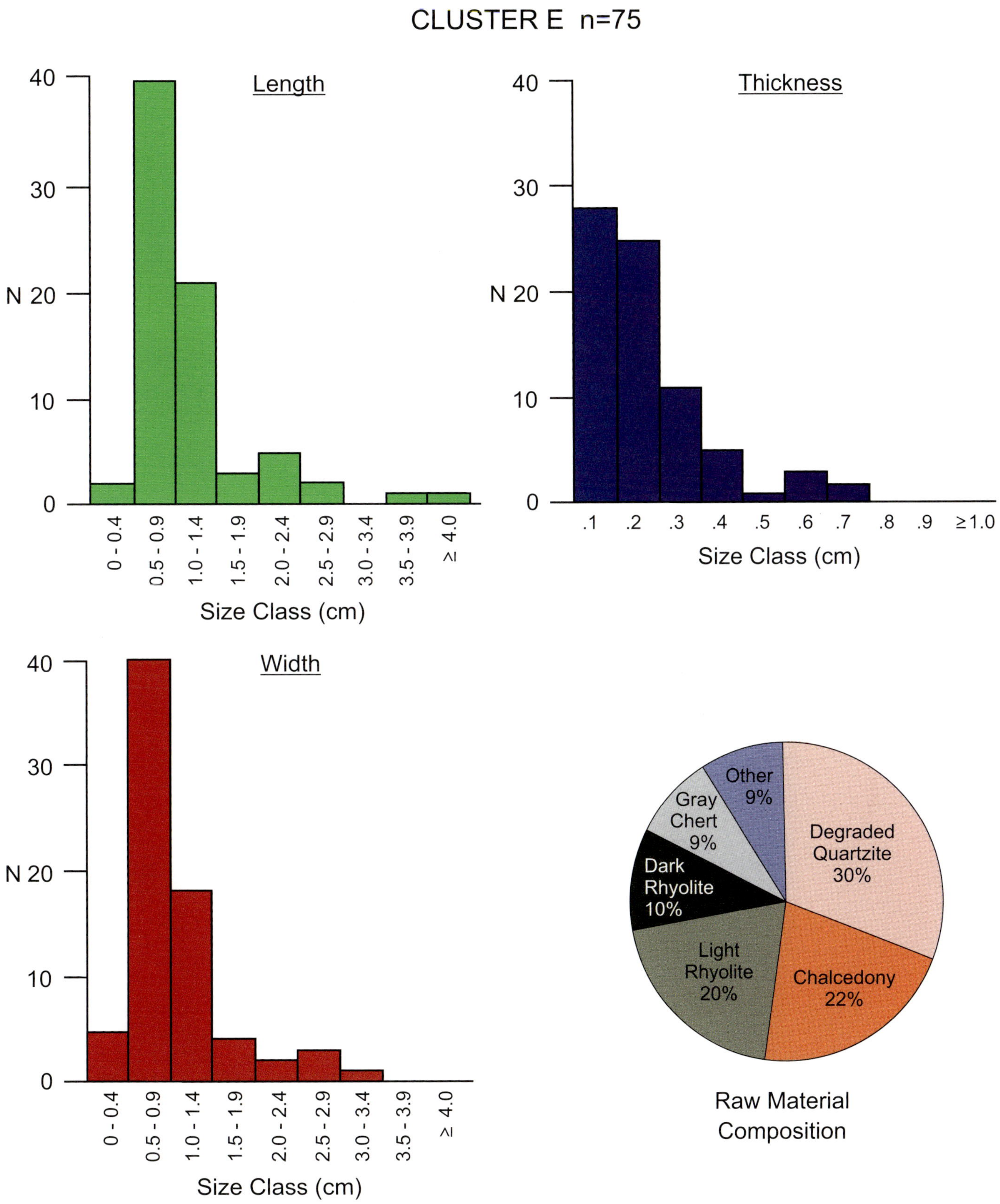

FIGURE 5.10. Raw material composition and flake sample dimensions for Cluster E, Component II.

TABLE 5.8. Cluster E: Tools and large flakes

Catalog No.	Description	Material	Dimensions (cm)
77-1879	Triangular biface	Chalcedony	3.3 × 2.1 × 0.6
77-2840	Point tip	Chalcedony	0.9 × 0.8 × 0.2
77-2219	Biface	Dark rhyolite	17.6 × 7.2 × 4.8
77-1880	Large flake	Degraded quartzite	13.6 × 6.3 × 2.5
77-2555	Large flake	Dark rhyolite	7.1 × 3.7 × 1.5
77-5007	Large flake	Degraded quartzite	7.3 × 4.7 × 1.3
77-2439	Large flake	Degraded quartzite	7.2 × 5.2 × 2.3

The large biface exhibits heavy use wear in the form of step-flaking along one side. The flake dimension profiles are strongly skewed toward the smaller end of the size spectrum, the modal length and width categories being 0.5–0.9 cm. The waste flakes appear to fall into two groups: (1) very small flakes (< 1.0 cm), frequently of chalcedony (and some of which possess the characteristics of bifacial sharpening flakes), and (2) slightly larger flakes (> 1.0 cm), often of degraded quartzite, displaying heavy step-flaking. The latter include two gray chert flakes that bear extremely heavy step-flaking retouch along portions characterized by steep edge angles. There are several large flakes of coarse-grained material that lack discernible wear. Faunal remains are present but unidentifiable.

Cluster F

Cluster F contains more than 680 flakes (table 5.9; figure 5.11). Raw materials of poor quality predominate (degraded quartzite and quartzite), but light rhyolite and chert are also present. Microblades, burins, and other evidence of their production are absent, except for one possible burin spall. Four small, finely worked bifacial knives and two knife fragments (chert and rhyolite) were recovered, and there are two larger bifacial tools (diabase and degraded quartzite) of cruder manufacture as well. Four scraping tools were found, and on the southern periphery, there are six chalcedony core-scrapers. In addition, there is a chalcedony scraping tool and one utilized flake.

The scraping tools all possess sharp edge angles (< 45°), and three of these manifest moderate damage in the form of edge crushing. Five of the steep-edged core-scrapers bear heavy step-flaking retouch. The chalcedony scraping tool exhibits edge polish; several of the bifaces and scrapers exhibit grinding along certain edges. The flakes are generally small and might display the characteristics of bifacial and scraping tool-sharpening flakes. Some of the larger flakes of coarse-grained material bear step-flaking retouch. Faunal remains are present but unidentifiable.

Cluster G

This cluster (northernmost portion unexcavated) is composed of approximately 2,800 flakes (table 5.10; figure 5.12). The most important raw materials are gray chert and chalcedony, although a significant quantity of degraded quartzite is also present. There are no less than 13 microblade cores and more than 760 microblades and microblade fragments. Cluster G also contains 5 burins and 22 core-scrapers. In addition to this, there are 2 denticulate tools and

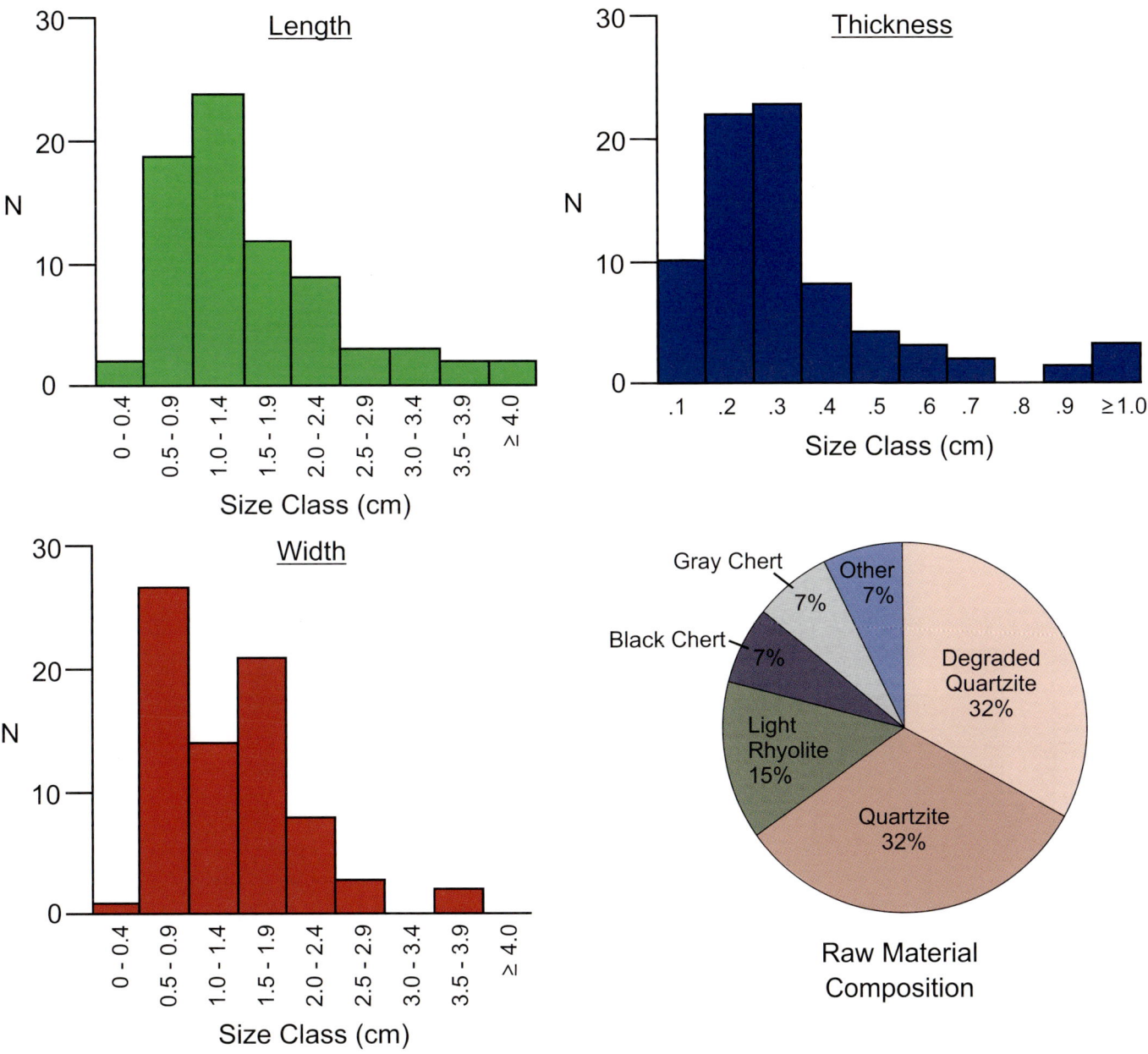

FIGURE 5.11. Raw material composition and flake sample dimensions for Cluster F, Component II.

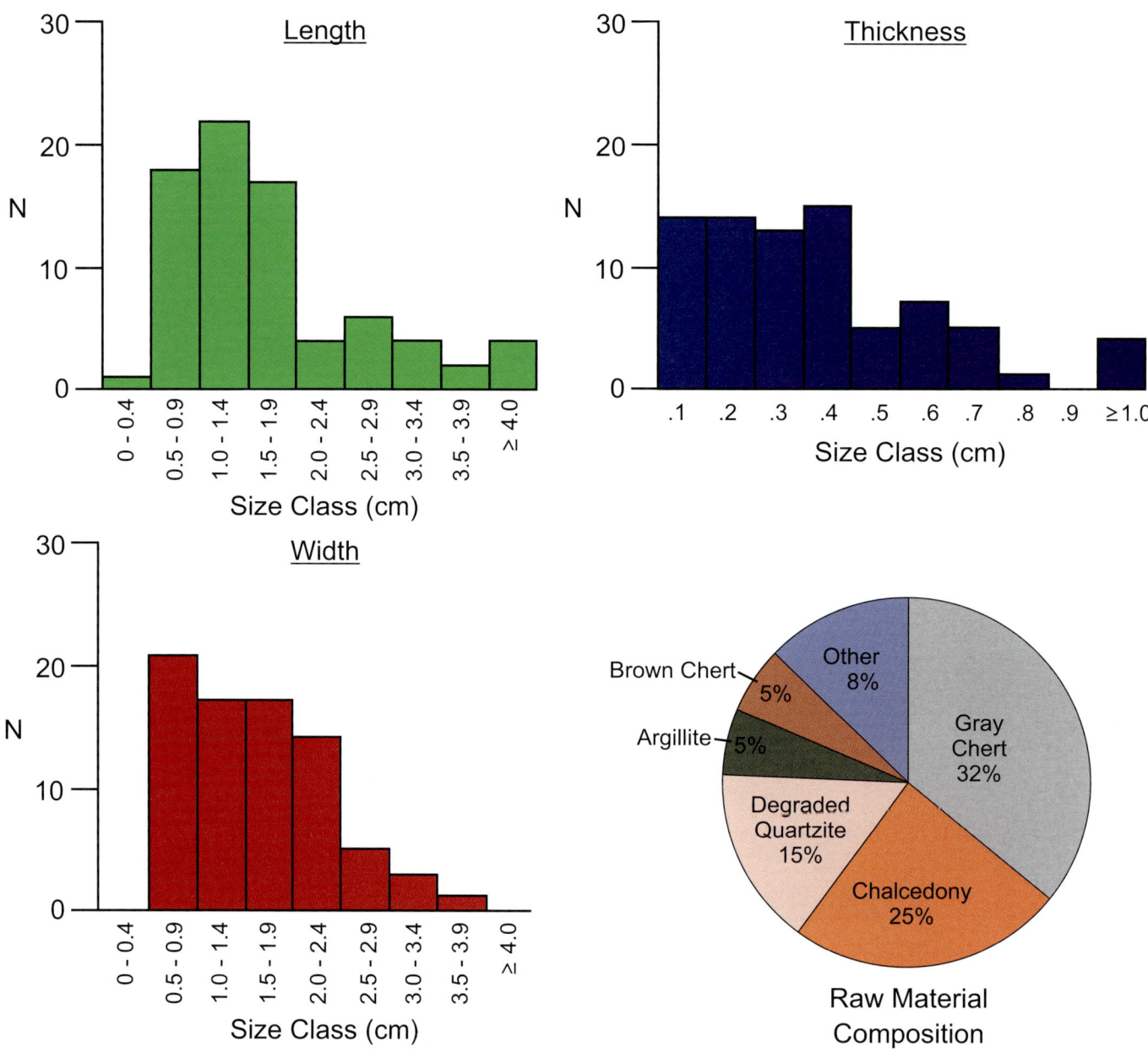

FIGURE 5.12. Raw material composition and flake sample dimensions for Cluster G, Component II.

TABLE 5.9. Cluster F: Tools and large flakes

Catalog No.	Description	Material	Dimensions (cm)
77-213	Knife	Light rhyolite	5.5 × 2.1 × 0.8
77-210	Knife	Light rhyolite	4.6 × 2.4 × 0.8
77-2959	Knife	Dark rhyolite	5.4 × 2.5 × 0.9
77-1976	Knife	Gray chert	5.0 × 2.6 × 0.9
77-3179	Knife fragment	Light rhyolite	1.6 × 1.6 × 0.6
77-1847	Knife fragment	Gray chert	1.6 × 1.7 × 0.5
77-269	Biface	Degraded quartzite	18.2 × 6.5 × 2.5
77-215	Biface	Diabase	8.8 × 5.4 × 1.8
74-81	Scraper	Degraded quartzite	9.9 × 8.5 × 2.5
74-82	Scraper	Degraded quartzite	6.0 × 9.3 × 1.9
77-2659	Scraper	Light rhyolite	5.2 × 8.7 × 0.9
76-5244	Scraper	Light rhyolite	9.5 × 4.8 × 0.8
77-2386	Scraper	Chalcedony	5.0 × 3.8 × 1.1
76-5606	Core-scraper	Chalcedony	5.9 × 3.2 × 1.8
77-2384	Core-scraper	Chalcedony	4.4 × 4.3 × 1.1
77-2387	Core-scraper	Chalcedony	4.8 × 3.5 × 1.8
77-2388	Core-scraper	Chalcedony	4.9 × 3.4 × 2.3
77-209	Core-scraper	Chalcedony	4.6 × 4.0 × 1.5
77-2385	Core-scraper	Chalcedony	4.3 × 3.3 × 1.3
76-5248	Large flake	Degraded quartzite	7.4 × 4.7 × 0.7
77-2244	Utilized flake	Degraded quartzite	8.0 × 4.1 × 0.7

2 large, crude biface fragments (argillite), as well as 2 smaller bifaces (chert) and a battered cobble. Also present are 2 fragments of small bifacial knives, 2 utilized flakes, and 1 retouched flake.

The utilized flakes bear signs of use in the form of edge crushing, and one of the denticulates is heavily worn (step-flaking) around the notch. According to Del Bene (unpublished notes), edge-wear patterns on the retouched flake suggest that it was used to scrape yielding material, and polish on the tips of several pointed microblades suggests the scraping or cutting of soft material. The flake dimension profiles indicate an unusual degree of thickness relative to the length and width, which reflects a generally small size. Many of these thick flakes exhibit heavy step-flaking retouch, as do the core-scrapers. Some possible bifacial trimming flakes can be distinguished. There is a large amount of faunal material present, but none of it is identifiable.

Cluster H

This cluster (partially unexcavated) contains more than 860 flakes (table 5.11; figure 5.13). Degraded quartzite was the chief raw material used here, although some chert and rhyolite of high quality was also used. There are no microblade cores or core parts, but one gray chert microblade fragment was recovered. Burins and burin spalls are also absent. The tools are simple and crude, consisting of three bifaces and four utilized flakes. There is a very large

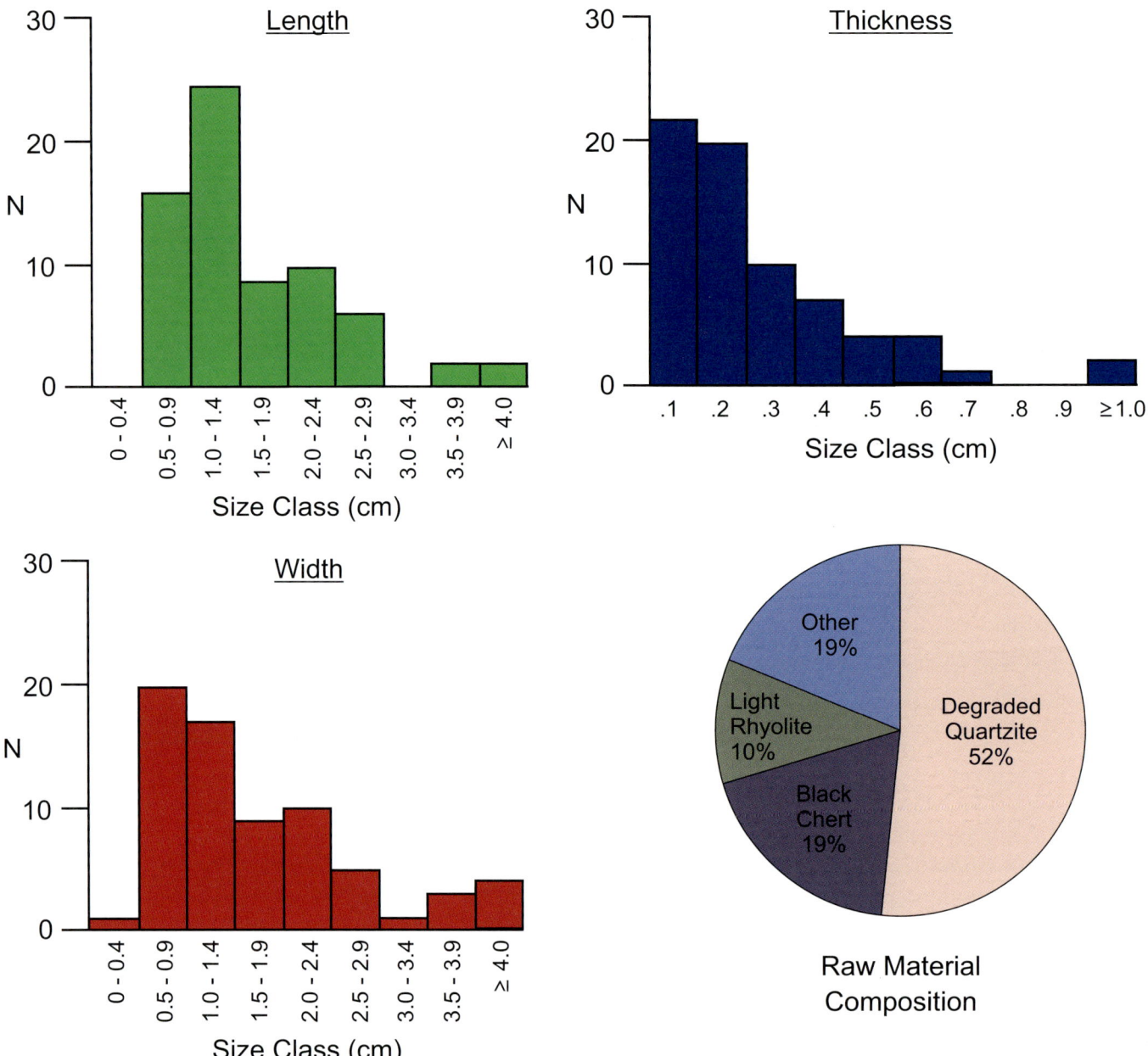

FIGURE 5.13. Raw material composition and flake sample dimensions for Cluster H, Component II.

TABLE 5.10. Cluster G: Cores, tools, and large flakes

Catalog No.	Description	Material	Dimensions (cm)
76-273	Microblade core	Light rhyolite	2.8 × 1.3 × 3.1
76-588	Microblade core	Chalcedony	2.9 × 1.7 × 2.9
76-474	Microblade core	Jasper	3.3 × 1.4 × 2.5
76-4058	Microblade core	Brown chert	4.6 × 1.5 × 2.4
76-4097	Microblade core	Black chert	3.2 × 1.0 × 2.5
76-278	Microblade core	Chalcedony	3.3 × 1.4 × 2.5
76-4518	Microblade core	Chalcedony	3.0 × 1.4 × 2.7
76-241	Microblade core	Chalcedony	2.5 × 1.4 × 2.3
76-764	Microblade core	Chalcedony	2.3 × 1.3 × 2.1
76-731	Microblade core	Chalcedony	2.4 × 1.0 × 2.6
76-757	Microblade core	Chalcedony	2.6 × 1.2 × 2.3
76-587	Microblade core	Chalcedony	3.4 × 1.5 × 2.4
76-278	Microblade core	Chalcedony	3.5 × 1.8 × 2.3
76-832	Burin	Chalcedony	2.4 × 2.2 × 0.5
76-4135	Burin	Chalcedony	2.1 × 2.0 × 0.9
76-274	Burin	Light rhyolite	4.1 × 2.8 × 0.9
76-501	Burin	Light rhyolite	2.5 × 2.1 × 0.6
76-230	Burin	Gray chert	2.5 × 1.2 × 0.7
76-775	Denticulate	Chalcedony	5.2 × 2.9 × 1.4
76-148	Denticulate	Gray chert	3.6 × 2.2 × 0.6
—	Core-scrapers (22)	Chalcedony	Mean = 4.9 × 3.4 × 1.9
76-4103	Knife fragment	Light rhyolite	2.2 × 2.0 × 0.5
76-4475	Knife fragment	Light rhyolite	1.9 × 2.3 × 0.7
76-4400	Biface	Black chert	6.5 × 4.1 × 1.6
76-4047	Biface	Gray chert	5.1 × 4.2 × 1.6
76-259	Biface fragment	Argillite	10.3 × 6.6 × 2.1
76-90	Biface fragment	Argillite	7.7 × 3.4 × 1.8
76-4384	Utilized flake	Black chert	5.8 × 5.5 × 1.0
76-656	Retouched flake	Light rhyolite	5.8 × 5.8 × 0.6
76-4067	Retouched flake	Gray chert	4.6 × 2.4 × 0.7
76-762	Large flake	Degraded quartzite	9.2 × 4.7 × 2.5
76-4049	Large flake	Sandstone	6.6 × 6.0 × 1.0
76-997	Large flake	Degraded quartzite	8.9 × 8.8 × 1.7
76-624	Utilized cobble	Sandstone	19.2 × 10.0 × 2.7

diabase artifact that may be classified as a core and/or chopper and two large flakes of the same material. One core-scraper of black chert is present.

The steep-edged core-scraper bears step-flaking retouch. The flakes are within the small size range for the site. The utilized flakes exhibit considerable nicking along their sharp edges

TABLE 5.11. Cluster H: Cores, tools, and large flakes

Catalog No.	Description	Material	Dimensions (cm)
76-4035	Biface	Dark rhyolite	9.9 × 5.3 × 1.9
74-258	Biface	Degraded quartzite	14.5 × 7.6 × 4.7
74-266	Biface	Degraded quartzite	11.2 × 5.8 × 2.2
74-289	Core-scraper	Black chert	5.8 × 5.3 × 2.5
74-264	Utilized flake	Black chert	5.8 × 4.7 × 1.0
74-267	Utilized flake	Black chert	6.7 × 4.9 × 1.8
76-4039	Utilized flake	Black chert	6.0 × 3.4 × 0.6
74-265	Utilized flake	Gray chert	7.1 × 6.2 × 1.1
76-4092	Core	Diabase	17.5 × 14.4 × 6.4
76-24	Large flake	Diabase	4.8 × 8.3 × 2.2
76-51	Large flake	Diabase	10.4 × 9.9 × 3.9
76-26	Large flake	Sandstone	3.8 × 7.4 × 1.4
76-4036	Large flake	Quartzite	8.5 × 4.4 × 2.3

(35°–45°), while some of the waste flakes possess step-flaking retouch. Other waste flakes have faceted platforms, suggesting bifacial resharpening. A number of small retouch flakes of light rhyolite and gray chert were recovered, although tools of these materials are not present. Faunal remains are present but unidentifiable.

Cluster I

Approximately 880 flakes were recovered from this (partially unexcavated) cluster (table 5.12; figure 5.14). A good deal of chert was used here, much of it of medium quality and presumably of local origin. No microblades, microblade cores, or burins are present. The tools consist of one crude biface, two biface fragments, a core-scraper, and a unifacially retouched tool classified as a scraper. In addition to this, there are a number of large flakes and one obsidian-utilized flake. A finely worked black chert knife lies on the border shared by this cluster and Cluster J.

The core-scraper is typically steep edged (75°–90°), while the scraper has a low edge angle (35°–45°). The former exhibits moderate step-flaking retouch and, according to Del Bene (unpublished notes), the obsidian flake (35°–45° edge angle) was used to scrape a hard surface. The chert knife lacks any clear traces of use or hafting. The flakes lie in the medium range for the site as a whole with respect to size. No faunal remains are present.

Cluster J

This is a large cluster with approximately 2,200 flakes (table 5.13; figure 5.15). Raw material types of poor quality predominate: degraded quartzite, local chert, and even some volcanic pumice. Some rhyolite was used also, however. Microblades, microblade cores, and burins are absent. Tools include three medium-sized bifaces and one fragment. Evidence of projectile point technology is represented by a thin, parallel-sided biface tip and a possible point base. Two unifacially retouched pieces have been classified as scrapers, and a small retouched flake is also present.

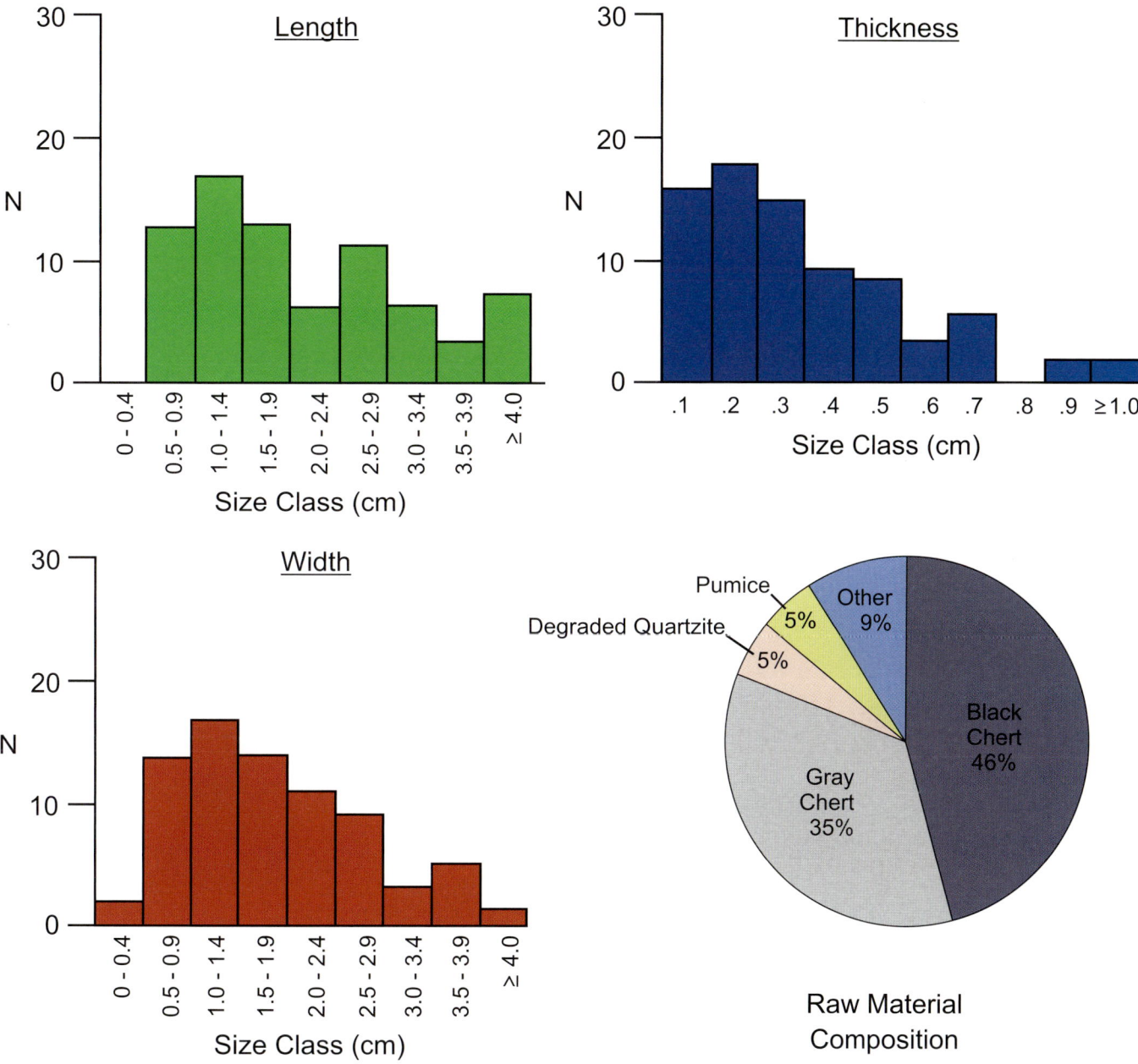

FIGURE 5.14. Raw material composition and flake sample dimensions for Cluster I, Component II.

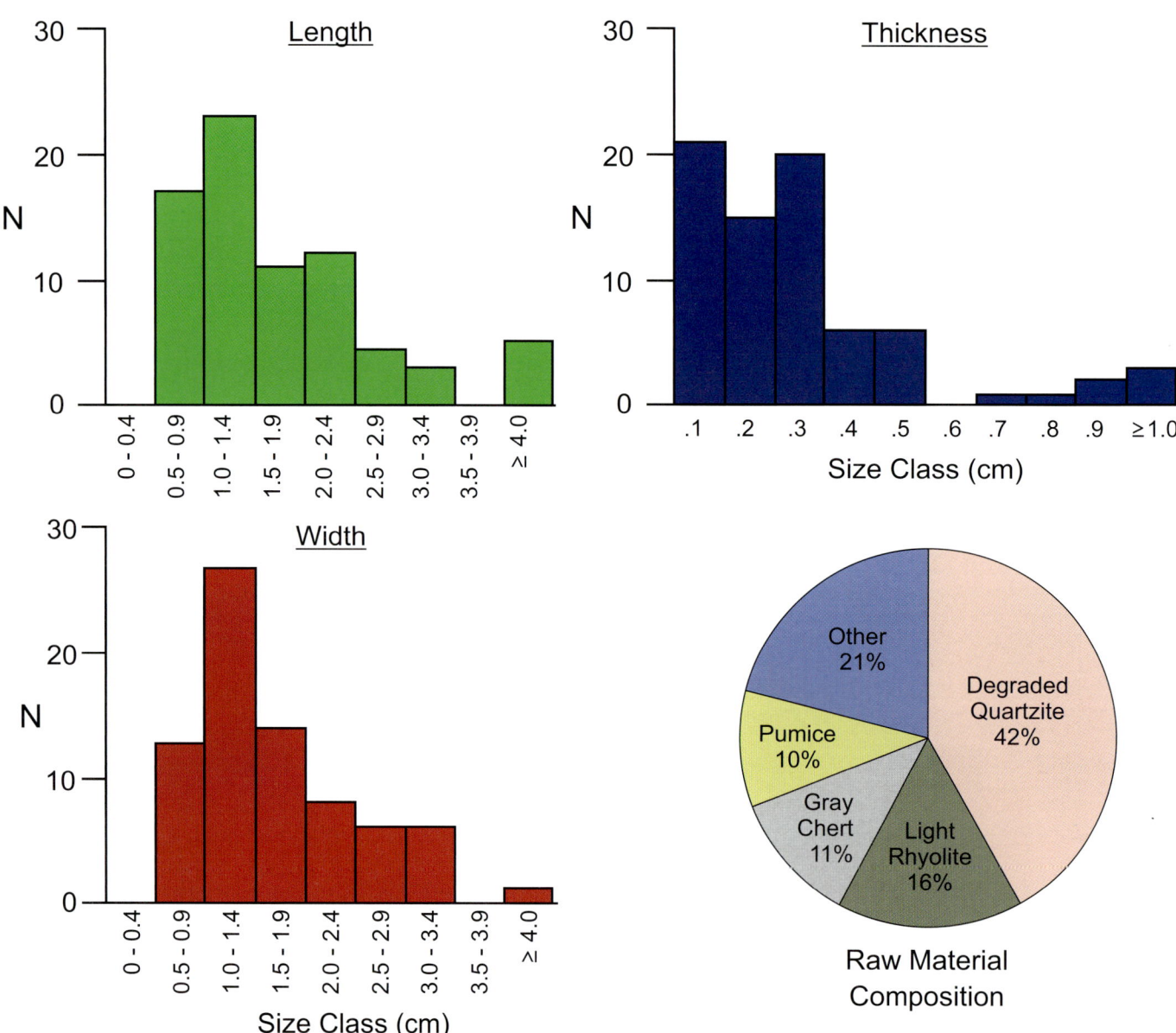

FIGURE 5.15. Raw material composition and flake sample dimensions for Cluster J, Component II.

TABLE 5.12. Cluster I: Cores, tools, and large flakes

Catalog No.	Description	Material	Dimensions (cm)
77-745	Biface	Light rhyolite	9.0 × 5.0 × 3.6
77-746	Biface fragment	Pumice	2.8 × 3.2 × 0.9
77-449	Biface fragment	Dark rhyolite	6.1 × 5.2 × 2.6
77-1591	Knife	Black chert	8.6 × 2.5 × 1.2
77-454	Scraper	Degraded quartzite	6.8 × 4.2 × 1.0
77-525	Core-scraper	Gray chert	6.5 × 4.0 × 2.7
77-524	Utilized flake	Obsidian	4.9 × 2.1 × 0.5
77-1684	Core-scraper	Gray chert	10.3 × 9.3 × 7.5
77-744	Large flake	Degraded quartzite	6.2 × 9.5 × 2.0
77-1959	Large flake	Degraded quartzite	7.5 × 8.0 × 2.8
77-741	Large flake	Gray chert	6.1 × 6.1 × 1.2
77-1708	Large flake	Gray chert	7.7 × 5.4 × 2.1
77-523	Large flake	Degraded quartzite	6.2 × 4.3 × 0.8
77-1683	Large flake	Gray chert	10.6 × 5.7 × 3.5

TABLE 5.13. Cluster J: Tools and large flakes

Catalog No.	Description	Material	Dimensions (cm)
77-1593	Biface	Dark rhyolite	9.9 × 5.1 × 3.0
77-1902	Biface	Gray chert	7.9 × 4.8 × 1.6
77-931	Biface	Degraded quartzite	12.0 × 7.0 × 2.7
77-930	Biface fragment	Degraded quartzite	5.9 × 4.6 × 1.2
77-929	Biface fragment	Degraded quartzite	7.7 × 2.9 × 0.9
77-2009	Possible point base	Degraded quartzite	1.9 × 1.8 × 0.8
77-2013	Scraper	Siltstone	9.1 × 6.2 × 1.5
77-1505	Scraper	Degraded quartzite	7.2 × 3.4 × 0.6
77-2248	Retouched flake	Black chert	4.0 × 1.9 × 0.4
77-2010	Large flake	Gray chert	8.8 × 4.7 × 3.0
77-999	Large flake	Gray chert	11.6 × 6.4 × 3.2

The bifaces all possess sharp edge angles (35°–45°), and two of them exhibit a moderate amount of edge crushing. Del Bene (unpublished notes) believes that the point tip was used to cut resistant material. The "scrapers" also have sharp edge angles and display heavy wear in the form of crushing and nicking; both of these tools exhibit edge grinding. Flake size is generally small, as illustrated by the dimension profiles. The waste flakes include large forms with step-flaking retouch and smaller delicate forms that appear to be the possible product of pressure-flaking. Faunal remains were recovered and tooth fragments were identified as *Bison priscus*. A group of fossil gastroliths is also associated with this cluster.

Cluster K

More than 1,660 flakes were recovered from Cluster K (table 5.14; figure 5.16). Raw materials of fairly high quality predominate, including light rhyolite and chert. Microblades and microblade cores are lacking. However, four square-based point fragments are present (three of light rhyolite and one of chalcedony). In addition to these, there is a small, parallel-sided chert biface (classified as a knife), a simple burin on-a-snap, a retouched flake, and a utilized flake. Beyond the western periphery of the cluster lie a dark rhyolite biface and a brown chert biface fragment.

Two of the point fragments are broken in such a way as to suggest possible impact fracture. The chert knife has a rounded tip and full-length unifacial flute. According to Del Bene (unpublished notes), edge wear on this tool indicates that it might have been hafted and used to cut soft material. The utilized flake possesses a steep edge (90°), which exhibits some crushing and some polish. The retouched flake manifests a narrower edge angle (about 45°) but no visible wear. The flake dimension profiles are skewed toward the small end of the spectrum. Waste flakes belong to the small, possibly pressure-flaked variety but also include examples with thick striking platforms perpendicular to the length axis and heavy step-flaking retouch along the dorsal-proximal edge. Some faunal remains were found in association but are unidentifiable; fossil gastroliths were also recovered.

Cluster L

More than 760 flakes were recovered from this cluster (table 5.15; figure 5.17). Degraded quartzite was virtually the only raw material used; a few pumice flakes were also found. No microblades, microblade cores, or burins are present, and the tools are confined to large, crude bifacial implements. Two of these are true bifaces (one is broken), but the remaining four are relatively amorphous pieces that might be cores. There are two medium, somewhat more finely worked biface fragments, one of which is made of light rhyolite.

It is difficult to distinguish edge-wear patterns on such coarse-grained, irregularly worked material, and the only damage visible is on the proximal end of some of the flakes. There is

TABLE 5.14. Cluster K: Tools

Catalog No.	Description	Material	Dimensions (cm)
77-1375	Point fragment	Light rhyolite	2.1 × 2.3 × 0.8
77-2318	Point fragment	Light rhyolite	4.8 × 1.7 × 0.9
77-3325	Point fragment	Light rhyolite	3.4 × 2.3 × 0.8
77-1578	Point fragment	Chalcedony	3.3 × 2.7 × 0.8
77-939	Knife	Black chert	6.5 × 2.5 × 0.9
77-1570	Burin	Gray chert	2.8 × 3.1 × 1.0
77-3317	Utilized flake	Gray chert	4.3 × 2.5 × 0.9
77-1361	Retouched flake	Light rhyolite	5.5 × 3.2 × 0.8
77-1999	Biface	Dark rhyolite	12.7 × 6.8 × 3.1
77-479	Biface fragment	Brown chert	5.8 × 7.3 × 2.1
77-3804	Core tool	Sandstone	12.5 × 9.8 × 6.8

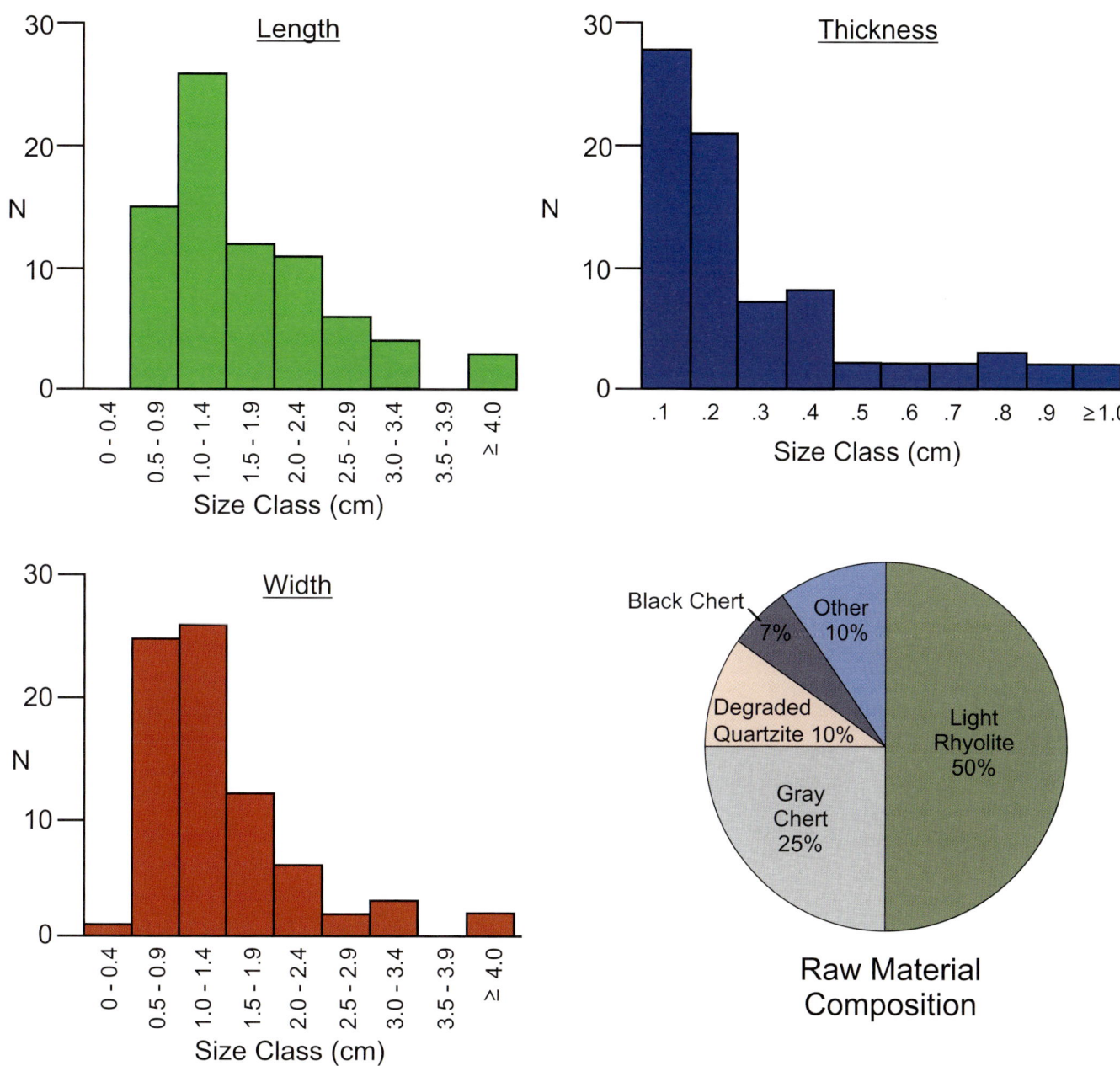

FIGURE 5.16. Raw material composition and flake sample dimensions for Cluster K, Component II.

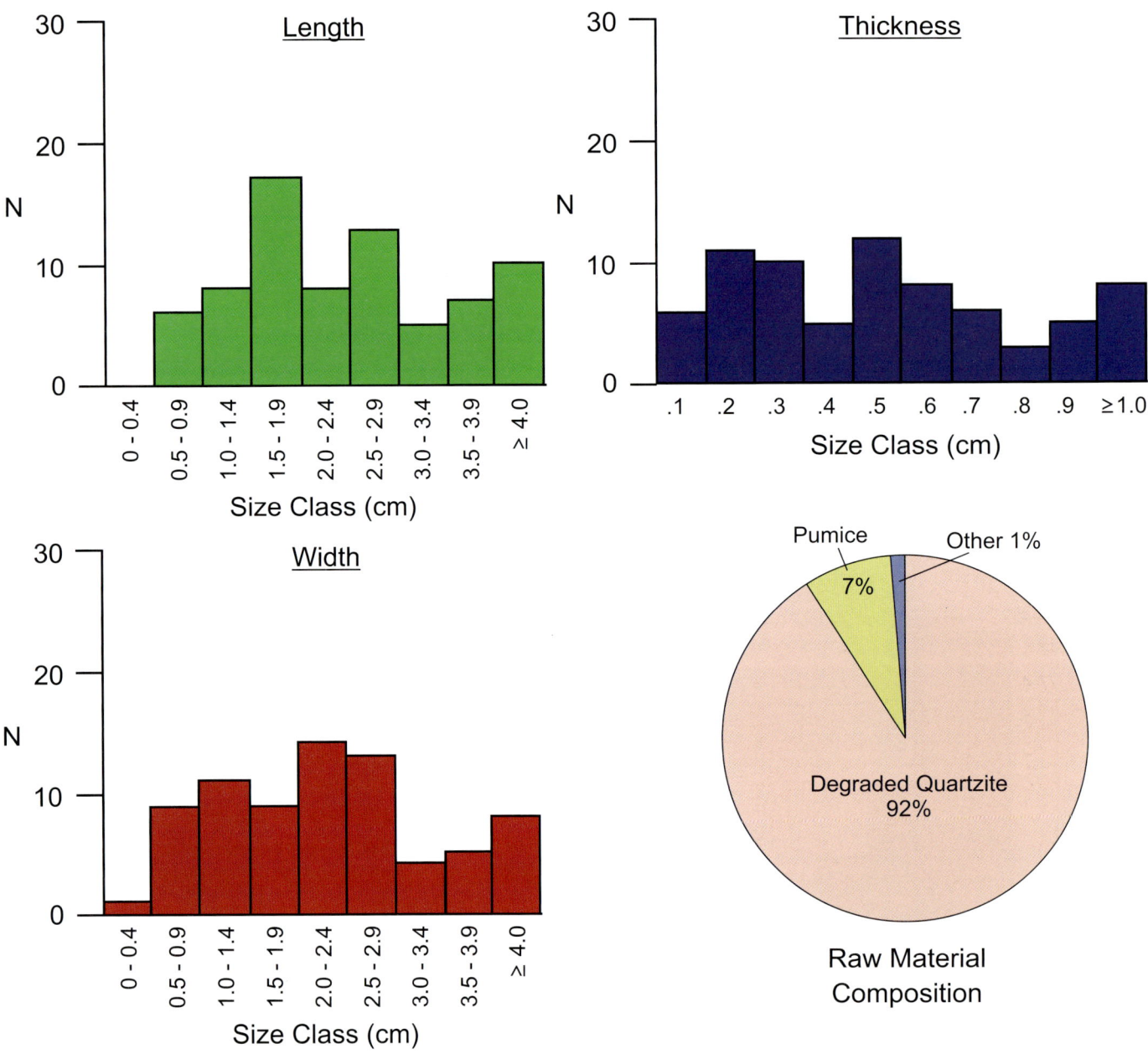

FIGURE 5.17. Raw material composition and flake sample dimensions for Cluster L, Component II.

TABLE 5.15. Cluster L: Tools and large flakes

Catalog No.	Description	Material	Dimensions (cm)
77-4920	Biface	Degraded quartzite	25.0 × 12.3 × 3.9
77-5104	Biface	Degraded quartzite	15.0 × 10.5 × 3.7
77-5129	Biface	Degraded quartzite	14.8 × 8.9 × 4.9
77-4026	Biface fragment	Degraded quartzite	10.1 × 4.9 × 2.5
77-5127	Biface fragment	Light rhyolite	4.7 × 5.0 × 1.6
77-5124	Percussion tool	Degraded quartzite	16.0 × 10.5 × 7.5
77-5129	Percussion tool	Degraded quartzite	14.7 × 9.5 × 7.3
77-5129	Percussion tool	Degraded quartzite	17.0 × 10.3 × 7.1
77-5129	Percussion tool	Sandstone	14.2 × 11.1 × 6.8
77-4820	Hammerstone	Degraded quartzite	20.2 × 5.2 × 2.8
77-5126	Large flake	Degraded quartzite	8.7 × 7.4 × 1.8
77-5125	Large flake	Degraded quartzite	3.9 × 7.4 × 1.7
77-4326	Large flake	Degraded quartzite	6.6 × 5.2 × 2.1
77-4260	Large flake	Degraded quartzite	7.3 × 6.8 × 1.0
77-4999	Large flake	Degraded quartzite	4.8 × 9.2 × 2.2
77-4266	Large flake	Degraded quartzite	7.8 × 4.2 × 1.4
77-4321	Large flake	Degraded quartzite	8.0 × 4.1 × 1.4
77-4306	Large flake	Degraded quartzite	5.1 × 7.3 × 1.5
77 4900	Large flake	Degraded quartzite	4.2 × 6.7 × 1.1
77-5115	Large flake	Degraded quartzite	8.0 × 3.0 × 1.0
77-4317	Large flake	Degraded quartzite	4.8 × 9.7 × 0.8
77-4323	Large flake fragment	Degraded quartzite	7.2 × 6.4 × 1.6
77-3018	Large flake	Sandstone	6.7 × 2.9 × 1.0
77-3381	Large flake	Dark rhyolite	4.4 × 8.3 × 2.0

no sign of retouch. The flakes are unusually large and thick and do not exhibit any of the characteristics of resharpening flakes. Faunal remains are present but unidentifiable.

Cluster M

Cluster M is a large concentration (although a significant portion of it might remain unexcavated), containing more than 2,050 flakes (table 5.16; figure 5.18). Coarse-grained materials predominate, especially degraded quartzite, sandstone, and a low-quality brown chert, presumably of local origin. Microblades and burins are absent, although a light rhyolite microblade core lies within the cluster. The tools include 29 large, crude percussion implements and 2 large flake tools classified as scrapers.

Neither the heavy percussion tools nor the flake scrapers (the latter possessing sharp edge angles of 35°–45°) exhibit detectable wear. However, many of the large flakes (which are usually large and thick) do bear heavy step-flaking retouch on their proximal ends. Faunal remains were recovered from the western periphery of the cluster but are unidentifiable.

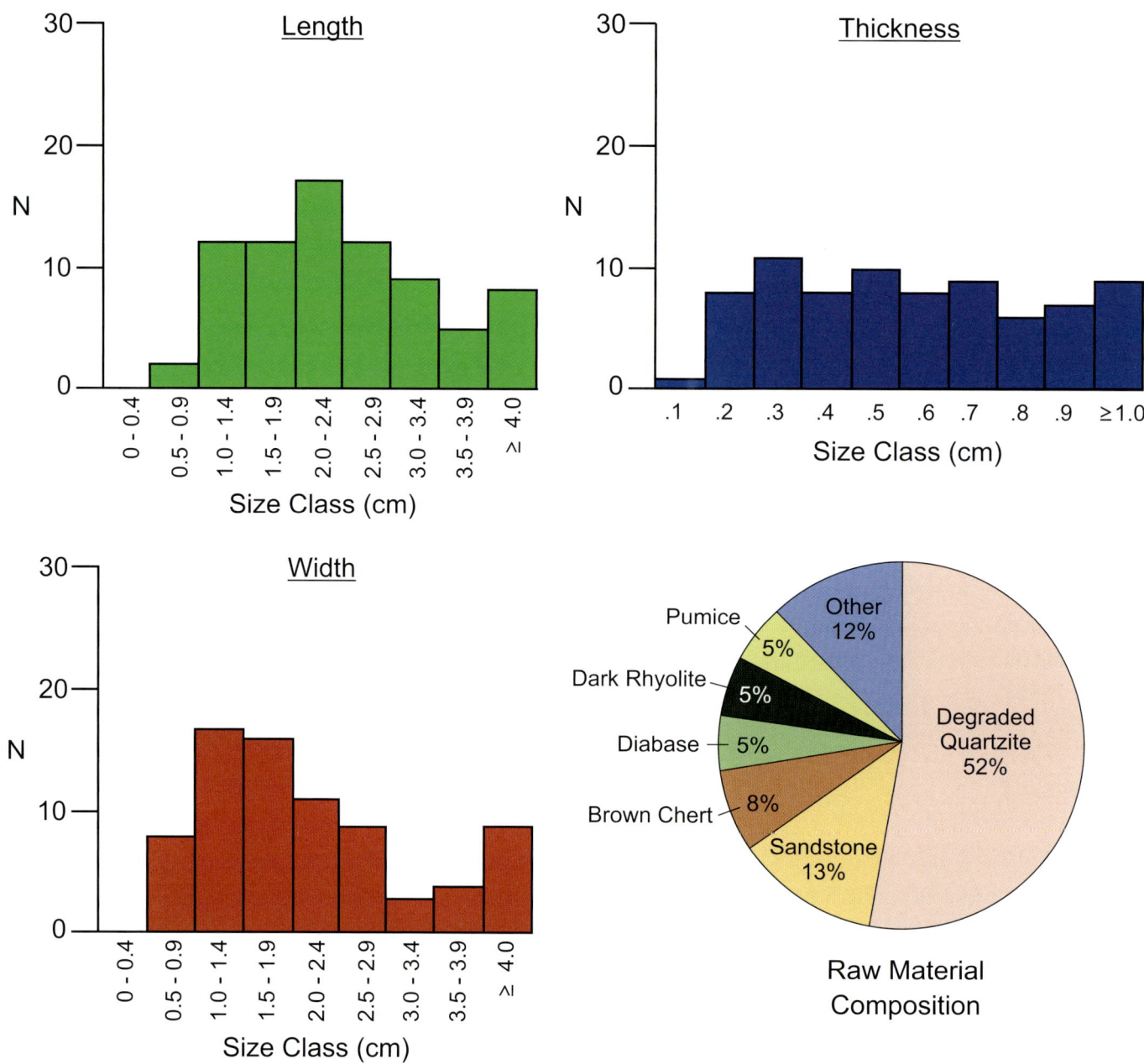

FIGURE 5.18. Raw material composition and flake sample dimensions for Cluster M, Component II.

TABLE 5.16. Cluster M: Cores, tools, and large flakes

Catalog No.	Description	Material	Dimensions (cm)
77-5131(a)	Core	Degraded quartzite	13.0 × 7.2 × 10.8
77-5131(b)	Core/biface	Degraded quartzite	15.2 × 9.4 × 7.2
77-5131(c)	Percussion tool	Sandstone	15.5 × 10.0 × 7.0
77-5131(d)	Percussion tool	Degraded quartzite	13.9 × 9.9 × 3.3
77-5131(e)	Percussion tool	Siltstone	12.8 × 5.4 × 3.0
77-5131(f)	Percussion tool	Degraded quartzite	13.0 × 10.5 × 7.1
77-5130(a)	Percussion tool	Degraded quartzite	16.6 × 9.3 × 7.0
77-5130(b)	Percussion tool	Sandstone	2.7 × 9.3 × 6.7
77-5130(c)	Percussion tool	Degraded quartzite	16.9 × 7.8 × 7.2
77-5130(d)	Percussion tool	Degraded quartzite	18.7 × 10.4 × 7.4
77-5130(e)	Percussion tool	Degraded quartzite	17.6 × 7.6 × 4.3
77-5132(a)	Percussion tool	Degraded quartzite	13.9 × 11.1 × 6.5
77-5132(b)	Percussion tool	Sandstone	15.5 × 10.4 × 5.1
77-5132(c)	Percussion tool	Degraded quartzite	16.5 × 10.6 × 5.9
77-5132(d)	Percussion tool	Sandstone	13.9 × 10.4 × 8.4
77-5132(e)	Percussion tool	Degraded quartzite	15.5 × 9.3 × 6.5
77-5132(f)	Percussion tool	Brown chert	15.7 × 10.5 × 5.8
77-5135(a)	Percussion tool	Degraded quartzite	11.4 × 12.1 × 7.7
77-5135(b)	Scraper	Degraded quartzite	14.4 × 10.0 × 4.1
77-5135(c)	Scraper	Degraded quartzite	10.5 × 9.4 × 2.5
77-5135(d)	Percussion tool	Siltstone	16.0 × 10.2 × 5.0
77-5135(e)	Percussion tool	Degraded quartzite	15.6 × 9.4 × 5.5
77-5135	Percussion tool	Degraded quartzite	15.0 × 10.5 × 5.5
77-5153	Core	Diabase	13.5 × 11.0 × 17.4
77-5141	Percussion tool	Brown chert	13.8 × 7.9 × 3.8
77-5152	Percussion tool	Degraded quartzite	12.2 × 9.6 × 3.0
77-3309	Percussion tool	Degraded quartzite	16.5 × 12.2 × 7.8
77-5199	Percussion tool	Dark rhyolite	20.3 × 8.5 × 8.2
77-5137	Percussion tool	Degraded quartzite	17.5 × 9.2 × 6.1
77-5151	Large flake	Degraded quartzite	6.0 × 6.4 × 2.0
77-3771	Large flake	Brown chert	3.7 × 7.3 × 1.9
77-3782	Large flake	Sandstone	8.4 × 4.5 × 1.9
77-5151	Large flake	Degraded quartzite	7.9 × 5.5 × 1.8
77-3292	Large flake	Sandstone	6.9 × 5.3 × 1.2
77-2330	Large flake	Quartzite	7.2 × 5.7 × 1.8
77-5144	Large flake	Sandstone	5.2 × 9.5 × 1.3
77-5136	Large flake	Degraded quartzite	4.8 × 10.0 × 1.8
77-5138	Large flake	Degraded quartzite	5.5 × 10.0 × 2.5
77-3768	Large flake	Degraded quartzite	6.7 × 4.6 × 1.1

(*continued*)

TABLE 5.16. Cluster M: Cores, tools, and large flakes (continued)

Catalog No.	Description	Material	Dimensions (cm)
77-5136	Large flake	Dark rhyolite	5.4 × 8.6 × 2.4
77-5136	Large flake	Degraded quartzite	10.7 × 5.5 × 2.4
77-5130	Large flake	Dark rhyolite	12.4 × 6.6 × 1.7
77-2031	Large flake	Dark rhyolite	10.1 × 6.7 × 1.7
77-5136	Large flake	Quartzite	7.5 × 7.1 × 1.1
77-5136	Large flake	Diabase	5.6 × 7.6 × 1.6
77-2356	Large flake	Light rhyolite	3.9 × 7.1 × 1.4
77-2031	Large flake	Dark rhyolite	4.6 × 6.6 × 1.0
77-2337	Large flake	Sandstone	5.9 × 6.9 × 1.1
77-3307	Large flake	Sandstone	7.4 × 3.7 × 1.1
77-3296	Large flake	Degraded quartzite	3.8 × 9.9 × 1.6
77-2032	Large flake	Sandstone	7.8 × 6.5 × 1.0
77-4774	Large flake	Sandstone	8.1 × 4.6 × 1.2
77-4872	Large flake	Degraded quartzite	7.0 × 5.3 × 1.6
77-3311	Large flake	Dark rhyolite	8.0 × 7.2 × 2.1

Cluster N

Cluster N lies on the western margin of the excavation, where the complex stratigraphy of the site is highly compressed. Assignment of artifacts to their proper stratigraphic context was difficult, and this problem was further compounded by high frozen-ground levels during excavation. Many of the artifacts might in fact belong to Paleosol 2 and constitute part of a separate, younger component. In this analysis, these materials have been lumped together because of the difficulties of isolating the potentially younger artifacts.

More than 1,190 flakes were recovered, and the raw material composition includes both low-quality local rock (degraded quartzite and sandstone) and high-quality imported rock (light rhyolite and gray chert; table 5.17; figure 5.19). Two gray chert microblade cores are present, associated with a total of 385 microblades and microblade fragments, mostly of gray chert. The microcores are unusually wide relative to their length. Nine burins were also recovered, along with several burin spalls. Bifacial tools are absent.

Although flake size is generally small, there are a number of larger crude flakes of coarse-grained material. These lack discernible signs of wear. It is difficult to characterize most of the waste flakes except to note that they do not suggest bifacial work. Faunal remains were found on the eastern periphery of the cluster but are unidentifiable.

Other Areas

In the central portion of the excavated area, between clusters C/D and F (N5-7 E17-18) and between clusters F/D and G (N8-10 E18-20), there are several small flake concentrations. These include a cluster of 231 degraded quartzite flakes in N5-6 E17-18 and a more dispersed scatter of light rhyolite flakes around N8-10 E18-20. No faunal remains or charcoal fragments were found in association with these artifact concentrations.

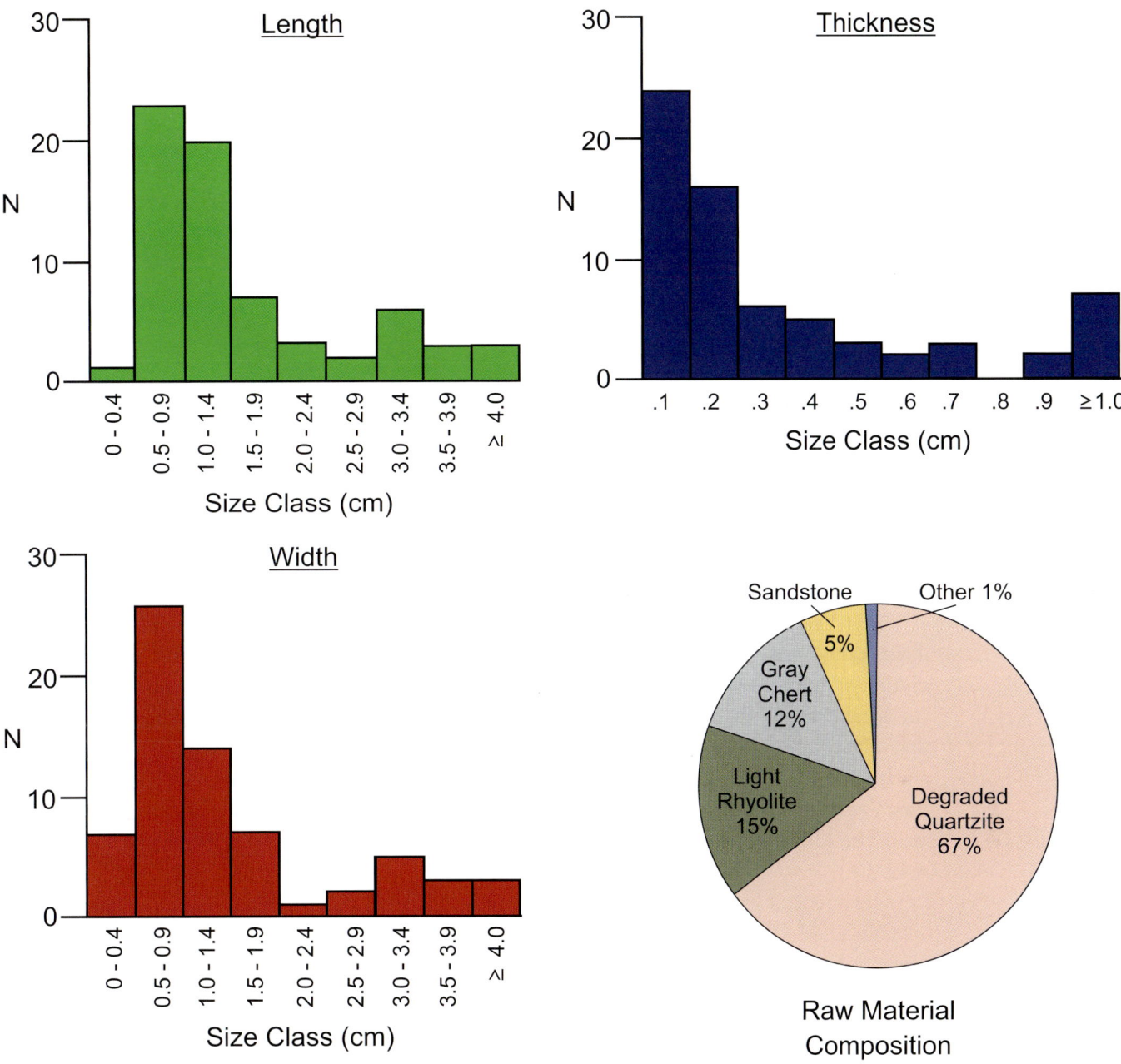

FIGURE 5.19. Raw material composition and flake sample dimensions for Cluster N, Component II.

TABLE 5.17. Cluster N: Cores, tools, and large flakes

Catalog No.	Description	Material	Dimensions (cm)
76-5058	Microblade core	Gray chert	1.7 × 1.5 × 2.2
76-1787	Microblade core	Gray chert	2.0 × 1.3 × 2.2
77-2777	Microblade core*	Light rhyolite	2.3 × 1.7 × 2.8
76-2346	Burin	Gray chert	2.0 × 1.7 × 0.6
76-2366	Burin	Gray chert	1.8 × 2.0 × 0.4
76-2023	Burin	Gray chert	2.2 × 1.9 × 0.4
76-2017	Burin	Chalcedony	2.7 × 1.4 × 0.4
76-2030	Burin	Gray chert	4.2 × 1.2 × 0.8
76-2081	Burin	Gray chert	3.5 × 1.9 × 0.5
76-2016	Burin	Light rhyolite	2.2 × 1.6 × 0.5
76-2125	Burin	Black chert	2.7 × 2.3 × 0.5
76-1785	Burin	Gray chert	2.3 × 2.0 × 0.6
76-1725	Large flake	Degraded quartzite	8.1 × 8.5 × 0.8
76-1831	Large flake	Dark rhyolite	10.3 × 4.3 × 2.2
76-1950	Large flake	Dark rhyolite	8.1 × 5.4 × 2.2
76-5035	Large flake	Sandstone	10.4 × 8.8 × 1.8

*Located in Cluster M.

Small flake clusters also occur in the central part of the excavation within N10-14 E9-14. Four scatters of avian gastroliths were recovered from this area, and a relatively large concentration of faunal debris was found at N13-14 E13. Small clusters of black and brown chert are located within N12-13 E11-13, while degraded quartzite flakes are distributed more sparsely over a larger area.

In the northern portion of the excavated area, west of Cluster K (N17-20 E10-13), small scatters of light rhyolite (chiefly concentrated in N17-19 E11-12) and degraded quartzite (diffuse distribution) were observed. Faunal debris was recovered from N17-18 E12-13.

More densely concentrated artifacts are present in the western part of the site, adjacent to Cluster M (B8-10 E0-W2). Although no faunal remains or charcoal were recovered from this area, a mass of degraded quartzite flakes (> 1,000) occur here (chiefly within N8-9 W0-1), along with smaller concentrations of light rhyolite (> 400 flakes) and diabase (ca. 40 flakes).

Analytical Conclusions

What conclusions can and cannot be drawn from this analysis and other supporting data concerning the Dry Creek occupation floors?

The site was occupied at least twice (accounting for the two separate occupation levels) and was probably occupied many more times. The quantity and variety of remains on the Component II floor might reflect a particularly complex history of occupation, given the contrast among some of the debris clusters. These differences could be a function of multiple site functions (i.e., different uses of the site by the same people, perhaps during different seasons) or of site use by different people (i.e., of differing ethnic affiliation). None of these alternatives can be eliminated at present. It should be kept in mind that despite the quantity of debris in Component II, it cannot be assumed that the site was used more intensively during this period, as rates of loess deposition appear to have been lower (indicated by the presence of numerous thin paleosol horizons in Loess 3). A significantly greater time depth might be represented by Component II compared to Component I. Unfortunately, the problems of site disturbance discussed earlier seem likely to have compromised microstratigraphic distinctions among clusters within Component II.

The occupants brought a wide variety of stone material types to the site, which may be broadly divided into materials of poor to moderate quality (probably obtained locally) and materials of high quality (apparently imported from some distance). The high-quality stones (e.g., high-quality chert, obsidian, rhyolite) were used for certain artifact types such as points, burins, and microblades, while poor-quality materials (e.g., degraded quartzite, sandstone) were typically used for the larger tools (e.g., large bifaces, "percussion" tools) and large unretouched flakes. Smith's (1985:7–8) reconstruction of reduction sequences for Component I confirms this relationship between the finished products and debitage in the earlier occupation level. In both components, flake size among the better-quality raw materials is consistently smaller than among the poor-quality materials.

Remains of large mammals were also probably brought to the site. Both the topographic context of the site and the quantity and character of debris deposited there appear inconsistent with the pattern of kill sites (Isaac 1971), and it may be assumed that the faunal remains were imported to the site. The remains (comprising teeth and bones) are poorly preserved, and it is probably significant that the identifiable elements are highly resistant body parts (i.e., cheek teeth; Brain 1976, 1981; Shipman 1981). It should be kept in mind that the original faunal assemblage might have been much larger and more diverse, including a wide range of represented species and body parts.

Fires were built at the site by its occupants. The low visibility of former hearths (represented by scattered charcoal fragments) might partly reflect wind erosion and partly reflect a pattern of relatively ephemeral hearths with limited baking of surrounding sediment and little accumulation of charcoal and ash. The extent to which wood was available for fuel is uncertain; pretreatment of charcoal samples for radiocarbon dating revealed them to be chiefly composed of humic material (Thorson and Hamilton 1977:167). Wood fragments appear to be absent or rare in pre-Holocene alluvium in the Nenana Valley (Thorson and Hamilton 1977:169).

Most of the stone tool debris is tightly concentrated in clusters, which seem unlikely to have been formed by natural processes (e.g., frost action) or human disturbance and appear to represent "working areas." Lithic reduction sequences, reassembled by Smith (1985) for

Component I, confirm that the concentrations represent debris from multiple reduction events, involving production of flakes and blades and manufacture and modification of tools. Limited reassembly of Component II sequences confirms that microblade core reduction and burin manufacture were occurring on this level; other tool types and unmodified flakes were also being produced in the Component II working areas, but reduction sequences have yet to be reconstructed for these. Yellen (1977:134) and Gould (1980:197–99), drawing on ethnoarchaeological observations, have cautioned against the assumption that associated remains are functionally related.

The clusters of debris vary markedly in terms of size, raw material type composition, number and size of artifacts, and associated finished implements. On both occupation levels, concentrations may be divided into those in which high-quality raw material and small-medium flake size predominate and those in which poor-quality raw material and medium-large flake size are dominant; however, some clusters do not fall clearly into one group or the other. On the Component II occupation floor, a sharp contrast is evident between concentrations containing the products and debitage of microblade and burin manufacture and those lacking such remains. No significant patterns were observed with respect to variation in area, number of artifacts, presence of charcoal, associated faunal debris (identifiable or unidentifiable), or associated avian gastroliths; in the case of the last two items, the lack of pattern recognition could be due to small sample size.

There is no evidence of postmolds, tent rings, or other indications of former structures. However, it should be noted that given the apparent lack of wood in the valley during the period of occupation (ca. 11,000–10,000 ^{14}C yr BP [13,000–12,000 cal yr BP]), tent poles might have been transported up the valley and subsequently carried to another site. Sources of wood were probably present in the Tanana Lowland at this time (Ager 1975:87).

The occupants of the site engaged in both the manufacture and the curation of stone implements in the working areas. In the Component I working areas, bifaces and possibly scrapers were produced and/or modified, and cobble cores of poor-quality raw material were used to produce large flakes (Smith 1985). Microblades and burins were produced in Component II; manufacture and/or modification of points and other implements might have taken place on this occupation level, but this has yet to be demonstrated.

It is not clear if any butchering or meat-processing activities took place at the site. The mammalian faunal remains could have been brought to the site for use as tools or raw materials for tools, and fresh bone might have been used as fuel (a small quantity of burned bone was recovered). Poor preservation of the remains precludes observations concerning surficial damage or breakage patterns, and other than teeth and mandibles, no specific body parts can be identified.

The season of occupation cannot be reliably determined on the basis of existing data. Although the morphology of the avian gastroliths recovered on the Component II occupation floor indicates probable late summer–early winter deposition (see chapter 6), their relationship to the artifacts and features remains problematic; gastroliths were found in archaeologically sterile layers as well.

What contrasts can be drawn between the two occupation floors?

The occupants of the upper occupation level (Component II) engaged in the manufacture of microblades and burins; microblade technology is completely absent, and burin technology is rare in Component I. This is the most important difference between the two occupation levels and does not appear to be a case of sampling error. Not only have large samples been collected from the two levels at Dry Creek, but a similar pattern is evident at other sites in the Nenana Valley. Assemblages dated to the Component I time range at Moose Creek and Walker Road (both located on the east side of the valley) also lack microblade technology (possible burin spalls have been recovered from Walker Road; Hoffecker 1985; Powers and Hoffecker 1989[3]). The contrast could potentially be explained in functional and/or cultural terms.

The upper level contains a significantly larger collection of artifacts and associated debris. Because equal areas of both levels were excavated, this is also unlikely to be a product of sampling error. However, this does not necessarily indicate a more active period of site use but might simply be a function of reduced rates of loess deposition (noted previously).

The upper level contains a significantly more diverse collection of artifact types. Not only microblades but large bifaces, heavy percussion tools, spokeshaves, and various other tool types are lacking in Component I. The meaning of this is not clear. It could reflect a wider range of site activities during the later period of occupation, use of the site by different peoples, or both.

Faunal remains vary between the two levels in terms of represented species, but this seems likely to be a product of sampling. The total number of identifiable remains from each component is very small. Remains of Dall sheep (*Ovis dalli*) and wapiti (*Cervus*) were recovered from Component I; remains of sheep and steppe bison (*Bison priscus*) were recovered from Component II.

Speculations Concerning Site Function

Despite (or perhaps because of) a growing body of ethnoarchaeological information (e.g., Binford 1983; Gould 1980; Yellen 1977) on the organization and function of hunter-gatherer sites, interpretation of archaeological data in these terms retains many uncertainties. Having previously presented a series of conclusions that are firmly grounded on systematic analysis of the available data, I will now turn to the more speculative domain of how the site functioned within the larger settlement and subsistence system of its inhabitants.

One of the more fundamental indices of site function is the diversity of artifact types. "Limited activity sites" such as kill localities typically contain little artifact diversity, while at the other end of the scale, the most complex "multiple activity sites" (usually equated with base camps) reveal a highly diverse assemblage of tool types (Wilmsen 1970; Wood 1978). Assemblages from short-term camps (e.g., temporary hunting or collection camps) apparently lie between these extremes. Although late Pleistocene sites in Alaska provide a limited basis for comparison, Paleoindian sites in midlatitude North America yield a broad range of examples, from kill sites (e.g., Casper [Frison 1974], Colby [Frison and Todd 1986]) to base camps (e.g., Lindenmeier [Wilmsen and Roberts 1984]).

Within this frame of reference, the Dry Creek Component I assemblage lies at the less diverse end of the spectrum, while the Component II assemblage lies at the more diverse end of the scale. Numbers of tool classes (excluding nonstone tools) from the Dry Creek assemblages and selected Paleoindian sites are given in table 5.18. (It should be noted that I have combined and divided classes used by different investigators in order to achieve greater classificatory consistency among sites.)

However, because the sharp contrasts among the artifact concentrations on the Component II floor could reflect different uses of the site at different times, and possibly by different peoples, it cannot be assumed that all types of activities represented in the working areas occurred during the same occupation episode. The remains in Component II could have been deposited by the same people using the site for different purposes, different people (i.e., more than one ethnic group) using the site for the same purposes, or different people using the site for different purposes. Thus, although the Component I assemblage conforms to the more limited activity site pattern, the Component II assemblage, when its various possible subsets are considered, occupied a broad portion of the spectrum.

On the basis of its topographic context, the site seems likely to have been used as a hunting overlook (Thorson and Hamilton 1977:174). A similar function has been hypothesized for other late Pleistocene sites in comparable topographic positions in the Nenana Valley (e.g., Moose Creek [Hoffecker 1985:44]). Located along the edge of the Healy terrace, approximately 100 m above what was, at the time of occupation, the Nenana River floodplain (see chapter 3), the Dry Creek site would have afforded a broad vista of the southern end of the valley and adjoining mountain front. The site also offers a well-drained surface for habitation and immediate access to a side-valley stream (although it should be noted that the modern stream flows infrequently and carries a heavy sediment load).

TABLE 5.18. Number of tool classes represented at Dry Creek and midlatitude North American Paleoindian sites

Site	Tool Classes	Area Excavated (m²)	Source
Colby	5	> 500	Frison (1986:91–103)
Olsen-Chubbuck	6	> 75	Wheat (1972:127, table 20)
Dry Creek (I)	8	347	(based on chapter 4)
Dry Creek (II)	11	347	(based on chapter 4)
Lindenmeier	> 12	> 1,800	Wilmsen and Roberts (1984:62–134)
Blackwater Draw No. 1	13	ca. 1,000	Hester (1972:97–118)

Game observation in itself has no archaeological visibility. However, sites used as hunting overlooks are often the locus of other activities, some of which might be related to the hunt (Binford 1978:330–32, 1983). The presence of whole and fragmentary projectile points in both components at Dry Creek suggests that hunting weapons might have been manufactured and/or repaired at the site. Although some debitage exhibit the characteristics of bifacial thinning or resharpening flakes (Bordes 1961:6–7; Frison 1968:149–50), no point reduction sequences have been reconstructed.

Microblades were produced in significant quantities on the Component II occupation floor, and if used in a similar fashion to microblades in Siberian Paleolithic sites like Kokorevo I on the Yenisei River (Abramova 1979:106–8; Chard 1974:31–32), these might have been set into laterally grooved bone or antler points. The burins (and burin spalls) so closely associated with the microblade technology could have played an integral role in the manufacture of such composite points; these steep-edged tools are especially suitable for working resistant materials such as bone and antler (Semenov 1964:94–100; Wilmsen 1968:157, 1974:91–92). Other types of steep-edged tools (e.g., core-scrapers) are present in both levels, and many exhibit wear patterns (e.g., step-flaking wear [Keeley 1980:24, figure 10]) typical of hard-material working (Hester et al. 1973:92–93).

Some butchering or processing of large mammal carcasses might have been undertaken at Dry Creek. Given the topographic position of the site, it is likely that most if not all of the faunal remains were retrieved from other localities; unfortunately, poor preservation conditions have destroyed the most useful sources of evidence concerning activities related to these remains (e.g., wear and breakage patterns on the bones themselves, represented body parts; Klein 1980; Shipman 1981). Faunal debris might have been brought to the site primarily as raw material for artifact manufacture, as described previously. Although implements appropriate for butchering activities—that is, large sharp flakes, heavy tools (Hammat 1970:150–51)—are present on both levels, and some concentrations reflect a predominance of these elements (e.g., Cluster L on the Component II floor), these tools are relatively unspecialized and could have been made and used for alternative purposes.

A variety of other activities might have been performed by the occupants of the site, and it should be noted that a few artifacts exhibiting wear patterns consistent with soft-material (e.g., skin) working, including polish, are present in Component II. On the other hand, some of the artifact types found at suspected base camp sites such as Lindenmeier (e.g., perforators, drills, notches [Wilmsen and Roberts 1984]) are either rare or absent at Dry Creek.

In sum, the site appears likely to have been used as a hunting overlook and locus of related activities that probably included some production and/or repair of hunting weapons and might have included some processing of large mammal carcasses. The occupants might have engaged in other activities (e.g., skin working) that were only indirectly related to the hunt. Dry Creek probably functioned therefore as a short-term hunting camp, its location having been chosen not only on the basis of its advantages as an observation post but also on the basis of its suitability as a campsite (e.g., proximity to water, fuel; see Jochim [1976], Wood [1978], and others for discussion of site location decisions). Within the larger settlement/subsistence system of its occupants, the site would probably have been linked to one or more long-term base camps (perhaps situated in the Tanana Lowland) and to multiple kill localities (probably within a radius of a few kilometers in the Nenana Valley or along the mountain front; Binford 1983; Butzer 1982; Dennel 1983).

Drawing further on ethnographic comparisons, I would suggest that Dry Creek was probably occupied for relatively brief periods (several days?), possibly by special task groups (e.g., a hunting party of several adult males). The lack of well-developed former hearths and traces of long-term shelters (e.g., house depressions) is consistent with this line of speculation; the multiplicity of working areas probably reflects repeated use of the site for the same purposes. Although the season of occupation cannot be reliably determined at present, a late summer/fall exploitation of upland areas is not uncommon among recent hunter-gatherers (including the aboriginal inhabitants of the Nenana Valley [Plaskett 1977]) and has been hypothesized for Dry Creek by Guthrie on the basis of the site-fauna composition (see chapter 6).

Significant changes in site function might have occurred over time, and this could account in part for the differences between Component I and Component II. Specifically, the broader range of artifact types in the upper level suggests that a wider variety of activities took place during the later period of occupation. On the other hand, some of the artifact variability might be a product of ethnic differences; for example, the hunting weapons that were probably manufactured with either projectile points or microblades might have served the same function. In this case, the Component II occupation floor might reflect use by different peoples but for essentially the same purposes.

Notes

1. The author is grateful to Timothy A. Smith and Terry A. Del Bene, who provided unpublished data. Earlier drafts of this chapter were reviewed by Karl W. Butzer, E. James Dixon, David M. Hopkins, Richard G. Klein, Richard E. Morlan, W. Roger Powers, and T. A. Smith. Their criticisms and comments were extremely helpful in preparing the final draft, although they should not be held responsible for its remaining shortcomings.
2. In the tables and pie charts, artifacts originally attributed to 'degraded quartzite' appear to actually represent basalt, according to more recent analyses (chapter 8; Gore and Graf in press)
3. In Hoffecker's last revised version of this chapter, dated 1987, this reference (Powers and Hoffecker 1989) was cited as an unpublished manuscript.

CHAPTER SIX

Paleoecology of the Dry Creek Site and Its Implications for Early Hunters

R. DALE GUTHRIE

Paleoecological Significance of the Site

At about 11,500 ^{14}C yr BP (13,500 cal yr BP), plus or minus a few tens of human generations, monumental changes culminated throughout North America and Eurasia. Jet stream patterns had shifted, climatic fronts had moved, and the large ice sheets were rapidly retreating. Entire plant biomes began to undergo restructuring and redistribution. Large grazing ungulates and many other animal species underwent a time of rapid extinctions; among surviving species, some experienced withering declines while still others increased and flourished. Among the latter were humans. The ensuing millennium saw the explosion of early Paleoindian peoples in North America and in South America as well.

Quaternary researchers disagree about many aspects of this critical time. The exact nature of climatic changes is controversial, as are the megafaunal extinctions and their causes. The origins of Paleoindian peoples and their effects on the plant and animal communities, the causes underlying their population distribution, and prey shifts during this time are even more controversial (Haynes 1982; Martin 1982; Morlan and Cinq-Mars 1982).

There are a number of New World archaeological sites dating from this period, but few contain substantial artifactual and paleoecological information. More well-dated sites are necessary to allow us to reconstruct what occurred during that critical millennium. Alaska might play a unique role in disclosing the story, as it was an important focus of faunal interchange, including the Paleoindian peoples themselves.

The Dry Creek site is the first multicomponent, deeply stratified site in Alaska that dates from this critical millennium and contains megafaunal remains in association with tens of thousands of lithic artifacts. Though the bones and teeth are poorly preserved and are not in the profusion found at mass kill sites, they provide us with the following information:

1. The two oldest components of the site mark the end of the faunal changes associated with a shift from steppe to woodland. Though the major extinctions probably have occurred within the preceding centuries, the megafaunal community in that area was still composed of grazers.
2. Holocene dwarfing of the megafauna at the site had just begun or had not yet begun—that is, the sheep, bison, and wapiti in the site are as large as the Pleistocene forms.
3. The mammalian fossils were poorly preserved, and only weathered enamel was identified to genus (by species, based on context only). However, from Component I, I could identify several specimens of *Ovis* and *Cervus*; and from Component II, several of *Ovis* and *Bison*.
4. While the lithic assemblage is different between the two lower cultural layers in numerous tool patterns and suggestive of a cultural and activity shift, the fauna still seems to suggest a similar pattern of ecological exploitation.
5. Though very incomplete, the Dry Creek occupation patterns, taken together with other information, contribute additional perspectives on early northern hunting strategies and community organization.
6. Although much the same kinds of large mammals were hunted, hunting patterns and perhaps group organization seem to be different from those that produced the mass kills on the Great Plains. The Dry Creek site was neither a kill site nor a base camp site. It was probably a temporary seasonal hunting "spike" camp or processing station, where kills were brought to process the meat and hides for transport elsewhere. Other game could be spotted from the site while the Dry Creek people were engaged in repair or replacement of their hunting and processing equipment.

Although the Dry Creek site by itself does not settle many questions now in dispute about the Pleistocene-Holocene transition, it does add important information and clarity to the paleoecology of that eventful period and at the same time raises more questions and poses more puzzles.

Paleoecology of the Fossil Ungulates at the Dry Creek Site

We know that the Dry Creek site people hunted at least bison (*Bison*), wapiti (*Cervus*), and mountain sheep (*Ovis*). The fossil preservation at the site is poor, and there are many bone fragments in addition to the few that are identifiable, so it is quite possible that there were other species taken as well. The presence of the aforementioned species does allow us to explore the chronology of habitat occupation and hunting strategy. In this respect, Dry Creek is almost unique among early northern archaeological sites studied thus far.

In this chapter, I examine the ecology of the living counterparts of the species found at the site for clues to similarities in the paleoecology 11,000–10,000 ^{14}C yr BP (13,000–11,500 cal yr BP). There are potential hazards to using such indirect and circumstantial evidence inherent in making *neo-* and *paleo-* comparisons. However, I am familiar with the nature of at least some of the risks involved and hope that an informed and judicious use of such analogies can enrich our reconstruction of the lives and environment of these hunting peoples. In the case of the animals found in the site at Dry Creek, we can show that, in fact, the fossil ungulates were not living lives identical to their modern counterparts, and these differences allow us to show how the environment at Dry Creek was different from the environment in which these species live today.

Before discussing the paleobiology of each species, there are two general points that should be made. The first pertains to the special ecological adaptations of present-day northern ungulate species to their seasonally harsh environment; those same adaptations were undoubtedly also characteristic of Pleistocene ungulates in the Far North. The second point concerns the character of the environment itself. Ungulates in the Far North experience a seasonal boom-bust economy. This seasonality involves divergent swings in food availability and quality, predation exposure, and physical characteristics of the environment such as snow depth and terrain access. There are a number of ways northern ungulates adapt to these seasonal shifts. They are not territorial, thus migration and mobility are a central part of their behavior. This nomadism is not random but tends to follow seasonally favorable habitats.

Winter forage is extremely low in quality. Pastureland is made inaccessible by deep snow, and in wind-cleared areas available for grazing, nutrients are leached through exposure. Northern ungulates do not simply stop growing in winter; they actually decline in weight and physiological condition. Curiously enough, this decline is not only environmentally induced by lack of food but also intrinsic. Unlike domestic animals or some southern species, they do not have the ability to grow during this winter dormancy period. Raised on a high plane of winter dietary supplement, they still continue to decline (e.g., Norden et al. 1968). Winter adaptations are all geared toward a "get-by" survival strategy.

However, summer strategies are the opposite. Klein (1965, 1970) has shown that arctic herbaceous plants and many woody species eaten by northern ungulates are of exceptionally high nutritional quality. Thus, in contrast to their winter dormancy, these same northern ungulates have special growth abilities that allow them to grow and rear young during the brief seasonal flush of high-quality herbage. This rapid growth potential is also true of some northern small mammals, such as ground squirrels (Levenson 1979).

In addition to numerous physiological specializations, northern ungulates have behavioral characteristics that also encourage full utilization of summer greenery. These species follow the early growth stages of plants over the countryside as the phenological wave sweeps across different latitudes, land contours, and altitudes. The early plant growth is highest in nitrogen and phosphorus and lowest in antigrazing compounds, fiber, and phytoliths. Different landscapes thus tend to attract these ungulates in a seasonally specific manner, and these seasonal uses are usually repeated from year to year in traditional patterns of use—that is, northern ungulates can be expected to occur on particular landscapes with a fairly high fidelity for a given season. This phenomenon will play a critical role in the interpretation of the seasonal use of the Dry Creek site.

Although we have a fairly firm grasp of the environmental demands of these past ungulates (by analogy to the modern species, which are moderately well understood), we know very little detail about the past environment in terms of plant species or plant community composition that formed the rangeland. To settle the larger questions about the Pleistocene mammoth steppe (Guthrie 1982, 1985), we have to know a lot more than we do now, but we can say something about the ungulate use of the Nenana Valley at the close of the Wisconsinan Glaciation. Without knowing the past plant species, but knowing the present patterns of ungulate use in that area and in similar areas, we can reconstruct with some confidence the pattern of the most intensive seasonal use of the range and how this use was partitioned among the grazers. This, in turn, can tell us something about the strategies of the human hunters.

Somewhere between 14,000 and 11,000 ^{14}C yr BP (17,000 and 13,000 cal yr BP; Ager 1975; Hopkins et al. 1982), we know that mesic shrubs began to replace the xeric herbaceous vegetation of the glacial communities. Exactly how and why it happened we are not certain, but it is quite probable that these windy outwash areas from the mountain passes were the last holdouts of the mammoth steppe, and the ungulates living there were confined (seasonally at least) to the wind-cleared grasslands. The Dry Creek ungulates may then be looked upon as the late stage in the unraveling of the mammoth-steppe fauna.

Judging from the diverse mammoth fauna that lived nearby in the Fairbanks area during the last glacial period (Guthrie 1968, 1985), the major late Pleistocene megafaunal extinctions had already occurred when the Dry Creek peoples were living at the lowest levels of the site (Components I and II). Yet the *regional* extinctions of wapiti (*Cervus*) and bison (*Bison*) had not yet occurred. These two species are not native to Alaska today, though some small herds have been reintroduced from the south and now live in marginal habitats.

Thus the Dry Creek fauna occurs at a fascinating time for a paleoecologist. The species-wide extinctions have occurred, but the ecological changes of the Pleistocene-Holocene transition are not yet complete. The postglacial dwarfing of bison and sheep has not commenced, at least in a noticeably dramatic form, and the range contradictions southward are probably just in process. Thus I have taken the approach that the fauna from the Dry Creek site, though poorly preserved, deserves detailed paleoecological work in a setting of the larger questions in order to maximize its contribution to the problem of early people in the New World and their changing environments.

To do this requires a thorough examination of the biology of the living ungulate counterparts, an examination of the biology of species that are not represented in the site, and a look at some general phenomena of Holocene dwarfing (and hence Pleistocene gigantism) and the problem of late Pleistocene megafaunal extinctions.

Now I would like to discuss the paleoecological implications of the presence of sheep, bison, and wapiti singly and then look at the combined pattern they present. Dall sheep will be discussed first. They are the best base from which to work, as they exist in most mountainous areas of Alaska today and still occur near the Dry Creek site within easily huntable distances.

Dall Sheep, *Ovis dalli*

Though our knowledge of Dall sheep biology is far from complete, we do know many aspects of their ecology. My graduate students and I have worked with the extant wild sheep in the Healy–McKinley Park area (immediately adjacent to the Dry Creek site) for a number of years, and I have a captive flock of sheep taken from this same area as part of an ongoing study of their ecology, growth, and behavior. In addition to my professional experience with Dall sheep, I have also hunted them almost every autumn for the last twenty years.

The bones and teeth at the site were treated in situ with an acetate-acetone mixture and then removed with the silt matrix. With few exceptions the material was fragmentary and poorly preserved. Many gray "smears" of bone were not collected from the site, as they were almost totally decomposed. These diffuse smears sometimes had a few particles of the bone remaining. Those that contained any solid forms of the bone pattern were collected. The osteological materials were taken to the laboratory for further preparation before they could be analyzed. Most remained unidentifiable. Several pieces of bone showed the silvery calcined white-and-black coloration characteristic of fire-charring.

The teeth consisted only of the more durable enamel (figure 6.1); the cementum and dentine were missing. Also, only one side of the teeth (lingual or labial) was usually present, though in some the entire enamel casing was preserved.

Of the 14 different specimens of sheep teeth that were identified, some consisted of a few fragments and others the entire alveolar row. The large specimens identifiable to specific teeth are listed in table 6.1. It is noteworthy that all but one were uppers. These occur in both the earlier levels (Components I and II).

The sheep are contextually referred to the species *Ovis dalli* with some confidence because of other zoogeographic information about these sheep. They occupied the unglaciated Alaskan–Yukon Territory refugium during the Wisconsinan Glaciation and continue to exist in that same area. As the Dry Creek site is in the middle of that region, it is unlikely that the sheep could be anything other than *Ovis dalli* in the general time period of 11,000–10,000 ^{14}C yr BP (13,000–11,500 cal yr BP).

The Dry Creek sheep are near modern sheep in tooth shape, except they are somewhat larger. This is a general pattern over much of the range of American sheep; the glacial forms were larger than their modern counterparts (Guthrie 1985). At this point in the discussion, it is most important to note that the sheep at the Dry Creek site were morphologically more akin to glacial than to postglacial forms.

The first sheep into North America (sometime in mid-Rancholabrean) were large, long-legged, foothill-adapted forms. In Central Eurasia, sheep (*Ovis*) occupy the rolling foothills and mountain plateaus, and goats (*Capra*) occupy the steep terrain. As a result, their body construction is quite different. Goats are stocky with short, stout limbs. The forelimb sets back farther into the body. Sheep are lankier, with the long forelimbs set forward, as in gazelles, for rapid running rather than for climbing (Geist 1971). These early sheep in the New World were thus quite large and lanky and are even given a specific status (*Ovis catclawensis*) by some. Geist proposed, however, that on entering Beringia and the New World, the *Capra* niche was unfilled, and so sheep underwent an evolutionary shift more toward the *Capra* body form and habitat preference. This seems to be well supported by the paleontological data.

Sheep were much more widespread throughout the Pleistocene in Beringia than at present. Fossils have been found in areas now unoccupied by sheep. Presumably the lowered treeline and the increased herbaceous cover allowed them to expand to precipitous areas that furnished adequate escape terrain but are now wooded. Of course, the more foothill-adapted sheep colonists would have facilitated this dispersal before they became more alpine adapted, probably the result of a rising treeline during past periods of treeline shifts. It appears that this alpine *Capra* adaptation in the New World occurred earlier in Alaska than farther south, as Wisconsinan-aged sheep in Wyoming and farther south still exhibit the long-legged foothill adaptation.

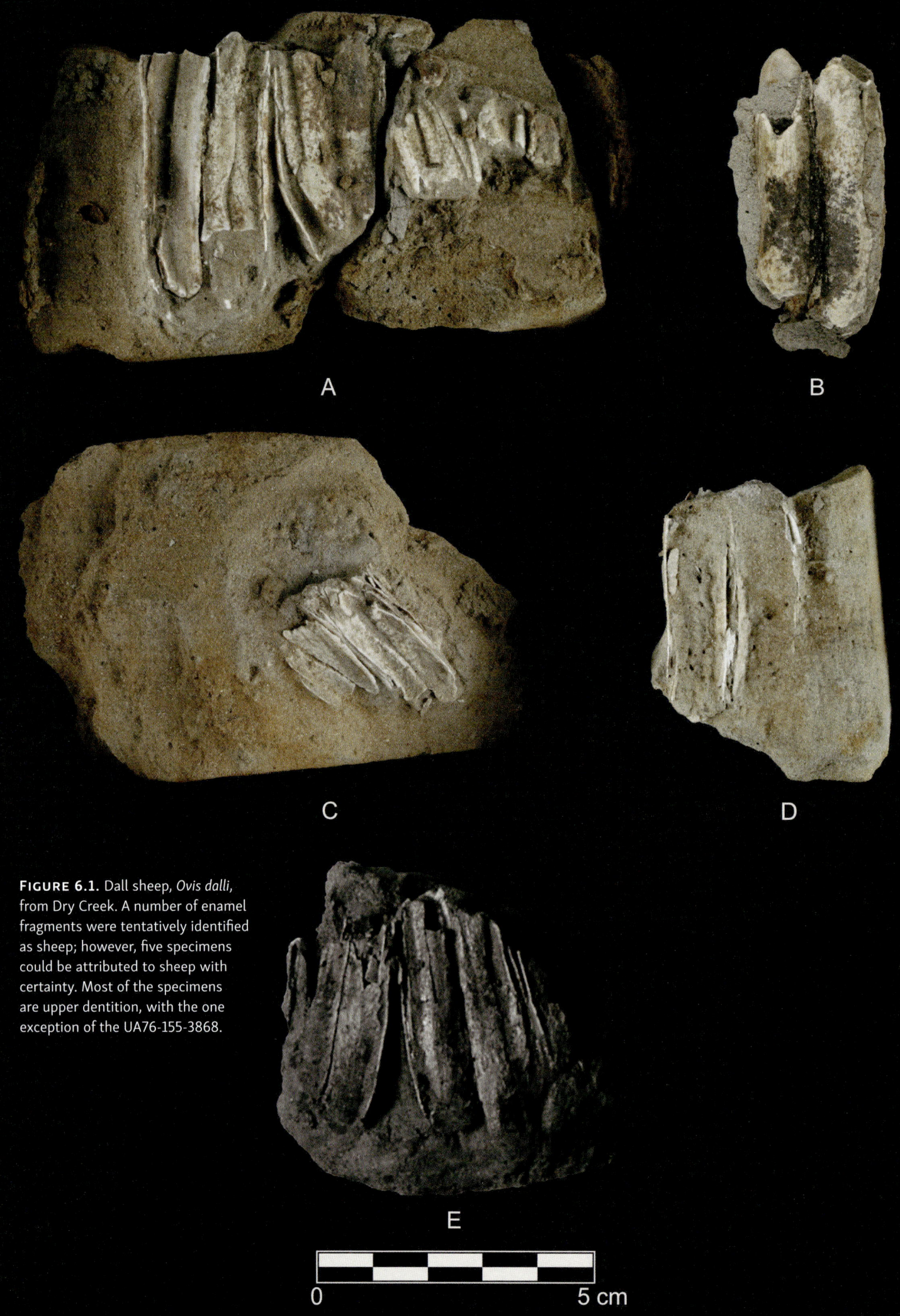

Figure 6.1. Dall sheep, *Ovis dalli*, from Dry Creek. A number of enamel fragments were tentatively identified as sheep; however, five specimens could be attributed to sheep with certainty. Most of the specimens are upper dentition, with the one exception of the UA76-155-3868.

TABLE 6.1. Sheep (*Ovis dalli*) teeth

Number	Tooth	Length (mm)	Estimated Age
UA 77-44-2740	Right M2	22.7	Juvenile
UA 77-44-4347	Right M2	21.2	Postjuvenile adult
UA 77-44-4399	Right M2	Incomplete	Middle aged
UA 77-44-1300	Right M3	Incomplete	Older adult innerface of tooth broken away looking at the labial side of the outer portion of the tooth row
	M2	Incomplete	
	M1	Incomplete	
	P4	Incomplete	
UA 76-155-3868	Right M3	Incomplete	Juvenile to young adult
	M2		

The factors affecting mountain sheep local distribution in the north have been known for some time and have been reviewed by Summerfield (1974). The major factor is escape terrain—that is, suitable topography—which also provides access to the food to which sheep are adapted. They rely neither on outrunning their predators on a straight chase (like *Antilocapra*, the pronghorn antelope) nor on climbing into inaccessible rocky cliffs (like *Oreamnos*, the Rocky Mountain goat); rather, they rely on escaping predators in an uphill chase. Thus American mountain sheep can never venture far from relatively steep slopes for long without being vulnerable to predation.

Because of the rugged character of the mountain slopes, where sheep can escape predators, the vegetation is not abundant. In these alpine areas, plants are usually thinly scattered and low growing. Mountain sheep are well adapted to grazing close to the ground, using rapid bites (more than 4,500 bites per hour [Horejsi 1976]). Mountain sheep are among the most selective feeders ever studied, specializing in the young leaves and fruiting parts of the plants that are the most digestible and nutritious. The summer range must have enough chronological diversity in plant phenology to cover several months of the year, allowing sheep sufficient time to grow and nurse lambs. Most mountainous uplands seem to fulfill this summer range requirement; it is the winter range that is usually the limiting bottleneck in sheep distribution.

Winter range not only has to have plant species that provide summer nutrients; the plant species must also retain some of their energy and nutritive quality aboveground and have gone relatively ungrazed during the summer. These plants must be located on escape terrain and, most importantly, must not be covered by crusted or deep snow. This latter factor is often critical because sheep do not have the ability to dig through deep snow for food, or at least it is energetically expensive for them to do so (Petocz 1973).

In most mountain systems, there are few areas that meet all these criteria for winter range. They are usually on exposed mountain range fronts or in windy pass areas where valleys widen from the mountains into the foothills or other major valleys.

There is another possible factor contributing to sheep distribution that is mentioned by Whitten (1975) but seldom mentioned by others, and that is the presence of salt licks.

New growth herbaceous material is high in potassium but low in sodium. The potassium causes the animal to excrete more sodium, thus causing sodium deficiencies. As a consequence, many animals turn to salt licks to replenish their sodium. Sheep normally concentrate at these natural licks in the spring just after returning from lambing range sometime in June. These mineral licks are traditional, and sheep return to them with considerable fidelity (Heimer 1973). Wherever one finds sheep, there are usually mineral licks within their home range. How limiting salt licks are to sheep distribution is not known.

Sheep movements can be divided into those that are intrinsic and seem to be independent of direct environmental control and those that are environmentally related and extrinsically controlled. The latter includes changes in temperature, predator harassment, snow cover, and most importantly, changes in plant phenology. Sheep follow variations in plant quality over the landscape. Movements to rutting and/or winter range seem to be more intrinsically controlled (Geist 1971; Whitten 1975).

Sheep then have several different traditional ranges that they use for different times of the year and for different reasons. The number and kind have been shown to differ between populations and to relate to several variables. What is important for our purposes, as seen from the perspective of early human hunters, is only that these ranges are predictable and that winter ranges tend to be in special environmental situations. I propose that the mountain pass area of the Nenana River, which is now a major Dall sheep winter range, would have also been winter range in the past. Furthermore, it is this behavior of traditionality that makes sheep (an otherwise difficult prey species) vulnerable to human hunting.

There are several things that we must know to make an evaluation of how and when Dall sheep could have been hunted by the Dry Creek peoples: When are Dall sheep concentrated and accessible from the campsite? At what altitude and areas do they occur seasonally? When do they move to winter range?

Summerfield (1974) reviewed the time of movement to winter range in different populations and subspecies of North American mountain sheep. I have presented his results in figure 6.2 and added the sheep from the Nenana Valley (my observations on the east side of the Nenana River and Whitten's [1975] on the west side). There is a clinal gradient from northern latitudes to southern ones. It might have been somewhat different 10,000 ^{14}C yr BP (11,500 cal yr BP) than it is today, but it is doubtful that the difference would have been in the order of several weeks. Pleistocene sheep would have moved onto winter range at least before the rutting season (late November to early December).

As we noted in chapter 2, the katabatic winds generated in the Nenana Valley are greater than virtually any other area in the interior of Alaska. As a result, the snow cover in the area is very light. This is true in the high mountainous borders as well as in the valley bottoms. Sheep move into the area in late September or early October and do not leave until May. Some rams usually linger until early June on the west side of the valley. Sheep use the spur ridges, which are snow-bare, or the interconnecting basins, which have only a light snow covering. The sheep remain high except for rare movements to adjacent highlands, where they must cross low-lying areas. On the east side of the Nenana River, the sheep move down onto the high terrace edges, which are also snow-free. The number of sheep wintering in the area has had a long, variable history (Murie 1944; Murphy 1974) but today is in the range of 200–300 animals on the west side of the valley and about 100 animals on the east side.

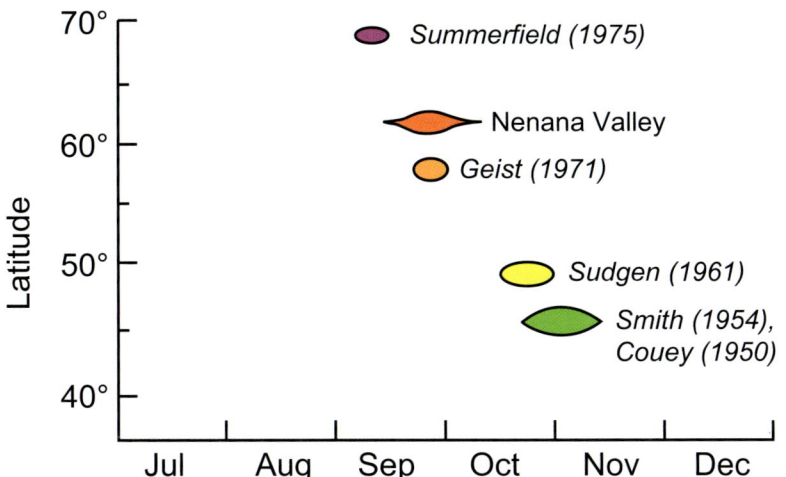

FIGURE 6.2. A graphic comparison of the time North American mountain sheep move onto winter range. There appears to be a latitudinal gradient. The recent sheep near the Dry Creek site move into the Nenana Valley onto winter range in late September and early October.

There is some very limited movement across the frozen Nenana River in the gorge between the populations on the east and west side; otherwise, they are separated. Several biologists have studied the populations on the west side. These sheep, which winter nearest the Dry Creek site, spend their summers and autumns in Denali National Park. They are thus unhunted and form a natural age group distribution. Murie (1944) studied a number of basic biological parameters in his classic study of the wolf and sheep. Murphy (1974) updated and reevaluated Murie's original population data and the history of the park's sheep movements and diet in a thorough way, and his work is most germane to our interests.

The Alaska Range is divided into two tectonic arcs set against one another. The higher southern arc is the Main or Inner Range, and the lower, northern-most arc is called the Outer Range. The sheep winter on the north face of the Outer Range (figures 2.2, 2.4), where despite the constant shadows during the short, subarctic winter days when the sun barely appears above the horizon, forage is available. The herbage is exposed by wind in numerous places along this Outer Range, and nowhere is this more marked than in the Nenana Valley.

The biology of these Nenana Valley sheep can be summarized graphically in a seasonal chart (figure 6.3) illustrating altitude, food, range use, time of movement, and so on. These data come mostly from Whitten (1975) and some of my own observations, and other elements are taken from other Dall sheep studies (Hoefs 1974; Summerfield 1974).

We can reconstruct the optimum time of the year that Dall sheep could have been hunted using knowledge of their behavior and seasonal uses of the landscape. These are presented in figure 6.4.

During the late spring or early summer, sheep frequently converge in large numbers from many kilometers to mineral licks (Heimer 1973). In so doing, they use well-worn traditional trails. They return to the lick repeatedly despite being flushed away. These lick areas would have been unusually good hunting locations, especially for prefirearm hunters. By June (the earliest time that licks are used by large numbers of sheep), there are few sheep remaining in the Nenana River areas accessible from the Dry Creek hunting site. It is likely that people using the Dry Creek campsite hunted sheep only in winter—that is, from late September to

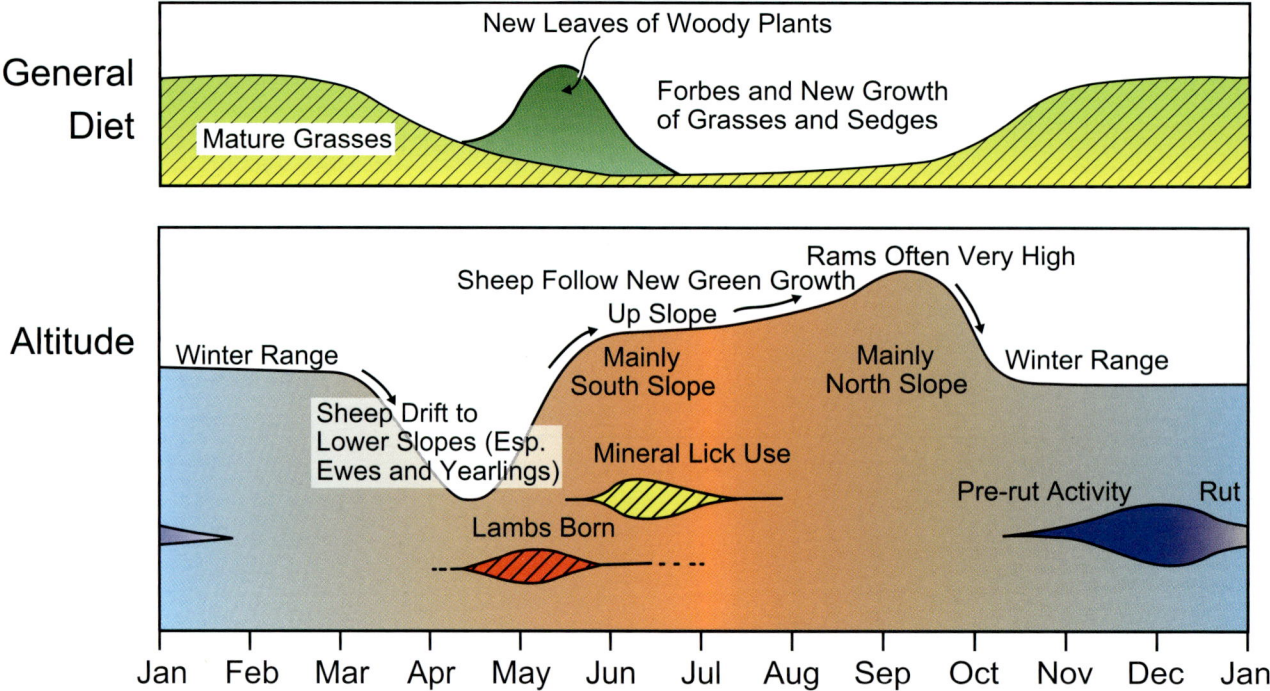

FIGURE 6.3. An encapsulation of the annual cycle of living Dall sheep, *Ovis dalli*. The timing of the depicted events varies somewhat from area to area, as does the diet, but these are the usual patterns. There are regular seasonal shifts in altitude and diet. Mineral licks are likely to be visited in late June after lambing. Rut peaks in early December, although rut activity commences well before that time. As discussed in the text, these seasonal changes make sheep especially vulnerable to human hunters at specific times and places.

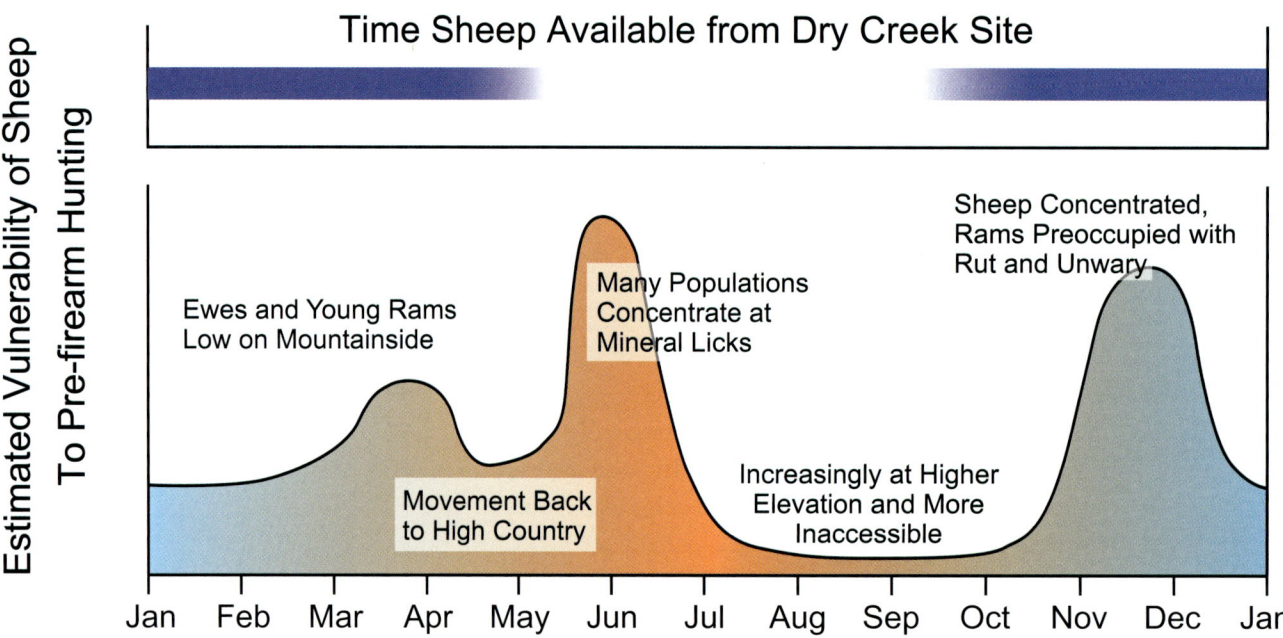

FIGURE 6.4. An illustration of sheep vulnerability to prefirearm hunters, with an indication of the time of the year that sheep now use the area adjacent to the Dry Creek site. During the early winter, sheep are concentrated in the period of rut. Rutting areas, which sheep use from one year to the next with high fidelity, could be seasonally used by human hunters.

late April. They might have hunted sheep at other times but probably not while camped at the Dry Creek site.

It is 2 km from the Dry Creek site to suitable escape terrain, which is today relatively snow-free as a result of winter winds. These areas were probably winter ranges during the late Pleistocene just as they are today. To reach sheep-summering areas, one has to go considerably farther and climb high, steep, rocky terrain. It is unlikely that the Dry Creek peoples would have done so, at least from that camp, particularly when one realizes that the skull bones of a large sheep (maxillae with teeth) were found at the site.

Sheep on winter range spend most of the few daylight hours feeding with heads down. They are frequently dispersed over the ridges and hillsides, presenting more varied opportunities for stalking or driving in predictable directions. During the first part of the winter, rams are in the rut and quite preoccupied with the ewes. This would probably be the most opportune time of year to use a decoy strategy. On the whole, the winter range confines the sheep to a small part of the countryside, allowing humans to study natural movements and escape routes. Sheep tend to use the same trails, shortcuts through passes, and detours around rocky outcrops. They become vulnerable to surprise or snaring. When feeding in a given area, they can be driven along predictable routes and can, of course, be confined to these routes by the use of drive fences, thus allowing other hunters to lie in ambush concealed next to the escape route or to place snares connected to drags prior to the drive. Corrals at the end of the fence funnel are another possibility (Frison 1978).

Although the collections from Dry Creek are meager, one can see that several sheep were killed, most in the early adult age range. Animals in this age range normally suggest a hunt by stealth or decoy on unwary, naïve animals as opposed to the trail ambush system. Old males tend to lead the line of sheep moving up the trail (Geist 1971) and would be killed first; however, this does not always happen if sheep are moving rapidly away from danger—for example, if a sheep dog is pursuing them. At Jaguar Cave in Idaho both wild mountain sheep and dogs occur in the same horizon of 11,000 ^{14}C yr BP (13,000 cal yr BP; Sadek-Koores 1966), about the same age as the Dry Creek site. Whether the Dry Creek site peoples had dogs is unknown.

A late prehistoric site (about 400–300 ^{14}C yr BP [500–300 cal yr BP]) in the Nenana Canyon also contained Dall sheep material (Plaskett 1977). These sheep were identified as autumn kills by the growth stage of the dental cement. Despite the much closer proximity to sheep range (the Nenana River Gorge site is at the base of Mount Healy, where many sheep now overwinter), there are skull and tooth remains and yet no evidence of sheep upper limb bones (scapula, pelvis, humerus, femur). The deboning of the meat was a common practice of early hunters to lighten the load for transport to camp (Frison 1974). The Carlo Creek site, also in the Nenana Canyon south of the Dry Creek site, dated at least 8,500 ^{14}C yr BP (9,500 cal yr BP; Bowers 1980), contains sheep, caribou, and ground squirrel. Bowers also concludes that it is probably an autumn processing-butchering station. Information from the prehistoric use of the Nenana River Gorge site corroborates our model of the chronological uses of the Dry Creek site for sheep hunting in fall to early winter, with sheep heads being carried back to camp for the valuable raw material of the horn sheath and perhaps the choicest food items of tongue, brains, and orbital fat (see Speiss 1979).

Like most other ungulates, there is a spatial sexual separation in sheep during most seasons except the rut. However, ewes and rams can be in the same general area during parts of the year. The greatest concentrations occur during June when the maternal bands flock together with their lambs and during October and November on rutting range. Unlike bison, where old males are difficult to drive, rams do frequently occur in modest-sized flocks (of around ten) and are drivable.

Sheep grow quite fat in the late fall, up to 20 percent body weight (Hoefs 1974). Sheep fat is highly polysaturated, more so than most ungulates, and the rendered lard becomes quite solid when cooled. It is thus excellent for making pemmican and for the treatment of leather for winter garments. The small amount of carbohydrates in the diet of northern peoples makes fat an extremely important commodity. I will return to this point later.

There are several interesting questions raised by the presence of sheep in the site: How important were sheep in the Dry Creek hunters' diet? At what intensity would they have had to hunt sheep for them to be a major part of their diet? A sample of sheep with a bias skewed toward adult rams would average about 100 kg apiece. Of this 100 kg, a maximum of 50 kg would have been usable for food (this includes lean meat, fat, liver, kidneys, some marrow, pancreas, etc.). Boning-out more than two dozen large, mature rams has shown that this figure is liberal. Large rams boned-out provide only 30–35 kg of muscle. Wheat (1972), in a thorough review of the literature on historical observations, arrived at the figure of about 5 kg as the average fresh meat consumption per day of Great Plains hunters. Being conservative and using half of his estimate (2–5 kg), one sheep would then furnish enough food for 20 person-days (the 5-kg estimate was an average for all people in the camp). To be liberal, we could say that this figure includes some meat loss to storage and scavengers.

The number of people occupying the Dry Creek campsite at any one time is unknown. Estimates on the Great Plains groups from about the same time period run between 100 and 200 with a theoretical fraction of 20 percent grown males in the populations to carry out the hunting forays. Northern hunting groups might not have been that large. A hypothetical group of 25 would have somewhere between 4 and 8 grown men. A population that size eating only sheep would require 9 sheep a week, or 38 sheep a month. If they were to live solely on sheep for the winter (October to April), a total of more than 700 sheep would have been required. This figure should be modified downward somewhat, especially if a substantial part of the fat was used for food or if they lived, seasonally, on a suboptimal caloric diet.

A sheep population can tolerate varying degrees of harvest depending on a number of factors. Let us say that, considering a balance of both favorable and unfavorable production years and some density-dependent increases in production with hunting, an average sheep population can sustain 15 percent annual adult harvest at its maximum sustained yield. The huntable sheep population from the Dry Creek site is not less than 300, and even if one doubled it to 600 for Pleistocene numbers, less than 100 sheep could be taken on average (less than 3 months' food resource for a group of 25 people). This figure does not include dogs that could have been present. Nor does it include wounded sheep that escaped and died later and were not found. I have performed this exercise to emphasize the following: (1) the difficulties (or virtual impossibilities) in relying solely on mountain sheep for a long period of time, (2) the difficulties for hunters in ever-establishing stores by depending solely on these medium-sized animals, (3) the varied benefits of hunting bison- to mammoth-sized prey characteristic of the Paleolithic-Paleoindian specializations, (4) the people who hunted deer- to sheep-sized animals did so mainly as a protein supplement, (5) the large role women traditionally played in gathering and in small mammal trapping and hunting, and (6) the significance of carbohydrates in reducing hunting intensity.

The northern peoples have been hunting specialists because there are virtually no plants that can contribute to large carbohydrate stores. The tubers and fruits in the north are small and sparsely scattered. Likewise, part of the seed dispersal strategy of most northern plants is to produce small seeds. The vegetative parts of northern plants are at best only edible in their early growth stage. All this means that the Paleolithic diet in the north was mainly animal tissue with some seasonal plant garnish. One can readily understand the adaptations to the use of coastal invertebrates and anadromous fish after the demise of the mammoth fauna.

Steppe Bison, *Bison priscus*

The taxonomic status of Alaskan Pleistocene bison is in dispute, but they are closely related to the ubiquitous European, Asiatic, and Beringian *Bison priscus*, and for contextual reasons, these fossils will be tentatively assigned to that species.

Bison do not seem to have played as large a role in the Eurasian Paleolithic hunter's diet as in North America. Although bison are common in camp refuse in the Eurasian Paleolithic, most concentrations seem to be of horse, reindeer, red deer, or mammoth. It is possible that bison were simply never as abundant as they were in the New World, at least in the large herds, or were more thinly dispersed though fairly ubiquitous. The exception might be the area of the Caucasus. There are several archaeological sites in the Caucasus that look similar to the bison drive sites on the American Great Plains (Vereshchagin 1967). One can safely say that there are more bison found among the post–11,000 ^{14}C yr BP (post–13,000 cal yr BP) Paleoindian sites in North America than any other large mammal species. The popularity of bison as a prey species was undoubtedly due to its large numbers, the quantity of meat it provided, and perhaps its behavioral susceptibility to mass-kill hunting techniques. Whatever the reason, there are numerous bison kill sites scattered across North America, concentrated in a belt from west Texas up through Alberta (Guthrie 1980). They cover a time period of somewhere around 11,000 ^{14}C yr BP (13,000 cal yr BP) to the decline of the bison herds in the late AD 1800s. There are several common features that most of these bison kill sites share, as outlined in the following list:

1. They are mainly autumn sites (Ewers 1955; Frison 1974). Peak bison hunting prior to the horse seems to have been in late summer, fall, and early winter. Seasonality of the kill is determined mainly by the age of young animals.
2. Much of the bison meat was preserved for stores to be eaten at a later date, unlike the strategies of many modern hunting societies (Lee and Devore 1968). This is determined mainly by the butchering techniques and quantity of meat removed from the site.
3. More female bison were taken than males. It had been suggested (Frison 1974) that this is due to the herding behavior of females, whereas males are in small groups and thus not easily driven toward hunting ambushes.
4. Some special topographic features of the landscape were used to concentrate and prevent escape while hunters dispatched animals. This was either an impoundment, arroyo, embankment, sand dune, or something similar. The geology of the site usually provides this kind of information.

5. Camp was not moved to the kill site; rather, the meat was taken to camp along with certain bones and parts of the bones, some of which were used to carry the meat. The reconstructed butchering techniques reveal an absence of these elements at the site. Often the yearlings are missing, and it is thought that they, too, were taken whole to camp.
6. At least in the autumn, and probably in the winter, human hunting groups tended to be moderately large (not simply a nuclear family group). This has been surmised from the large number of people required to perform a major drive-kill, the number of tools present, and the disassembly line technique of butchering (removed limbs were taken to specific areas and the bone disposed of in stacks).
7. Though they were large mammal hunting specialists, other supplemental game species were widely used by these early hunters. Early campsites reveal a diverse assortment of fauna. Among modern hunting societies, the women often provide the greater portion of the calories to the diet in the form of small game and plants in season (Lee and DeVore 1968).
8. Frison (1974:110) concludes, "Some means of distribution of exotic materials for stone-flaking purposes existed during the Paleoindian period on the Plains."

It is curious that the Dry Creek site, several thousands of kilometers from the Great Plains, in what is now a quite different environment, should be so similar to that of the Paleoindian bison hunters from Montana to the Llano Estacado. With the exception of the Asiatic microblade technology and the permafrost, they could be almost identical in fauna and similar in lithic technology. There is ample reason to suspect that the overall ecological texture of Beringia during the late Wisconsinan glacial period shared features with that of the early Holocene American Great Plains.

With the recession of the Wisconsinan ice sheets, the Beringian bison moved into the Great Plains, probably bringing the bison hunters with them. This seems to be the correct picture, but there is a heated controversy about this issue (Guthrie 1980; McDonald 1981).[1] Sometime before that, however, bison had begun to decrease in size from the ubiquitous *B. priscus* that occupied the Far North from western Europe across Beringia, at least during the Illinoian-Wisconsinan (Riss-Wurm) interval. Throughout its range, bison began to decrease in size during the late Wisconsinan. In Alaska and northern Canada, it became known as *Bison occidentalis* (*B. bison occidentalis*), which might have colonized the northern plains and might have continued to decline in size to become *B. bison* (*B. bison athabascae*). During the early expansion of the Great Plains in the late Wisconsinan, a large bison known as *Bison antiquus figginsi* also continued to decline in size to *B. antiquus antiquus*. Hence *Bison occidentalis* met *B. antiquus*. According to Wilson, they were exposed to one another on the open plains with no artificial barriers and, being closely related, they probably interbred (Wilson 1975; see McDonald [1981] for a different interpretation). The body size and character of the horn core in the resulting Plains bison (*B. bison bison*) underwent further size reduction but, for the most part, retained the style of horn shape of *B. occidentalis* from the north—that is, the horns were directed posteriorly with respect to the skull, and the tips were posteriorly twisted and pointed. The more southern form had horn cores extending at right angles from the skull with little or no posterior twist. Unfortunately, no bison horn cores are present in the Dry Creek site, but *B. priscus* fossils with horn cores are known from other areas in Alaska.

The identifiable bison remains in the Dry Creek site are all teeth. Bison (and all the Bovinae—*Bos, Bubulus, Syncerus*) have very hypsodont teeth, more complex than most bovids. This high-crowned characteristic is an adaptation to grass and grasslike plants, which use silicious phytoliths as an antiherbivore defense (McNaughton et al. 1985). As a result of this increased rate of wear, high-crowned molars evolved in *Bison* and most other grazers. Many grazers do not use much grass in summer (see our discussion of sheep diets), and many ungulates with brachyodont or low-crowned teeth consume a lot of grasses or sedges when these plants are young and have not yet produced their antiherbivore defenses. Extant plants bison consume include some forbs and leaves of woody plants in the summer, but they eat mainly short grasses or the regrowth of grazed tall grasses. In fact, they are one of the least-selective ruminant grazers studied (Peden et al. 1974). Their winter diet, like that of most other grazers, consists almost entirely of grasslike plants. Unlike moose, bison do not have the capacity to rely solely on woody twigs in winter and thrive. Judging from the low rate of enamel wear among browsers, the lignin, hemi-cellulose, and cellulose is quite coarse and fibrous but not very abrasive. The rate of bison-tooth wear can probably serve as an index to the proportion of leaves of mature grass and grasslike plants in the diet (Haynes 1984).

Pleistocene bison from Alaska had a high rate of tooth wear (Guthrie 1990).[2] Introduced bison in Alaska today live in *marginal habitats* for bison along mountain passes, where windblown "deltas" exist relatively free of snow. In those areas, they must rely on browse and forbs (legumes), as there are insufficient grasses present for either summer or winter range (Miquelle 1985).

Unfortunately, bison herds in the wild were reduced and confined before any studies were done on their seasonal movements. McHugh (1972) reviews the migration-nomadism controversy. If we use the fragmented modern bison population and are justified in drawing analogies from other plains grazers such as wildebeest (*Connochaetes*), we may infer that bison followed the moisture-nutrient gradient around the countryside, probably with some traditional year-to-year regularity. Unlike the situation we observe in sheep, which live amid considerable topographic relief, bison tend to be foothill or plains dwellers. They do, however, occasionally graze high into the mountains (McHugh 1972). But even for bison, the lowland vegetation is not phenologically or nutritionally homogeneous in a fine-grained pattern. Local variations in rainfall, past fire history, differences in past grazing intensity, and proximity to open water create subtle variations in vegetational communities over the plains.

Bison sexes are separate for most of the year. Cows and calves are in larger bands, and the bulls are in small groups. They come together in the midsummer for rut (rut lasts, in most modern herds, from late June through September, peaking in late July or early August).

With this brief review of bison life history, we can now look at the significance of bison in the Dry Creek site. There is evidence of bison persisting in small, fragmented populations in Alaska and northern Canada through historical times (Guthrie 1990; Holmes and Bacon 1982). Adequate wintering areas seem to be the limiting bottleneck. The deep, dry powder snows of the interior make access to winter range difficult. Where snow-free areas exist, forage of sufficient quality might not. The most likely wintering ranges are glacial outwash surfaces in major mountain passes. The bison that have been restocked in Alaska are in these special areas. Their numbers are small, and the Alaska Department of Fish and Game manages them on a carrying capacity basis. The upper Nenana Valley could perhaps overwinter a small wild herd of bison today. Several bison actually were introduced there in the 1960s but were subsequently killed by a train while grazing on the railroad tracks in a narrow pass. Several horses have overwintered without supplemental feed on the revegetated strip mine areas on the east side of the Nenana River, where grass is exposed by winter winds from the pass. Tame bison now kept on a local ranch are hay fed during winter.

The terraces, foothills, and valley bottoms of the upper Nenana River were probably bison winter range for reasons that have already been discussed. As we know very little about the actual snow cover in the interior of Alaska during the Wisconsinan Glaciation—other than that it was light (Guthrie 1968, 1982, 1985)—it is difficult to determine the extent of winter range. The wind and precipitation shadow effect near the mountains in the area of the Nenana River must have produced stretches of snow-free winter pastures. As the mean snow cover began to deepen near the end of the Wisconsinan, these valley outwash fans would have been among the last remaining winter ranges to consistently overwinter bison.

There are numerous columnar tooth sliver fragments throughout the lower levels of the site that have the same thickness and character as bison teeth. One quite weathered cluster of extremely long enamel fragments in the original 1974 test trench was tentatively identified by me as *Equus* sp. (Thorson and Hamilton 1977). Subsequent study has shown that the unworn M3 of bison can also produce enamel fragments of that length. Though several subsequently excavated specimens are definitely identifiable as bison, only a few are complete enough to determine individual age and take reliable measurements (table 6.2). All consist of lower dentition and all occur stratigraphically in Component II.

The fact that these bison teeth are all mandibular teeth (no uppers were found [figure 6.5]) might suggest that no skulls were brought back to camp, only lower jaws.

Using Reher's (1974) indices of *Bison antiquus* individual age, the stage of tooth eruption of one specimen was in the middle of its fourth year of life. Another was between nine and

TABLE 6.2. Bison (*Bison priscus*) teeth

Number	Tooth	Length (mm)	Estimated Age
UA77-44-120	(Left) M3	52	Juvenile, hardly worn
	M2	36	
	M1	28	
	P4	25.5	
	P3	23.5	
UA77-44-1422	(Right) M3	46	Very old
	M2	33	
UA74-22-180	(Right) M3 fragment	Incomplete	Peak growth
UA74-22-181	(Right) M3	46.5	Quite young
	M2	Incomplete	
UA77-44-121	Unerupted crown	Incomplete	Immature

FIGURE 6.5. Steppe bison, *Bison priscus*, from Dry Creek. The best preserved large mammal specimen was a complete left lower tooth row (enamel only) of *Bison* (UA77-44-120). Only two other specimens were sufficiently well preserved to be measurable (UA77-44-1422 and UA-74-22-181). All the *Bison* and *Cervus* teeth were identified as lowers and all but one of the *Ovis* as uppers, suggesting that the sheep skulls were being brought back to camp, perhaps for their horns as tool material. The heavy bison and wapiti skulls were perhaps left behind at the kill and only the lower jaw brought back, perhaps for the mandibular marrow. The lower jaw also makes a suitable handhold with which to carry the tongue, cheek muscle, and neck meat back to camp.

ten, as determined by metaconid height. One of the best indicators of sex is depth of the lower jaw, but the jaws were not preserved at the site. The teeth are robust compared to modern bison and are larger. In M1 comparisons among bison from a number of archaeological sites in the Great Plains (see Reher [1974]), the Dry Creek bison fall outside the size range of historic and prehistoric *Bison bison* and well up into the larger size range of the earlier bison, with an M1 length of about 29 mm and a width of about 18.5 mm.

Unlike many other northern grazing genera (such as mammoth, *Mammuthus*, or collared lemming, *Dicrostonyx*), Pleistocene bison have not undergone any rapid dental changes. The first identifiable fossil bison have teeth very similar to living bison. There is one exception that Skinner and Kaisen (1947) have pointed to as an indication of fossil versus present bison: the posterior expansion of the metaconid of P4 in the living bison results in a shortening of the depth of the lingual fold and produces a fossette on moderately worn teeth. The development of a metastylid also contributes to the width of the mesoconid in living bison. The teeth from the Dry Creek site are primitive in regard to these characters. This information is consistent with the radiocarbon dates for the site.

Though the information about bison is fragmentary, it seems certain that the peoples camped at Dry Creek were hunting and eating bison. Because of the reconstructed seasonal grazing range characteristics, we can conclude that bison were probably in this area mainly in fall or winter. The fact that long bone fragments and jaws are present indicates that the animals were probably not killed a great distance from the site, as the bones would have added considerably to the pack weight and are usually left behind at bison kills (Wheat 1972). Whatever hunting techniques were being used, they were sufficient to take adult bison. Judging from the lithic assemblage, the bison were probably dispatched with bifacially tipped javelins or by spears with composite bone and microblade heads. Normally, Paleoindian bone points are rare in midcontinent North America. They do, however, occur at Blackwater Draw and other scattered areas throughout North America. A bison scapula from Kokorevo I in Siberia has embedded in it a composite point minus the inset microblades (Abramova 1967).

An intensive autumn and early winter hunting effort aimed at larger animals like bison would make sense. The same limitations on fat availability in the fall would have pertained to bison as well as to sheep. Also, the bison concentrations in autumn are a practical time to set aside stores. Active nomadism of the entire camp during winter, when plant and microfaunal resources are more limited, would have meant a total reliance on the unpredictable large mammals. A more sedentary winter existence would have allowed the hunters to take advantage of the fall and early winter game concentrations and kill beyond their immediate needs. Frozen, dried, or smoked meat would not have to be transported. Camp could be made near running water (probably not a minor convenience in a preceramic era). Fires could be maintained; starting fires in the arctic winter is more difficult than at warmer temperatures. Warm, heavy skins for sleeping could be seasonally accumulated for comfort. Heavier, sturdy skin tents could be built in conjunction with establishing a base camp. A permanent fall and winter camp would have allowed mobile hunting parties to forage great distances, accumulating and adding stores.

Traditionally, moose-hunting Athabaskans have been sedentary in the summer (at fish camps), when it is difficult to travel cross-country over impenetrable tussock muskeg and through blackfly and mosquito swarms. Athabaskans traveled in winter using snowshoes and dogsleds over the frozen landscape (Vanstone 1974).

Autumn or early winter hunting intensity is indicated by the human megafaunal kill sites on the Great Plains of North America (Frison 1974) and the prevalence of winter Paleolithic campsites in Eurasia (Klein 1973). Both argue for an almost universal *seasonal* hunting strategy of northern early peoples that is also exemplified by the traditional inland Eskimo and Athabaskan caribou hunters.

Not only would the Dry Creek campsite have been in the path of autumn movements of large mammals along the front of the Alaska Range, as well as the major terrace systems and the river flats; it also would have been on the winter range of several large grazing mammals.

Wheat (1972) calculated (by using dietary and ethnographic data) that a single bison would furnish food for about 60 person-days. This would mean that a group of 100 people living on bison meat alone would have had to have 400 bison to last them from early October through May. A group of 25 people would have had to have about 100 bison for the same time period. Or being conservative by using half those requirements, they would have needed 50 bison.

Traditionally, meat was preserved on the Great Plains by drying it in thin strips into jerky. Often, this was then pounded into a white flaky powder and mixed with fat (and usually some berries) as pemmican. The Plains tribes then wrapped these into standardized, fat-sealed containers of untanned bison skin (rawhide), which were known as *perfleche* among the plainsmen. These containers were inaccessible to small vermin and preserved their contents for long periods of time. It is likely that the hunting people of the late Pleistocene in the north also had similar ways of storing and preserving meat. A small fire under the drying racks to smoke meat strips adds preservability and flavor. I have made smoked jerky from wapiti, caribou, and moose for a number of years and find that it is excellent trail food. The weight is reduced by about 70 percent from that of lean fresh meat depending on how dry it is prepared. The processing of dried meat as pemmican not only is a more convenient package but is quite concentrated in nutrients and calories.

The advantage of dried muscle as opposed to cooked meat is that the heat-labile nutrients are retained. The pemmican is also a convenient vehicle by which fat can be eaten. Fat is more difficult to digest and less palatable when eaten alone.

It should be emphasized that this process of cutting the meat into thin strips (around 0.5 cm), drying/smoking them, breaking and pounding each into small pieces, dicing the fat into small chunks (less than 0.5 cm), and constructing a container is a time-consuming task. Several dozen big-game animals represent many hours of preparation after the animals are hunted, killed, skinned, butchered, boned, and packed to camp.

The great reliance on meat in the north meant that if larger game like bison were not available, very great quantities of medium- and small-sized animals had to be harvested. There were, however, other large species available during the period 11,500–10,500 ^{14}C yr BP (13,350–12,480 cal yr BP), and one of these was wapiti.

Wapiti, *Cervus canadensis*

While the fossil *Cervus* teeth found in the site cannot be definitely assigned to a species by any inherent morphological characters, they are probably wapiti (*C. canadensis*).

The habitat of wapiti is more difficult to delineate than bison and sheep. They can occur in the grassy heath meadows, like the Scottish red deer; in the woodlands of the Appalachians; the plains of Illinois and Kansas; or high in the Rocky Mountains. Lewis and Clark (1893) reported that "they are common to every part of this country."

From Murie's (1951) classic study, one can conclude the following:

1. Wapiti follow the high-quality early plant growth stages, which usually means movements into the mountains in spring and summer (when sufficient topographic relief is present).
2. They characteristically move out into the windswept prairies, or plains, whenever snow in mountains or woodlands becomes limiting.
3. Like mountain sheep, they are animals of habit and tradition, making similar seasonal movements and using similar ranges from year to year.
4. Also, as in sheep, body and antler size are general indications of range quality.
5. They are very social, with sexes living separately through the year, except for the autumn.

On high-quality ranges where sheep numbers are below carrying capacity, it has been found that wapiti eat much the same plant species as sheep: forbs in summer and grasses in winter. If both sheep and wapiti are on the same range, there is direct competition, but because of different range-use patterns, this overlap does not occur often.

The cervids, or deer groups, have radiated as browsers in a deciduous woodland setting, specializing on young seral plant growth at the parkland edge. *Rangifer* and *Cervus* have moved away from this original focus, the former becoming a specialist on lichen for winter diet and the latter becoming more of a grazer in winter. Both have moved away from the woodland cover in parts of their range and have become relatively social, occurring in herds of large size in an un-deerlike fashion.

In the northern Rocky Mountains, the rut begins in late August and runs until mid-October (Murie 1951). The period of calving is from May 15 to June 15. Wapiti, like both mountain sheep and bison, give birth to a single calf and virtually never have twins. The males drop their antlers in February and commence to regrow them in April. In a number of ways, the life histories of sheep, bison, and wapiti are quite similar. They mature at somewhat similar rates, eat much the same foods, have similar mortality curves, have roughly similar social organizations, and have only one young per year, and the males are elaborately adorned with social paraphernalia. They have evolved toward similar aspects of the open grassland environment from different evolutionary lines and have experienced somewhat the same general selection pressures.

Wapiti were a common element of the mammoth steppe, which extended across Eurasia to Alaska (Guthrie 1966, 1968), and were generally well represented in Rancholabrean faunas throughout the Holarctic. The few ^{14}C dates from Beringia, however, cluster in the very late glacial or Holocene, suggesting a possible increase in numbers with the decline of the mammoth steppe. Straus (1977) concludes that wapiti is the most abundant megafaunal component throughout the Paleolithic sites of northern Spain. This is true for many northern European sites as well. *Cervus* is the most frequent species at Star-Carr (Clark 1954), a famous Mesolithic site in England. It is also present in several early sites in North America and in northern Asia. Wapiti never seems to have been a major early dietary item in the New World, at least it is uncommon in Paleoindian archaeological sites (Frison 1978), so it might not have as yet undergone its Holocene increase.[3]

We know very little of the early techniques used to hunt wapiti. As major predictable migrations occur when wapiti are in large herds, rock fences with snares (anchored to drags) across the openings could have been used in the same manner that the Kutchin hunted caribou (Roseneau 1977). We do know that *Cervus* was driven into impoundments in Medieval Europe. The construction of elaborate hunting facilities in the Nenana Valley would have meant a more traditional commitment to those hunting grounds. Other hunting techniques might have been used, of course. At Star-Carr, there are a number of *Cervus* skullcaps with antlers attached from young *Cervus* bulls. These have two holes for the attachment of a chin strap so that they can be worn on the head. Whether these were used in allowing the hunter to approach a herd or used in some ceremonial rite is unknown (Clark 1954).

The large wapiti antlers would have provided a valuable raw material for tools. Antler is less dense and easier to work than the diaphyses of long bones, and yet it is strong and durable. It is porous and works quite easily when wetted.

There are a number of specimens of enamel fragments from the site that could be *Cervus*, but only two specimens are definitely identifiable as such (table 6.3). Both are lower dentitions and both come from Component I (figure 6.6).

Although sex is not known, these appear to be large teeth consistent with the findings from other wapiti fossils in Alaska and the Yukon Territory (Guthrie 1966). These fossil Beringian wapiti were the largest group in the past and present range of *Cervus*. One can only conclude that they were on a high-quality range.

Reconstruction of the Late Glacial Megafaunal Community of Interior Alaska and Its Paleoecology

Because the Dry Creek megafauna are transitional between the full glacial and Holocene community, it would be worthwhile to first reconstruct the general features of these periods before analyzing and interpreting the Dry Creek fauna.

TABLE 6.3. Wapiti (*Cervus canadensis*) teeth

Number	Tooth	Length (mm)	Estimated Age
UA76-20-110	(Left) M2	Incomplete	Only medium tooth wear
	M3	Incomplete	
UA76-155-3891	(Left) M3	Incomplete	Very worn (+16)
	M2	30.5	
	M1	22.5	
	P4	Incomplete	

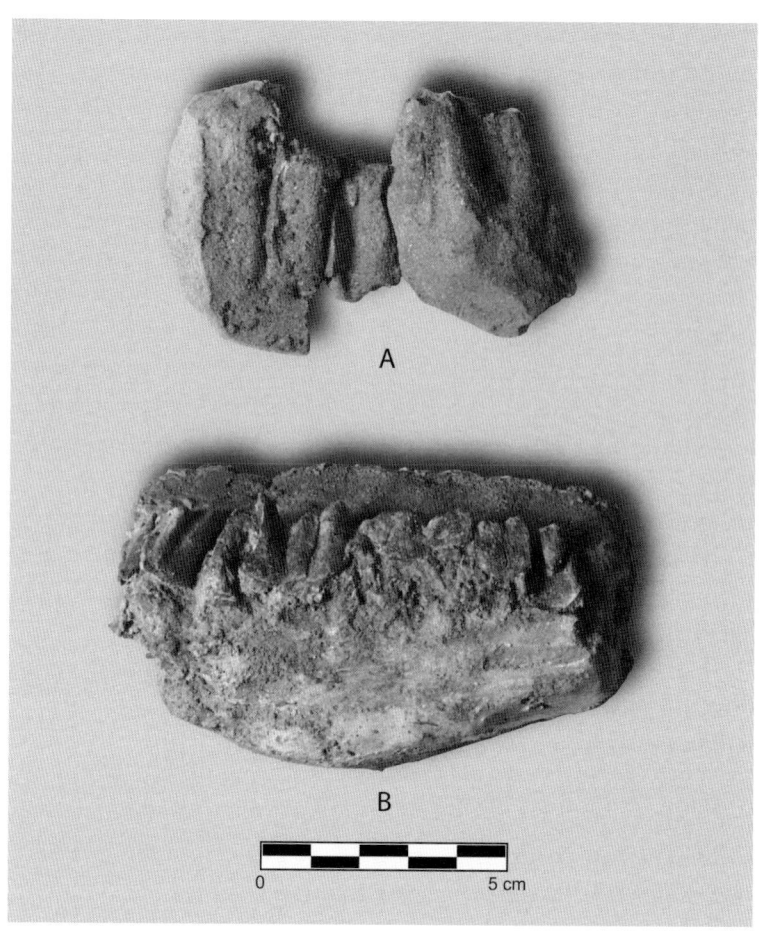

FIGURE 6.6. Wapiti, or elk, *Cervus canadensis*. Two specimens were identified as *Cervus*; both are left-lower dentitions from two different animals.

Although the palynological evidence suggests otherwise (Colinvaux 1980; Cwynar 1982; Cwynar and Ritchie 1980; Ritchie 1984; Ritchie and Cwynar 1982), the vertebrate fossils indicate a dry steppe environment during full glacials (Guthrie 1968, 1980, 1982, 1984a, 1984b, 1985). There is very little direct evidence as to exactly what that steppe was like. Indirectly, however, one can argue that the specialized adaptations seen in the morphology of bison, horse, ass, sheep, and mammoth indicate a dependence on grass, at least for winter forage. The great diversity within the faunal grazing community also points to a steppe flora of considerable spatial and growth-form diversity. I have referred to this general biotic province—which extended from England to the Yukon Territory, Canada—as the *mammoth steppe* (Guthrie 1980, 1982, 1985).

We know from the studies of ungulate grazing communities elsewhere that competition is avoided or reduced by niche specialization (Bell 1969). Some species even increase the volume of forage available for other species (e.g., equids removing grass tops, which improves access to midstem leaves for bovids). Also, when there is direct competition, grazers have often evolved a spatial separation. For example, sheep have undergone adaptations to climbing in order to escape predators on steep alpine slopes (Geist 1971), while saiga antelope (about the same size as sheep and with similar dietary preferences) use the flat lowlands and depend on speed to outdistance their predators.

Jarman (1974) and Bell (1969) have shown that in a grazing community, there is a species gradient in the ability to use fiber in the diet. Some species select for portions of the plant that are lower in antiherbivory devices (fiber, phytoliths, resins, alkaloids, etc.), while others are less selective, specializing in the abundant parts of plants that are high in fiber. Jarman and Bell were able to show that this use of fibrous plants relates to body size. The smaller ungulates require a larger percentage of digestible protein in their forage (lower quantity of fiber) than do larger-bodied species. Sheep and saiga-sized animals require relatively high concentrations of protein, hence they are extremely selective. They graze by selecting the best (least defended) plant parts (Whitten 1975). Elephants, at the other extreme, consume vast quantities of low-quality forage (Laws et al. 1975).

With these general principles in mind, there is much we can say about the ecology of the grazing ungulates that occupied the mammoth steppe even though we do not yet have a firm grasp of the exact character of the vegetation.

In an earlier publication (Guthrie 1968), I showed that bison increase in numbers relative to equids from the Tanana Valley toward the Tanana Uplands. Sheep bones are virtually unknown from the lowlands and, when found, usually show signs of stream transport from the uplands. One must be cautious, however, about reading these distributions as indicative of year-round occupation because of the seasonality of range use. But we do know from numerous wildlife management studies that most adult mortality occurs during late winter or early spring or, in the case of more temperate climates, during the dry season (Laws et al. 1975). Thus in Alaska, natural fossil ungulate assemblages might be general indicators of winter range (in combination with depositional biases). The upshot of all this is that ungulate winter use of the Tanana Valley during full glacial conditions can be reconstructed through a combination of fossil distributions, ecology of relict forms, current models of grazing ungulate habitat partitioning, and information about the dental morphology.

Modern Ungulate Community

Much ecological information now exists about the ungulates living in central Alaska today. For background for the discussions that follow, I would like to give a simple review of the salient features of ungulate ecology in the north. The species are in general order of abundance: caribou (*Rangifer tarandus*), moose (*Aces alces*), and Dall sheep (*Ovis dalli*). Caribou are usually found in the rolling, tundra-covered foothills, though they often frequent the thinly vegetated lowlands and higher alpine areas. Caribou bulls generally tend to be high in the mountains in the summer months. Moose are generally a lowland animal; however, willow fingers reaching into alpine areas are favored habitats, and like caribou, bulls will frequently be found at higher altitudes in summer than will cows. Sheep are exclusively alpine dwellers, though they will come down rather low at times if suitable escape terrain is available. Rams tend to graze higher in the summer than ewes.

There is considerable overlap in the summer diets of caribou, moose, and sheep. For example, young willow leaves play an important role in all their diets. In winter, however, there is virtually no dietary overlap. Caribou crater through the snow for lichens and a few herbs. Moose shift to woody twigs above the snow. Sheep inhabit alpine areas of low snow cover and concentrate mainly on the green bases of grasses and a few other herbs.

The climax spruce forest is inhabited regularly by virtually no large mammals and relatively few small mammals, mainly red squirrels (*Tamiasciurus*), flying squirrels (*Glaucomys*), and red-back voles (*Cleithrionomys*).

Snow is one of the major limiting factors for ungulate populations in the north. It increases the amount of energy required for moose to move among the lowland high shrub patches, making the subnivian plants less accessible to caribou and decreasing access to the sheep's winter alpine ranges.

Ungulate Winter Range in the Nenana Valley

The archaeological record of late Paleolithic and Paleoindian peoples usually reveals a hunting economy centered on large mammals. As the peoples at the Dry Creek site were evidently also big-game hunters, the concentrations and distributions of large mammals should be a central focus of the paleoecological reconstruction. I wish to propose that the Nenana Valley, which is an important winter range for ungulates today, was also prime winter range during the time of occupation of Components I and II of the Dry Creek site.

The chief bottleneck for northern ungulates is the winter (Klein 1970). While most wild northern ungulates are not stressed by cold temperatures, winter food quality is quite low, and when access is restricted by snow, it usually results in rapid depletion of fat reserves, debilitation, or exposure to predation. Winter ranges thus become critical in northern ungulate biology. These are usually quite restricted, and ungulates use them in a traditional fashion (Summerfield 1974).

The windswept slopes around Healy are traditional wintering ranges of two different large herds of Dall sheep (*Ovis dalli*) from either side of the canyon. This area was also an important winter range of the Delta caribou herd. When the deep snow accumulates in the upper drainages of several adjacent rivers and tributaries, moose also concentrate in the Nenana Valley in late winter.

Several local residents of Healy turn their horses loose during the winter to graze on the revegetated excavations from open-pit coal mining. There is also a maverick packhorse that has gone feral in the area and cannot be caught. The horse has escaped pursuers and survived for several years. A rancher, Beryl Mercer, has also kept bison in the area.

There are two factors that make the Nenana Valley a good ungulate winter range. The first is reduced snow cover; the second is vegetation characteristics produced directly and indirectly by the wind. Wind is most destructive to the climax spruce forests. Not only does it blow down the shallowly rooted northern conifers, dehydrate them, and abrade the wintergreen needles with tumbling ice crystals; it magnifies the effects of fire by causing hotter burns and carrying the burn over wider areas. The dry wind also makes for greater plant susceptibility to burning.

Thus most vegetation around the Nenana Valley is in the early stages of fire succession or aridity subclimax. There are many grass meadows and willow thickets. These are the plants now used by many ungulates for winter range.

During the last glacial and early postglacial, we know from other lines of evidence (see Péwé [1975] and Hopkins et al. [1982] for a review) that woody plants were uncommon in interior Alaska. Given a more herbaceous plant community at that time in the Nenana Valley, it would have been an even more favorable wintering area for ungulates. This assumes the same type of katabatic air flow. If anything, the winds would not have been diminished by increased mountain glaciation but actually increased. Sand dune activity and thick loess deposition farther to the north (Péwé 1975) indicate even greater wind speed through katabatic flow or other mechanisms during glacials.

Unfortunately, until now, most Quaternary geological investigations in interior Alaska have been devoted to the Tanana Uplands with their famous glacial-aged thick loess, rich in vertebrate fauna. As yet, there have been no deeply stratified sites of Wisconsinan age with fauna and flora studied in detail for the north Alaska Range, so we know relatively little about the vegetation and animal communities there during the full Wisconsinan Glaciation, say 15,000 ^{14}C yr BP (18,000 cal yr BP). However, the species of the mammoth-fauna so common around Fairbanks have also been found along the north front of the Alaska Range. Our archaeological surveys produced a mammoth lower jaw in the Teklanika Valley (a nearby drainage running parallel to the Nenana River), and Joseph Usibelli, owner of the coal mine at Healy, said several mammoth and bison bones had been unearthed in the process of removing the overburden sediments from the open-pit coal mine.

It is possible, however, that the periglacial conditions produced a harsh environment for ungulates, though modern periglacial analogues suggest conditions even more conducive to ungulate range than farther out in the plains (Geist 1978). But of course, there might be no appropriate modern analogues to the Pleistocene periglacial environments (Guthrie 1982).

It might be possible that the Dry Creek fauna represented a relict of the mammoth steppe, which we know occurred farther away from the glacial front throughout most of unglaciated Alaska, Beringia, and northern Eurasia. It would have been relict, because by 11,000–10,000 ^{14}C yr BP (13,000–11,500 cal yr BP), the shrublands and woodlands were probably well along in gaining dominance over the mammoth steppe (Ager 1975). These relict habitats probably existed only in scattered refugia with special windy conditions.

Megafaunal Analysis and Site Chronology

Wilson (1975) has illustrated the utility of bison size in dating archaeological sites in the northern Great Plains. Because the sequence of size reduction in Alaskan bison is not documented to such a fine degree as in the northern Great Plains, the same size-date equivalents cannot be as precise in Alaska. In addition, bison numbers were probably declining in the Far North at the beginning of the postglacial so that exact parallels between there and the expanding herds of the Great Plains might never be available.

Be this as it may, the Holocene bison in Alaska did undergo a rapid size reduction. Harington (1978) has several large male skulls from the midlower range of *B. priscus* at about 12,500 ^{14}C yr BP (14,800 cal yr BP) from the Yukon Territory, Canada. At the other end of the scale, an older male bison skull collected by Dr. Frederick Hadleigh-West from near Anchorage, Alaska, was dated at 500 ^{14}C yr BP (530 cal yr BP; Guthrie, unpublished). The latter is well within the size range of *Bison bison*, though near the high end.

On the basis of climatic expectation alone, one might expect that the bison in the Dry Creek site would be generally smaller than the Wisconsinan *B. priscus* mean. We know from pollen evidence (e.g., Ager 1975) that the climatic change resulting in woodlands along the Yukon drainage was well under way by 11,000–10,000 ^{14}C yr BP (13,000–11,500 cal yr BP). Thus the currently accepted paleobotanical view is a rapid reinvasion of the mammoth steppe by shrubs by 14,000 ^{14}C yr BP (17,000 cal yr BP) and conifer lowlands by 11,000–10,000 ^{14}C yr BP (13,000–11,500 cal yr BP). Given the 11,000–10,000 ^{14}C yr BP (13,000–11,500 cal yr BP) age of the Dry Creek site's early components, such large bison were unexpected. We can safely say that whatever changes resulted in the size reduction of bison in the Holocene, they had not yet appreciably affected the bison populations living in (or visiting) the Nenana Valley. In fact, the Dry Creek bison is not only as large as the Wisconsinan *B. priscus* from Fairbanks but appreciably larger than the mean. Exactly what this means paleoecologically is not clear, but it definitely has some major implications for our ideas about Pleistocene gigantism (Guthrie 1984b).

The relationship of this large bison to an independent evaluation of the site chronology is also unclear. Superficially, it would suggest an earlier age than the ^{14}C date indicates; however, Wilson (1975) has shown that while bison in the Great Plains began to undergo size reduction by 10,000 ^{14}C yr BP (11,500 cal yr BP), there were still moderately large bison in the Hawkin site dated at 6,270 ± 170 and 6,270 ± 140 ^{14}C yr BP (7,160 and 7,170 cal yr BP, respectively). The major reductions in bison size on the Great Plains thus seem to have taken place after 6,000 ^{14}C yr BP (6,840 cal yr BP; Wilson concludes that this might have been caused by the arid altithermal). However, there is no reason to suppose that the size reductions in the Alaskan and Siberian bison corresponded exactly to this same chronology, although they do seem to have been roughly comparable.

Nutritional Considerations at the Dry Creek Site

It is important to consider the nutritional dimension to the story of early peoples in the north because, ultimately, that is going to be the central factor in their population biology, seasonal strategies, and general success or failure. The sparseness of carbohydrate resources in the north cannot be overemphasized. There are a few wild greens, legume tubers are the size of one's finger top, and most berries are quite small (in fact, there are little or no ericaceous pollen in mid to late Wisconsinan–aged sediments in Alaska, indicating that no berries were available at that time). These plant resources are not available in all areas, and when they are, it is for a very brief season. This leaves animal tissue as the major food source. There is, as yet, no good evidence that Paleoindians used fish to any great extent. Early peoples were dependent mainly on birds and mammals. Birds can be difficult to catch and do not provide much meat unless dozens are killed. The volume of small mammals required to meet nutritional needs is also very great. Thus the selection in the Far North quickly narrows down to large mammals as the main Paleoindian food resource.

Viscera, heart, lungs, liver, spleen, and so on are high in calories and vitamins but do not keep well. Muscle is the best storable protein source. Body fat and marrow also store moderately well when properly treated.

Muscle, however, is not very calorie rich. As I already mentioned, researchers have proposed from ethnographic studies that the average person required around 5 kg of red meat per day, if meat was the only food available (Agenbroad 1978; Wheat 1972).

To check this calorically, I went to the nutrition literature and found that for every 5.65 kilocalories per gram of protein (on the average), 0.45 is lost in feces and 1.20 is lost in urine. This leaves 4.0 kcal/g metabolizable energy, or *physiological fuel value* (Pike and

Brown 1967). Raw meat from game animals normally averages about 26 g of protein per 100 g of meat. This results in 136 kcal gross energy per 100 g meat, or 104 *available* calories. The calculation ignores the digestible intracellular fat content of the meat, which varies depending on conditions of the animal. Speiss (1979) calculated 800–1,000 kcal/kg, or 450 kcal/lb, which agrees with these calculations.

The mean daily caloric requirements are difficult to calculate without some knowledge of average human weights, age, sex structure, and activity budgets (Pike and Brown 1967). As a rough estimate in all human subgroups, very active males might require more than 4,000 kcal but only use between 2,000 and 3,000 per day in more relaxed conditions. Women use less than men because of lower metabolic rates, averaging 15–20 percent less than males. A moderately active 120-pound woman requires about 2,000 kcal per day. Children require slightly less. Using an average equivalent of 3,000 kcal per person would mean only 3 kg of meat per person and about 4 kg for a hardworking male. Fat has more than twice the available energy as protein (9.0 kcal/g vs. 4.0 kcal/g). A daily supplement of only 50 g of fat per day would decrease the daily calories required from lean meat by 450 calories.

Thus without the carbohydrate supplements in the north, people require a considerable volume of protein and fat. The calculations here are somewhat below those for the Great Plains but are not greatly different. These amounts of meat seem incredibly great to those of us on modern, carbohydrate-rich diets and living rather sedentary lifestyles.

The concentration of large game hunting activity in the autumn or early winter in both the late Paleolithic of Eurasia (Klein 1973) and the New World (Frison 1978) has been explained by the need for winter food stores, and this is undoubtedly correct. But there was probably another major reason. Even when winter hunting was profitable, fall would have been the only season that fat was available in any quantity. Killing animals at this time of year maximizes available calories. Speiss (1979) calculates, for example, that an adult caribou yields 76 person-days of food in September but only 15 person-days of food in December; the difference is mainly one of fat content.

In addition to being calorically rich, fat makes for both efficient stores and compact trail food. But probably most of all, its rich, sweet, salty flavor breaks the monotony of lean, roasted meat. Wild meat is not marbled like grocery store domestic cuts, and unless it is cooked quite rare, it becomes dry because of the lack of interfasicular fat. When drying meat for storage, one must first remove the fat, thus jerky eaten alone is dry like bark. Esthetically and energetically, dry meat is best when combined with fat as pemmican. Fat also greatly increases the heat from woody twigs when added to a fire. But burning one's food is generally an uneconomic proposition in terrestrial habitats (sea mammals produce so much fat it is sometimes used for fuel by Eskimos). Fat is a long-burning source of light, the only one available in dark houses or tents during the subarctic winter. Fat is necessary for water- and snowproofing garments and for retaining leather suppleness.

Northern wild ungulates killed at any time except fall are generally without appreciable body fat. Thus these autumn harvests of large mammals take on considerable importance because the fat must last until next summer. The available small mammals (such as hares) and birds (ptarmigan, grouse) are equally lean during winter. So fat was undoubtedly a commodity of high worth. Rather than postpone harvest through the winter, these early hunters would have maximized available calories by killing heavily during the fall when fat was obtainable. This probably explains the general concentration in fall and early winter harvests in both the Old World and the New. They were after the fat. But this principle might also tell us why these earlier peoples created a relatively well-used hunting camp or processing station on Dry Creek.

We can do little more than guess at how many large mammals were taken from the Dry Creek campsite and how many people this resource could have supported. As part of this study, I made an attempt at using the microblade cores at the site to approximate the numbers of points that could have been produced. Also, using the literature and some of my own experiments, I made some rough estimates of how effective these points could have been before they were broken. These estimates were made with wide confidence limits. Accepting these, one could calculate how much game was killed and then how many person-days that game could support. The estimated range of person-days ranged from 9,000 to 150,000, which is to say that from those calculations, the site could have provided large mammal resources to sustain 25 people for between 1.0 and 16.0 years. I do not wish to imply that this was a continuous occupation—undoubtedly it was cumulative (remember these calculations only include the microblade projectile points from Component II).

Although, like the other calculations in this chapter, these are only an exercise to give one a more concrete feel for the ecology of large mammal hunters, they do show that not only did early northern hunters have to kill hundreds of large mammals to exist; they seem to have produced the tools at Dry Creek to make that quantity of kills.

The Development of Big-Game Hunting in North America

Though not a rich, well-preserved bone assemblage, the material in the Dry Creek site provides some critical clues to the nature of big-game hunting techniques and their development in the New World.

Despite the availability of hundreds of productive Paleolithic archaeological sites in Eurasia, there is no well-defined evidence of hunting technique. The most obvious interpretation of the faunal diversity in the Eurasian sites is that the hunters were skillfully opportunistic, using a variety of methods on a variety of game species. The species represented in site middens are heterogeneous between and often within sites.

This opportunistic type of hunting persisted for many tens of thousands of years over a vast continental area, most of the Eastern Hemisphere. The people of the early Holocene in North America, however, became specialists at one major kind of hunting technique, the drive.

Drive sites are uncommon in Eurasia. The cliffs of Solutré are a notable exception. Here the land lay was uniquely suited to herding horses over a precipitous cliff (remains of more than one hundred thousand horses are present). But indications are that drives were a small part of the hunting repertoire; there seem to have been relatively few drive specialists in the Eurasian Paleolithic when compared to the American Great Plains during the Holocene.

From studies of the osteological assemblage associated with the bison drive sites (Frison 1974; Kehoe 1973; Wheat 1972), the kinds of bone least likely to be taken back to the encampment have been shown to be mandibles and skulls. Yet teeth are among the more common skeletal elements in Eurasian sites and the Dry Creek site. In fact, like the Eurasian

Paleolithic sites, the assemblage at Dry Creek has all the earmarks of the employment of a more opportunistic, heterogeneous hunting strategy. At least three different species of large ungulates are present; each species lived in different environments and terrain and used different escape behavior; each demanded a different kind of hunting strategy.

Another characteristic of the bison remains found in the drive sites is the preponderance of females. Judging from the large size of the Dry Creek site specimens, it is possible that they were males. Ethnographic studies of bison use indicate that females were more easily herded and driven and that, in fact, the hides and meat of females were more highly prized (Kehoe 1973). My experience in hunting other ungulates confirms this. The meat and hides of females and young are preferable to that of bulls for almost every use.

The massive concentrations of bone associated with drive sites have not been found in Alaska. When bone accumulations have been discovered, they have not been in contexts that would indicate a drive-kill site (the caribou fences from a much later time are an exception).

Another indirect piece of evidence that suggests that the northern hunters were using more opportunistic strategies is the location of archaeological sites along lookout prominences. Most sites of the Denali Complex (West 1975) in the upper Delta River valley dating from about the same time as the Dry Creek site and later are along the tops of serpentine moraines and eskers and along the north front and within the Alaska Range. This is also true of a number of other sites along the north face of the Alaska Range. The Campus site is likewise up off the valley floor, affording a good view to someone searching for game. Until the Dry Creek site was analyzed, it could have been argued that these were sentinel stations from which the "decoy" could sight the herd and prepare to drive the animals into the trap. The Dry Creek site, however, allows us to see that these early people were taking a variety of game and probably looking for any likely huntable animals from these spike camps.

The tool kit remains at these sites seem to have been light in weight like those of backpack hunters in Alaska today. Binford (1983) describes Nunamiut hunting-pouch contents, which are generally similar to those artifacts found at the Dry Creek site: a small stone core, blanks, flaking gear, and individual flakes for knives. With these, the hunter could repair tools and weapons or make new ones, start fires, prepare field food, skin game, and prepare skins and meat for transport, in addition to a wide variety of other tasks.

Thus a pattern begins to emerge from the spectrum of sites across Eurasia, into Alaska, and on into the Great Plains. There was a shift from the generalist-opportunist hunters to the more specialized hunting industry of first mammoth hunting and then bison hunting. The reason behind this shift probably lies in the nature of the large animal community. The Eurasian faunas were heterogeneously mixed: cervids, equids, rhinos, bovids, and proboscidians. Much the same is also true of the late Pleistocene faunas of the Alaskan and Asian mammoth steppes (Guthrie 1982).

In the Great Plains, however, at least after 12,000 ^{14}C yr BP (13,800 cal yr BP), the major faunal elements were mammoth and bison. After 11,000 ^{14}C yr BP (13,000 cal yr BP), the major biomass component of the large animals seems to have been mainly bison. This was not the case in the more mountainous areas, where Paleoindians were still hunting a wide variety of game species (e.g., Jaguar Cave, Idaho, with *Equus*, *Ovis*, *Bison*, and *Cervus* [Kurtén and Anderson 1972]). In the open plains, however, people were becoming bison specialists.

Frison (1978) refers to this as the "buffalo procurement complex," and Kehoe (1973) calls it the "bison industry."

At first, these bison specialists seem to have used the older opportunistic strategy of the northern hunters but went a step further and herded bison over natural precipices or, more often, into narrow ravines, where many could be speared before getting away. Still, these were probably opportunities of circumstance rather than laid-out plans (Wheat 1972). The final stage of this trend was an impoundment area, where all bison herded into the enclosure could be systematically dispatched (Kehoe 1973). With the coming of this technique and the Iberian horse, bison became similar to "open range" stock, which could be harvested at appropriate times of the year.

From this perspective, we can begin to see these Paleolithic colonizers in a different light. The Great Plains mammoth and bison hunters were specialists. The Dry Creek peoples and their Eurasian counterparts probably were not. Although these differences are surely a matter of degree, they do indicate a notable difference and provide an important ingredient in a reconstruction of how these early people must have lived. Combined with several other bits of information, they can provide some clues about the environment as well.

The fact that all sites in the 11,000–10,000 ^{14}C yr BP (13,000–11,500 cal yr BP) range or earlier thus far excavated in Alaska show neither long-term sustained nor intense use (other than a profusion of flakes that can be generated in a short period of time) is consistent with the exploitation pattern at Dry Creek, where a variety of animals were being taken. People were taking every game animal they could get, not in organized, planned drives of a single game species, but as individual groups of hunters using a variety of hunting and possibly trapping methods.

This kind of exploitation will not support a population in the same area indefinitely. Even seasonal use in this fashion eventually reduces the food productivity of an area. Most of the early sites in Alaska suggest this pattern of use. There is intense use for a season or maybe even sporadically over several years followed by a sterile horizon before any other signs of use, and there is often no more artificial evidence in the rest of the strata.

The Orb Model of Hunting Camp Settlement

There has been considerable discussion in the literature on how one predicts land use and settlement patterns from archaeological sites. Also, there are a wide variety of terms used to describe these different cultural patterns. The Dry Creek encampment seems to have been different from any analogy we could draw from extant hunting societies. Inland Eskimo (Nunamiut) have what Binford (1983) referred to as blinds, hunting stands, field camps, stations, and game observation sites, in addition to residential camps. But his descriptions of these do not fit very well with the Dry Creek site contents, and the actual land use by inland Eskimos does not seem to appropriately fit the information from the Dry Creek and other early Beringian sites. Kung Bushmen have a residential camp and forage from it in different directions (in what Binford calls a "daisy" pattern) and sometimes have outlier camps (Yellen and Harpending 1972), but again, these do not seem to offer a good parallel of Dry Creek and similar sites in the north.

The character of the Dry Creek site has led us to conclude that it was a "spike camp," a secondary, short-term encampment used primarily to spot game (its position on an exposed prominence), to prepare game for retransport (the presence of knives, bones, and worn large flakes), and to manufacture and repair hunting tools (the presence of the complete microblade manufacture sequence and broken bifaces). The term "spike camp" can be defined as a small hunting camp in which virtually all materials have been carried there by the hunting party on their backs, in one trip, and where one would normally spend more than one overnight (this differs from a sleeping stop while on the trail). Meat is brought back to the spike camp for processing and retransport, though some is regularly eaten there—usually the boney parts most difficult to carry (like ribs) and those that do not preserve well (edible viscera). This kind of camp seems to best fit our reconstruction of the activities at Dry Creek. The "residential camp" aspect of the tool assemblage is poorly represented; polished end scrapers (see chapter 4) and leather working tools are rare. These are not elaborate habitation areas with indications of long-term use, and there are no house pits, tent-ring stones, center postholes, or the like, although the camp is positioned on an exposed windy prominence.

Interestingly, many of the early sites in Siberia are characterized by these same features: heterogeneous fauna, few end scrapers, a predominance of points and knifelike tools, and locations on river terrace or moraine prominences overlooking areas where big game frequently travel (Mochanov 1977). The Siberian and Alaskan ridge-crest sites have much the same overall character. Binford (1983:330) refers to these kinds of sites as "special purpose locations" that "are more discrete in their location and more redundant in their use and contents."

I would like to argue that these frequent Siberian camps—along with those of the Denali Complex, the Campus site, and most other northern sites in the 12,000–8,000 ^{14}C yr BP (13,800–8,900 cal yr BP) range—are part of a land-use system that involved a moderately stable base camp and numerous outlier spike camps. Hunters used these spike camps in a radiating pattern away from the main hub and more permanent base, like the web of an orb spider.

In addition to foraging out from residential camps in a daisy pattern, one can move laterally in an orbital pattern at a right angle to the "spokes" from spike camps if hunting success is low. The thin and less predictable distribution of game in the Far North usually requires that kind of flexibility.

The conceptual model would allow a more thorough use of a thinly distributed, only moderately predictable, big-game resource. There is a critical size for a hunting group, well above that of the nuclear family (between 25 and 100 is the generally agreed upon range). A small nomadic band of a family group would be extremely inefficient year-round hunters. The entire camp and equipment (it takes no mean amount of gear for anybody to live in the north) would travel slowly and noisily. Game would soon be flushed out of the general area. If these few hunters ran into a long period of poor hunting, they would be doomed with no backup reserve buffer. It is most difficult to imagine that people could survive in the north as truly nomadic hunters, always faced with the unknown or poorly known terrain.

A central camp exerting light hunting pressure scattered throughout a wide area of known conditions has a better fit with the data. A stationary base camp would allow larger stores to be accumulated for winter or lean times. Life in the north requires stores of some sort, as Binford (1983) described in his comparison of several varied societies—predicting that storage quantity is inversely correlated with the length of the growing season. Maintenance energy (preparation of garments, cooking facilities, etc.) would not be stressed by constant travel. Camps could be chosen for optimal locations of water availability, dust and wind protection, fuel access, substrate for tents or dugouts, and other things that would be compromised if campsites had to be chosen for direct mobile hunting purposes as well. The individuals who were less able to travel, but had nonetheless important contributions, could remain sedentary: pregnant women, small children and women with small children, the aged, and the sick or injured.

This base camp would allow very mobile bands (I suspect mostly adult males) to travel throughout a great distance, continually adding to the meat stores with big game already processed in a preliminary fashion for retransport. In this way, a familiarity with a large region and traditional patterns of game use could be achieved. This hunting pattern would, in general, result in a system moderately buffered against the vicissitudes of the big-game-hunting existence in the Far North. If game resources became depleted, the information would already exist as to which direction the main camp could be moved, thanks to the familiarity the widely ranging hunting bands would have with the expanse of their territory. The mention of territory raises another issue, which is intergroup aggression. The base camp "orb" hunting territory would allow the moderately rapid availability of the most people for the defense against invaders. A simple family band would be relatively defenseless in comparison.

The permanence of the base camp, however, might have only been seasonal. Athabaskans and Eskimos have traditionally had different campsites for winter and summer. One might imagine that winter would be the harsh season, where people (like Eskimos) would band together in the larger kinship units with food stores and continued organized hunting pressure over a larger region.

The circumstantial arguments for a stable base camp during winter are, to me, persuasive. The large amount of equipment required for winter living in the north (skin blankets, extra robes and clothing, various specialized tools, and a stock of raw materials) would be relatively immobilizing without dog teams and sleds for travel. But the most important reason to be sedentary is the quantity of food stores necessary to provide a buffer against the boom-bust conditions in the north. The ethnological literature about caribou-dependent northern tribes discusses the hundreds of animals taken seasonally for stores. Enough dried meat to last for a lean 2 months would alone weigh about 120 kg per person (2 kg per person for 60 days), or more than could be carried. Also, condiments like tubers and berries gathered in the summer could not be kept if one were mobile. In addition, most groups in the north at the time of European contact depended heavily on concentrated marine resources (sea mammals on the coast and salmon in the interior); few were exclusively big-game hunters, as the Dry Creek peoples might have been.

However, it would be difficult to lay up meat to last all winter; I imagine that stores were only meant to last for long hungry times and that hunting would have to have been continued throughout the winter. I believe that most of the early sites in the north were fall or winter spike camps of people residing away from the main camp with only the rudiments of necessities. Most of the tool manufacture or use is for traditional tools, weapons, and butchering tools. The few scrapers are probably to do preliminary work on the hides in order to dry them for transport. A green hide, with the subcutaneous muscles and fascia not removed, weighs several times that of a fleshed, dried hide. Also, if the hide is dried first, fleshing becomes very difficult.

The spike camps could serve as cache areas to store raw materials later used in tool and weapon manufacture and repair. Also, the select bones that were to be used for implements or even horn could be cached there. There is a large accumulation of stones at the Dry Creek site, though it is unclear why these particular stones were piled up.

It is likely that some preliminary meat preparation occurred at the site, judging from the presence of bones and bone fragments, knives, and the signs of sustained human activity (hearths and stoneworking remains). Given the amount of meat required to maintain a hunting group, large quantities would have had to be carried back to camp. As meat is more than 70 percent water, it can be dehydrated by drying (even in winter) for easier transport back to the base camp by cutting it into thin strips and hanging it in an exposed spot.

The bone fragments of skulls at the site make it unlikely that these were "trail food" bones, like the long bones and ribs attached to dried meat that Binford (1983) described among inland Eskimo. Dried meat is difficult to remove from between the complex boney processes of the skull. Unlike marrow (the reason Eskimos carry long bones and ribs on the trail), brains do not preserve well but petrify quickly by forming anaerobic conditions for decay.

A more likely explanation for the presence of skull bones at the Dry Creek site is that they were either a by-product of preliminary field processing or a food eaten in the field that would be difficult to carry back to the main residential camp, or both.

It is interesting to note that although the Dry Creek site was probably used by several different groups of people, at quite different times, hunting different species of animals, they seem to have been hunting in roughly the same manner—that is, they were camping on the bluff, using it as an observation prominence, making weapons or repairing them, retrieving their game back to the bluff, and processing it. Yet they left no sign of any substantial dwelling or general "household" activities. If I am interpreting the spike camp aspect of the site correctly, it would seem to be part of a larger pattern of land use that persisted in the north for a long time. Binford (1983:335) observed that "during general systems changes the role of a strategy in an overall adaptation may change but the strategy itself does not necessarily change."

This orb pattern would make a great deal of ecological sense. Thinly scattered resources would be tapped over a wide area. It also inherently creates a partitioning system of land use between groups; expansion could only be done in the direction of the new occupation or in abandoned areas. Out-group relations could be relatively stable for exogamous marital exchanges (biologically necessary in groups of only modest to small size) and trade goods (nonlocal, exotic lithic material is present in virtually all northern early archaeological sites, including the Dry Creek site).

Because of the large area probably required to support the central base camp, the record of these central camps would be rare; at present, there exists no good search model for the location. According to the model, the spike camp (processing station) sites should be common. Intensive surveying done on good lookout prominences or game crossing areas generally does uncover these latter kinds of sites. They are, however, usually not deeply stratified due to the erosional character of the prominence or terrace lip and, for somewhat the same reason, have seldom preserved the nonlithic material. The deep deposits at Dry Creek, with their spatial separation of artifact clusters, clear separation of vertical components, and preservation of large mammals combined with the lithic manufacturing sequences, provide some new information with which to view the paleoecology of other sites of similar age in Beringia that lack this information.

Traditionally, it has been assumed that these sites are products of mobile nuclear or extended families because, in general, there are none of the signs of a large, stable encampment. The orb model, however, provides an alternative explanation and, I think, a more effective way to live in the north as big-game hunters.

Dry Creek Bioliths: Gastro and Phyto

It was not until early in the last excavation season (1977) that one of the field school excavators brought in a handful of small pebbles that were recognized as bird gizzard stones. This was discussed with the rest of the crew, and throughout the following days, they began to be found by many people. Undoubtedly, they were present earlier but had not been a part of the crew's artifact "search image." All this was despite careful skim troweling. Our experience suggests that they are present in many archaeological sites but are overlooked. Gastroliths have been identified from other archaeological sites, but Bottema (1975) concluded that the variation was too great to be diagnostic. However, the roundness and polish indices can be used as a good seasonality indicator in the Far North. In some cases, even species or more general bird groups can be identified when accompanied by some knowledge of the environmental conditions. Contrary to Bottema's experience in parts of Europe, where large sand grains and gravel are common in archaeological deposits, the loess silt and small sand-size ranges of interior Alaska make gastroliths extremely important as seasonal indicators of site use.

Many bird groups use a gizzard to masticate food before it passes into the stomach. The food can be ground against the horny walls of the gizzard, or seeds can be used as a grinding compound, but usually the birds pick fine, hard stones as grit. In the Far North, these are seldom replaced during midwinter, even when opportunities are provided—though birds will actively seek out windswept slopes in early spring to renew the grit. Extant ptarmigan killed near Healy in early February, with continuous access to new winter grit on the windblown gravels, all had rounded and polished gastroliths. During the fall, well before the first snow, there is a concentration of grouse along river bars and road shoulders gathering new angular grit (usually quartz) that lasts them through the winter. These angular stones first begin to round, and as wear continues, sometime after the first of the year, they acquire a high-glass polish. Spring and early summer birds have mixed gizzard contents of rounded, polished, and angular unpolished stones.

In a more thorough discussion of the analysis techniques and the theory of identification (Hoskins et al. 1970), fossil gastrolith samples from the Fairbanks loess and a cluster taken from a peat sample were shown to be from winter-kill birds.

From what we can reconstruct of the upper Nenana Valley 11,000–10,000 ^{14}C yr BP (13,000–11,500 cal yr BP), the valley was not a good habitat for waterfowl (another important game bird with gastroliths); it was most likely a good grouse habitat.

There are a number of northern grouse. Of these, the sage grouse (*Centrocerus urophasianus*) does not have gastroliths. The blue grouse (*Dendragapus obscurus*) and spruce grouse (*Canachites canadensis*) are normally found in coniferous woodlands. The ruffed grouse

(*Bonasa umbellus*) is a deciduous woodland species. That leaves three species of ptarmigan (*Lagopus*) and the sharp-tailed grouse (*Pedioecetes phasianellus*). At this time, we have not been able to locate sharp-tailed specimens (during the study, grouse were in the low period of their ten-year cycle in interior Alaska).

Rock (*Lagopus mutus*) and white-tailed (*Lagopus leucurus*) ptarmigan are characteristically found in treeless tundra. Willow ptarmigan (*Lagopus lagopus*) are usually in the shrub zones. Sharp-tailed grouse occur on the open grasslands but more frequently in brushy stands in the grassland.

The mean size of the gizzard stones in the Dry Creek site is 2.14 mm for both the total mean and the mean of the different clusters. The distribution is unimodal (figures 6.7, 6.8). There could have been some sampling biases in the recovery of the gastroliths. Only the clusters were recognized from skim troweling. Wet-screen washing of back-dirt in fine mesh

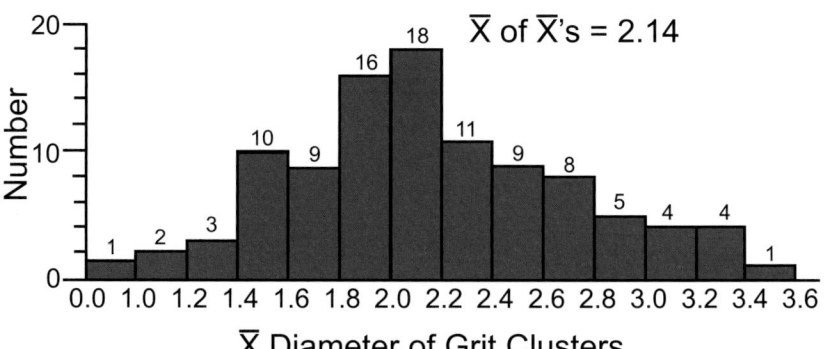

FIGURE 6.7. Distributions of means of the various clusters of gastroliths found at the Dry Creek site.

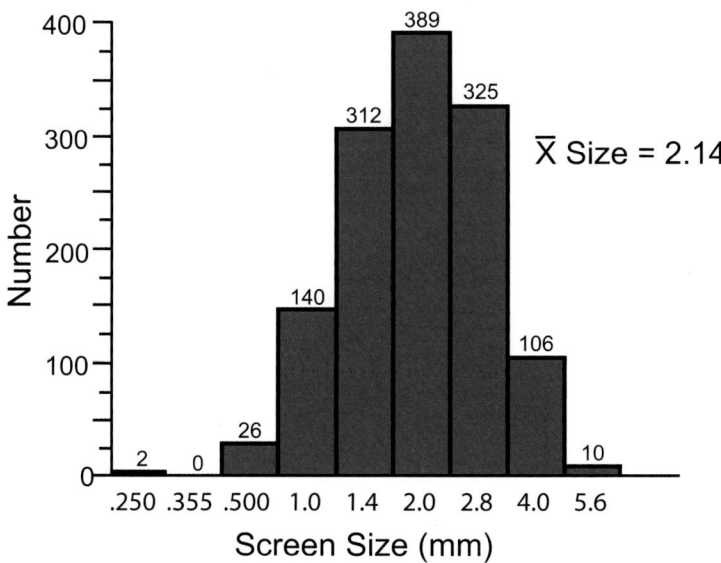

FIGURE 6.8. Comparisons of the gastroliths collected at the Dry Creek site by diameter, using sorting screens.

screens showed many gastroliths and fine fragments of worked stone missed by trowelers. Biases would likely be on the small end of the distribution tail, which would tend to lower the observed mean. When or if other portions of the site are excavated or sites nearby are worked, wet-screened techniques should be used in gastrolith sampling.

We can assume, for the time being, that the mean gastrolith size is 2.14 mm, or smaller, for the Dry Creek clusters. This is not a diagnostic figure, but it suggests birds in the ptarmigan range. Preliminary small samples from late winter birds showed very little size differences among Alaska Range ptarmigan species (Hoskins et al. 1970). Weeden (personal communication), however, gathered much larger samples and showed willow ptarmigan to have significantly larger grit (60 percent of their grit were 3 mm and larger as opposed to around 10 percent for rock and white-tailed ptarmigan). There were no observed differences in the grit size distribution between rock and white-tailed ptarmigan, except that white-tailed gizzards contained slightly more grit.

Judging from the reconstructed physical and vegetative environment, either rock or white-tail would be expected to be the most abundant grouse.

The Powers roundedness scale (Powers 1953) reveals a skewed distribution toward angularity (figure 6.9). Most of the stones show only slight signs of rounding, and there is no polish characteristic of grit from early spring birds. These, then, could represent birds from summer, fall, or early winter. During much of midsummer, ptarmigan are dispersed and very secretive while the brood is being hatched, and it is virtually impossible to locate them, let alone catch them. Once the birds are fledged and are of moderate size (usually late July to early August), the family broods coalesce into large flocks. Probably because these flocks are composed chiefly of young birds, they are not very wary of humans, and they are most vulnerable to hunting during this time period. It is even common to kill them with thrown stones. They turn white in the fall, but sometimes the synchronization of plumage change is poor, and the white birds stand out against the drab brown autumn vegetation. Alternatively, the early snows might reveal brown birds exposed on the open hillsides. So despite the fact that grouse, or ptarmigan, gastrolith angularity at an archaeological site could mean summer, fall, or early winter hunting, it is more likely an indication of fall or early winter use (September). Snares can be set in among the low shrubs with great success when ptarmigan are abundant. This is the traditional interior Eskimo method. Ptarmigan move to and from summer and winter ranges in great numbers (Weeden 1964), and major valley systems in the Alaska Range are often concentration areas during migrations.

Unfortunately for archaeological interests, most silt sediments in the interior of Alaska contain gastroliths, for every time a bird dies, these highly preservable parts remain. Thus in the noncultural sediments at Dry Creek, there are also gastroliths. Gastroliths in these sediments show the same wear stages as those from the archaeological Components I and II. Careful plotting of the gastrolith distribution at the site showed a concentration within the general stratigraphic units that contained the living areas and hearths, but they were not concentrated specifically with the living areas or with the hearths. Thus the question of grouse-ptarmigan use at the site will have to await further detailed studies. During the time of year ptarmigan are most vulnerable to being caught by humans, they are most vulnerable to other predators. Raptors, foxes, and so on might well have brought them to this same high rise to eat them as did humans. So, like a cave deposit where faunal elements in an archaeological

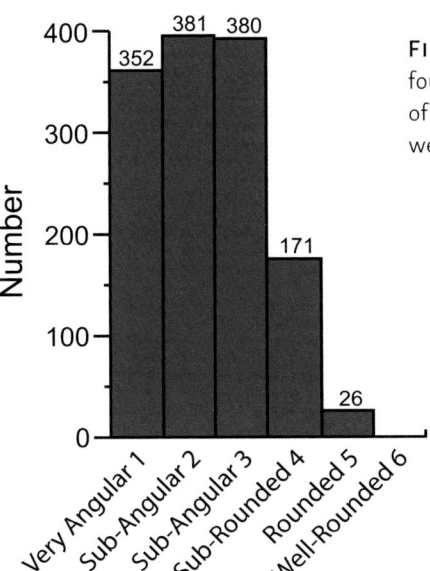

FIGURE 6.9. Distribution of roundedness of the gastroliths found at the Dry Creek site. Most cluster at the angular end of the spectrum, suggesting that the birds that were killed, or were brought to the bluff, died in the late summer or autumn.

assemblage must be distinguished from the natural paleontological background, gastroliths can only be judged an artifact by careful study of their distributions, concentrations, and associations. Fine wet-screening of these early northern sites in the future will be of utmost importance.

The role, if any, of ptarmigan in the diet was probably to add variety and garnish rather than to be a caloric staple. Ptarmigan do not accumulate fat like waterfowl but are lean throughout the year. There is an average of approximately 0.25 kg of meat per bird (including liver, heart, etc.). Assuming a 3-kg lean-meat requirement per person per day, approximately 12 ptarmigan are required per day per person, or 1,200 for 100 people per day, or 400 for 25 people per day. Ptarmigan do not occur in those numbers in any one area to sustain harvests at that level for more than a few days. This also assumes a superbly efficient hunting technique. Even today with a modern shotgun it is difficult to live on ptarmigan, particularly when the birds are in the low of their ten-year cycle.

The sediments of the site are rich in opaline phytoliths. These siliceous crystals are secreted by the plant and tend to have a characteristic shape for each plant group. They preserve well and will probably be an important tool for Quaternary research once identification guides become better developed and some of the dynamics of their preservation and deposition are better understood.

Presumably, phytoliths are part of the plants' antiherbivory defenses and are best developed in the grass leaves, which tend not to have elaborate chemical antiherbivory substances or much fibrous lignin. It is the phytoliths that seem to be responsible for the increased tooth wear of mammalian grazers, ultimately resulting in the characteristic high-crowned complex molars of these species (McNaughton et al. 1985).

Originally, a student had wished to study the Dry Creek site's phytoliths as a thesis topic but decided on another thesis instead. As there were no other people in Alaska working on phytoliths, we decided to run only preliminary analyses and save the in-depth analyses for the future. To do justice to a phytolith analysis, a thorough key to their shape and distribution in northern plants would have to be constructed.

Both James McCalpin, the geologist on the site in 1976, and Mary Calmes, a botanist technician, extracted phytoliths from the sediments. Calmes took her samples from the darkened areas interpreted as hearths. Both people found abundant phytoliths. Many could be identified as festucoid grasses.

Either the Dry Creek peoples were burning the dung of large mammal grazers where the phytoliths are concentrated or the grasses grew in abundance on the hearth areas after the people had abandoned the site, or both. One of the phytolith shapes that occurred in at least two hearths was a microscopic-sized "Clovis point." It was dart shaped with a thinned concave base. This shape was not found in any of the published guides to phytolith identification.

General Paleoecological Conclusions

There are several obvious conclusions that one can draw from the large mammals present at the Dry Creek site. I will first outline and then discuss each of these.

1. *During the general time range of 11,500–10,500 ^{14}C yr BP (13,350–12,480 cal yr BP), the large mammals predominating in the vicinity of Dry Creek still tended toward the "grazer" end of the herbivore spectrum.* At the beginning of the Holocene, there was a major shift from a dominance of grazers (mammoth [*Mammuthus*], bison [*Bison*], horse [*Equus*], etc.) to a dominance of cervids (moose [*Alces*] and caribou [*Rangifer*]). (Hunters in Alaska now take moose and caribou from the area of the site.) The exact chronology of this shift to cervids is still unclear and undoubtedly had some real variations. The point at issue is whether the Nenana Valley in the vicinity of Dry Creek was a relict of the northern mammoth steppe or represented a general pattern through Beringia. Unfortunately, at the time of this writing, there are no other early dated, large stratified, multicomponent sites containing megafauna in Beringia with which to make comparisons. The valley-pass outwash nature of the Nenana River valley would prolong the steppic character of the vegetation. There is, in fact, information that documents the existence of bison (Holmes and Bacon 1982) and wapiti (Guthrie, unpublished) from the Delta River valley outwash area. If the Nenana Valley did retain a relict habitat for large grazers during the postglacial, it was probably limited or intermittent.

2. *Though our Dry Creek sample is from two points in time, it is consistent with other data, which indicate that blanket megafaunal extinctions of mammoth and horse had probably already occurred but that the regional extinctions of bison and wapiti had not.* The reduction of species of megafaunal grazers in the north seems to have been a general phenomenon. The extinction of a number of species was accompanied by the retraction of the distributional range of several others southward. The Dry Creek site is very important from this angle, as it seems to be situated chronologically in the middle of these two events—adding important information about their timing and sequence.

3. *At the time the lower levels at Dry Creek were occupied, the local megafauna still exhibited a Pleistocene body size.* That is to say, they had not begun to undergo the almost universal postglacial megafaunal "dwarfing." I (Guthrie 1984b) reviewed this elsewhere in detail.

4. From what we can surmise about the current ecology of the area combined with information from the site, Dry Creek appears to be an autumn-winter hunting camp. The venturi funneling of the winds through the Nenana Gorge of the Outer Range creates snow-free autumn and winter pastures. The dehydration effect, wind disturbance of the soil, and abrasion of woody plants by windblown ice and snow crystals all create a more grassy landscape than in any area in the entire region. The combination of the presence of grass and its availability due to lack of snow cover creates a situation more conducive to winter grazing by large mammals than in any nearby area. Sheep populations now only use this area during fall and winter. It is not a particularly good spring range and, because of its altitude and position in the shaded side of the Alaska Range, green-up is two to three weeks later than in many other parts of interior Alaska.

5. Though it was not itself a kill site, the overlook at Dry Creek afforded an excellent opportunity to spot all three large mammal grazers (sheep, bison, and wapiti). From the high point at the site, one can see the mountains, foothills, broad terraces, and valley floor, covering a diversity of large mammal habitat. There is every reason to believe it was not a kill site. For one thing, sheep-escape terrain is too far away. There are no natural entrapment areas, steep overhanging cliff faces, and so on that would indicate that the site was a camp associated with a drive-kill site. It is obviously a lookout area to which game was brought from somewhere within the purview of the site.

6. The scattered bone fragments and teeth within the site that is located in an area of opportune hunting, the extremely exposed location of the site, and the general lack of permanent fire pits, dwelling dugouts, and tent perimeter stones all suggest that this was not a central base camp but a hunting spike camp and processing station. From this camp, meat could be roughly processed for easier transport to a more distant base camp, and at the same time, one could continue to search for more game. The presence of faunal remains at the Dry Creek site and their general character creates an entirely different view of early hunting strategies.

7. The diversity of large mammal species at the site suggests an opportunistic hunting strategy that selected for a wide range of age and, perhaps, sex classes of megafauna. This opportunistic technique seems to be more Pleistocene-like than the Paleoindian sites later than 11,000 ^{14}C yr BP (13,000 cal yr BP) on the Great Plains, which generally have remains of females and young of mainly one species of large mammal predominating.

8. The Dry Creek site in the Alaska Interior lay, ecologically, within the "Great Bison Belt." If one plots the distribution of fossil bison sites, or the area of highest bison density in historic times, they fall along a belt from Mexico through northern Canada in the rain shadow of the Rocky Mountains. It is also along this same belt that Paleoindian kill sites are the most common (Guthrie 1980). For a short time period around 12,000 to 11,000 ^{14}C yr BP (13,800–13,000 cal yr BP), the bison-dominated large-mammal community in Beringia connected with the bison-dominated communities of the Great Plains, forming a long, roughly continuous belt of bison habitat that ran the length of North America (Guthrie 1980). The fauna at Dry Creek is essentially a Great Plains fauna of sheep, wapiti, and bison. In other areas in the Alaskan interior, there were ground squirrels, grouse, badgers, ferrets, and horses. The Great Bison Belt at or shortly after 13,000 ^{14}C yr BP (15,500 cal yr BP) thus tied the waning arctic mammoth steppe to the waxing grasslands of the Great Plains. The Dry Creek peoples appear to be connected, at least ecologically, to the early plains hunters to the south.

The colonization of North America by the Beringian people was facilitated by the continuity of the "plains" habitat, to which they had already adapted in the North. They followed it from Alaska—along the grasslands that lay between the two receding continental ice sheets—on through to the expanding grasslands of the High Plains and farther south.

Note

1. For an upade on this debate, see Shapiro et al. (2004) and Heintzman et al. (2016).
2. In the final unpublished version of this chapter, Guthrie cited this reference as an unpublished manuscript.
3. Meiri et al. (2013) provide a current synthesis of wapiti dispersal and human interaction during the late Pleistocene.

CHAPTER SEVEN

Dry Creek and Its Place in the Early Archaeology of the North

W. ROGER POWERS AND R. DALE GUTHRIE

Specific Conclusions

Like finding another Clovis site in the Great Plains, Dry Creek is another Denali Complex site in interior Alaska. The site produced neither significantly new lithic "types" nor any radical chronological revision of these lithic types. Rather, the main importance of the Dry Creek site is within these already established time boundaries and typologies. It begins to address the next level of issues and raises an entirely new set of questions about early Beringians and their activities.

New information and insights gained from the excavation and analysis of the site are considerable; a few key items can be summarized as follows:

1. The site clearly documents, for the first time, discrete typological and temporal variations within an Alaskan site of this time range. The earliest Component I (c. 11,100 ^{14}C yr BP [13,000 cal yr BP]) does not exhibit blade and core technologies; rather, the projectile point being used was a basally thinned, triangular, stone point. The later Component II (c. 10,600 ^{14}C yr BP [12,600 cal yr BP]) has two types of lithic clusters. One kind contains microblades and microcores. The other has no microblade technology. These latter, instead, have lanceolate stone points with expanding bases.
2. The complete sequence of microblade production—from refined preform to microcore, through microblade extraction, on to the final completed use of the core—is all documented within lithic clusters. This sequence helps us gain some insights and raises new questions as to how the site was used and how weapons were manufactured.
3. The hearths, the distribution of lithic clusters, the remains of large mammals, and the location of the site, when compared with other Alaskan sites of similar age, allow us to propose a new model of large-mammal resource exploitation.

4. The presence of identifiable remains of large mammals conclusively associated with well-dated artifacts in the 11,000 ^{14}C yr BP (13,000 cal yr BP) range in Alaska thus far is unique. It allows us to see what large mammals were being hunted and eaten—that is, it allows us access to broader paleoecological interpretations.
5. The comparatively good data (we always hope for better) from the Dry Creek site allow us to see the site in the context of the overall use of the Nenana Valley by early peoples, answering questions as to why the site was located along the terrace edge in that particular location. From this vista, one can see the area of the pass kept free from deep snow by the ubiquitous winter winds. Because of the wind, large-mammal herbivores would have had winter access to grazing land, a rare occurrence in interior Alaska.
6. Combining the archaeological and paleoecological information from northern Eurasia, Beringia, and the Great Plains, one can envision the Dry Creek peoples as a connecting link between the hunting traditions of two continents. Thus the Pleistocene grassland belt of Eurasia was buckled to the Great Bison Belt of North America.

Implications for Further Research

These new data and reconstructions raise a new set of questions. The foremost of these questions is probably the issue of how to interpret the typological differences between Components I and II, particularly with regard to the major difference in projectile-point design. Thus most of this current section will stress the implications of that issue.

The concurrent use of two weapon systems—composite microblade inset and bifacial lithic projectile points—between 11,000 and 10,000 ^{14}C yr BP (13,000 and 11,500 cal yr BP) in Siberia and parts of Beringia contrasts with the emphasis on the bifacial fluted point and related forms, which are the hallmark of early Paleoindian assemblages. Apart from these, the remainder of the assemblages shows a considerable amount of technological and formal similarity over tremendous distance. Dry Creek possesses both of these weapon systems, with bifacial stone projectile points occurring alone in Component I and then with the microblade inset technique in Component II; however, in the latter instance, the two systems occur in spatially distinct clusters. Hence Dry Creek displays both temporal variability and a degree of continuity with the classic Siberian-Beringian microblade technologies to the west. Formal similarities with American bifacial point technologies can be demonstrated, although geographic continuity is more difficult. The site then appears to be intermediate. This is in part due to its geographical location and position within an ecological continuum running from the northern Eurasian and Beringian mammoth steppe to the evolving plains of interior America during the late Pleistocene and early Holocene. However, there are cultural-historical factors that have surely affected the development of the early Dry Creek technologies.

Before attempting broader comparisons and discussing the implications for the earliest components of the Dry Creek site, we should briefly summarize the economic and technological activities presented in the foregoing analyses and point out the problems this information creates in seeking these comparisons.

Component I

The earliest presence of human culture at Dry Creek is represented by a relatively small assemblage of lithic remains composed of bifacial knives and projectile points, side scrapers, transverse scrapers, end scrapers, burins, flake tools, and cobble cores and tools. The spatial analyses indicate that the primary activity carried out at the bluff at that time was meat processing and a minor amount of weapons maintenance. This contrasts somewhat with the richer and more complex remains of Component II.

At the time of the formation of Component II, a greater range of activity might have been conducted at the site. This resulted in the spatially separate patterns of activity that fall into two main categories: microblade and nonmicroblade clusters. The microblade clusters contain large numbers of microblades, microcores, and attendant production and maintenance parts, plus distinctive bifacial knives, core-scrapers, core-burins, bladelike flake and flake tools, burins, and burin spalls. In contrast to this set of associated implements, we find the nonmicroblade clusters, possible burin use, crude bifacial implements, shaped scrapers, and projectile-point bases as sets of implements. The two types of clusters are quite distinctive.

With respect to economic activity, we know that both *Cervus* and *Ovis* were hunted by the people who left Component I and that *Bison* and *Ovis* were procured during Component II times, though we doubt that these differences have any major significance due to the limited number of animals in our sample. Although occurring in spatially distinct clusters, both microblades and bifacial points are associated with *Bison* remains, hence we assume that both were involved in some way with the procurement or processing of this species. We must bear in mind, however, that the nature of this association is tenuous, because we cannot be certain as to why any particular faunal remains are where they are in the site. The nature of the data simply does not allow that level of refinement. The facts we can work with are that (1) *Bison* remains occur within spatially separate clusters, and (2) these clusters are composed of different sets of lithic remains, characterized by two methods of producing weapon systems.

The Component II *Ovis* remains are not associated with any particular cluster. Little should be made of an apparent temporal difference in the fauna between the two components or the two types of clusters in Component II. Poor preservation has surely affected the faunal representation within both components.

It can be suggested, however, that the earliest occupants at Dry Creek were engaged in the same economic activity—hunting large herbivores—and that this activity cross cuts both components and the two types of clusters within Component II.

With respect to these types of activity, it appears that the earliest levels of Dry Creek represent very similar site usage and extrasite activity, at least based on the data now available, and that these activities represent only a part of the total subsistence cycle of the populations that were utilizing the Nenana Valley 12,000–10,000 ^{14}C yr BP (14,000–11,500 cal yr BP). This fact will clearly affect the validity of any comparisons with other sites of comparable antiquity, especially because the information presently available from such sites cannot be evaluated in terms of seasonality of activity areas. It is simply impossible to develop a common comparative standard in the absence of this information.

There are several possible interpretations of the technological variability displayed at Dry Creek. It could be argued that the remains from Component I can be accounted for by resorting to inherent difficulties in the sample size and that the variations of clusters in Component II represent differences in seasonal or other ecologically related activity. This position has been implied by previous summary treatments of the material (Powers and Hamilton 1978; Thorson and Hamilton 1977), where both early components were simply lumped into one classification and its relationships traced to other early Alaskan sites as well as to the late Paleolithic period of Siberia, specifically the Diuktai Culture.

While being fully mindful of the possibility that such problems as sampling error and lack of better faunal preservation might well be affecting our interpretations, it is also necessary, in view of more detailed analyses of the material, to treat Components I and II separately and to recognize that the internal differences within Component II require special treatment.

It should be noted that spatial clustering can result in the isolation of artifact sets that might represent a special activity within a technological continuum or might be totally unrelated to other clusters lying nearby. Therefore, the definition of complexes or other classification in northern sites should be done cautiously, especially where less than ideal stratigraphic or spatial contexts are present.

Although we cannot offer any final solution that accounts for the differences displayed in the early component, we can suggest two alternate interpretations.

Within the Nenana Valley, Components I and II at Dry Creek are presently the best understood early occupations and still form the basis of the temporal and cultural framework for understanding the prehistory of the region. Component I can be interpreted to represent a premicroblade horizon in the Nenana Valley. Additional support for this position has been derived from test excavations at the newly discovered Moose Creek site on the east side of the Nenana Valley (Hoffecker 1978, 1982). Here, preliminary data suggest that a technology characterized by bifacial tool production (projectile points and knives) and a related flake industry occurred in a stratigraphic context associated with paleosols dating in the 12,000–8,000 ^{14}C yr BP (14,000–9,000 cal yr BP) range. The dated paleosols lie at the top of the artifact-bearing horizon. At the present time, it appears that this material is similar to the lowest cultural horizon at Dry Creek.[1] Unfortunately, no faunal remains have been encountered to date. A similar typological and stratigraphic situation possibly exists at the Usibelli site on Healy Creek, but no radiocarbon dates or fauna are yet available (Hoffecker 1980).[2]

Within the context of central Alaskan prehistory, the phase of culture history manifested by Dry Creek Component I—and Moose Creek and possibly Usibelli—might be present at the lowest levels of Healy Lake in the upper Tanana Valley. Here, the Chindadn Complex at the Village site dates between 11,000 and 10,000 ^{14}C yr BP (13,000–11,500 cal yr BP; Cook and McKennan 1970). There are some superficial similarities between the projectile points at Dry Creek and Chindadn points at the Village site, but the techniques of manufacturing are distinctive for the two areas. It is possible that a microblade technology occurs in the Chindadn Complex, which sets it apart from Dry Creek Component I in the Nenana Valley (although not from Component II if these clusters are considered as part of the same cultural entity). In addition, as mentioned earlier in this work, the Healy Lake material occurs in a compressed loess section where ample opportunity for mixing exists. West (1981) has thoroughly reviewed numerous sites in Alaska that are similar typologically to the Dry Creek site, and some of these contain triangular stone points similar to those in Component I at Dry Creek (West, personal communication).

In addition to the central Alaskan sites that might bear some relationship to Dry Creek, mention should be made of the fractured bison calcanei and horse scapula at the Trail Creek Caves on the Seward Peninsula, which have been dated to between 15,000 and 13,000 ^{14}C yr BP (18,000–15,500 cal yr BP; Larsen 1968). The oldest tool found here is a generalized bifacial point from the lowest level of Cave 2 that could be as old as the fractured bones.[3] These finds would precede the earliest occupations of the Nenana Valley. It is possible that a technology somewhat similar to the bottom of Dry Creek might have been present in western Alaska that would have been roughly synchronous with or slightly older than the Nenana sites.

In addition to the aforementioned sites, human activity has been revealed at the Bluefish Caves in the northern Yukon. The occupation here is proposed to have spanned about 6,000 ^{14}C years, from 16,000 to 10,000 ^{14}C yr BP (19,300–11,500 cal yr BP). While the earlier part of the record is more cryptic and has produced only altered bone and microchips (the artifactual nature of these is still open to discussion), a later occupation contains lithic artifacts (a burin spall, flakes, and possibly a microblade) that are probably consistent with microblade technology elsewhere in the Far Northwest (Cinq-Mars 1979).

Seeking a broader cultural/historical relationship with the Siberian late Paleolithic yields little that can shed light on the origins of the oldest Nenana Valley occupation (Component I). Whereas bifacial-point (not necessarily projectile-point) technology is well documented in the Siberian Diuktai Culture (Mochanov 1977) and in scattered occurrences across southern Siberia during very late Pleistocene/early Holocene times (Medvedev 1968; Powers 1973), for the most part, they are awash in a sea of microblade technology. While the vast majority of these sites display considerable similarity with the microblade-cluster assemblages in Component II at Dry Creek, they bear little resemblance to Component I. There are, however, a few exceptions, the most notable of which is Kukhtui III, located about 1.5 km from the coast of the Sea of Okhotsk. Here, underlying a Neolithic level dated to 4,700 ± 100 ^{14}C yr BP (LE-995; 5,430 cal yr BP) is a collection of lithic artifacts, which, while classified as Diuktai, contains no microblade technology (Mochanov 1977). The Kukhtui artifacts include discoidal cores, flake knives, broken spearpoints made on large blades, bifacial oval knives, a fragment of a bifacial knife or spearpoint, a blank for a bifacial knife or spearpoint, a blank for a bifacial knife, a bifacial spearpoint, and a wedge. The one complete specimen referred to as a bifacial spearpoint is very similar to the triangular points from Component I at Dry Creek. The remaining portions of both assemblages are similar in composition and typology. This site is the only northeast Siberian example that we presently know of that might represent an industry emphasizing a refined bifacial technology and lacking a microblade (wedge-shaped core) technology that could possibly relate to the origins of nonmicroblade late Pleistocene industries in northern North America. Unfortunately, no date is available for the Kukhtui material, and there are no faunal remains.

As there are little data with which to compare Component I at Dry Creek in Alaska with Siberia, it remains to point out another possible area of relationship—the North American Great Plains. This idea was broached only briefly in chapter 4, where it was suggested that the tool kit in Component I is broadly comparable to Paleoindian tool kits from the plains of interior North America. Specific resemblances cannot be drawn with any of the established projectile-point types of this region, and any other similarities between flake-tool or endscraper types may be viewed as fortuitous. The points of similarity lie in the emphasis on a lithic technology composed mainly of stone projectile points and knives with a variety of flake tools and a generalized core/flake or large blade technique. This set of tools contrasts strongly with those possessing a wedge-shaped core/microblade technology with attendant distinctive burin forms. The Component I tool kit is broadly concurrent with the Clovis Culture to the south but is probably not old enough to provide a technological base from which to derive this technology. However, it might represent the terminal phase of a tradition with greater time depth and geographical extent that did give rise to Clovis Culture, although just when, where, and in exactly what sequence is presently unknown.

Fluted points are present in Alaska (Clark 1978; Dumond 1977). The verifiable specimens are all located north of the Alaska Range, and the majority of these are north of the Yukon River (Clark 1978). Dixon (1976) has argued that these points are younger in Alaska than in the south and that the point of origin for fluted points must lie outside of Alaska, presumably in Canada or the continental United States. However, Clark (1978) is of the opinion that some of these fluted-point occurrences in Alaska could be as old as 11,000–9,000 ^{14}C yr BP (13,000–10,200 cal yr BP). Morlan (1977) further hypothesizes that fluted points will be found to be older in Beringia and that they spread south as part of a cultural response to the collapse of the northern steppe biome at the end of the Pleistocene. While this is speculative, it does provide a model to direct further research on the problem, and one that is, more or less, in agreement with the data presented herein.

Although the data are at best sketchy, it might be that there were at least two concurrent early geographical variants of point styles at about 11,000 ^{14}C yr BP (13,000 cal yr BP) in Alaska. One was centered more to the northern interior of Alaska and beyond, with Clovis points proper, and a second was distributed across the southern interior and characterized by the more generalized triangular point styles of Dry Creek. Also, both of these point variants could share a common ancestry in Alaska. Haynes (1978), in briefly reviewing the evidence from the lower components at Dry Creek, suggests that two concurrent traditions might have existed in Beringia and that the one lacking microblades might have given rise to Clovis.

Based on the Dry Creek evidence and the scattered data pertaining to fluted points (Clovis) in Alaska, it is possible that Clovis did develop in the north, possibly even in Alaska, from a technological base similar to that found in Component I. It is further possible that these events predate the appearance of microblade technology in Alaska.

Further work at the new Moose Creek site near Dry Creek might help clarify the situation. The site might prove to truly predate both microblade technology in the north and Clovis in the south. In the latter case, the temporal extension is near enough to provide a clear source for Clovis, but again it might be a later part of the tradition that might have been the developmental base for fluted point technology.

Clearly, by the time of Component I at Dry Creek, Clovis was well established on the American Plains, but both of these cultural developments represent steppe adaptations geared toward hunting large Pleistocene grazing mammals. At this time, the two points on a geographic spectrum (Alaska and the Great Plains) could be seen as representing two parts of a north-south technological gradient adapted to a vast sheet of grasslands that stretched from the Alaskan interior to northern Mexico and perhaps beyond, a zone that included the older, northern mammoth steppe through intermediate transition zones in the newly deglaciated Canadian regions and the evolving steppes of interior North America—the Great Bison Belt (chapter 6). The impetus for this movement—a drift to the south of formative or developed Clovis—could be seen in the terminal Pleistocene deterioration of the mammoth-steppe biome (Morlan 1977; chapter 6) and the development of a new, rich grassland ecosystem along the eastern margin of the Rocky Mountains.

Component II

Component II at Dry Creek comprises the next major occupation in the Nenana Valley. For the most part, Component II constitutes a Denali Complex assemblage (West 1967) that is widespread in the neighboring parts of the Alaska Range. While the age of this complex has been controversial, West (1981) has presented evidence—including the 10,690 ^{14}C yr BP (12,540 cal yr BP) date for Component II at Dry Creek—that the Beringian Tradition in Alaska (which includes the Denali Complex) dates to at least 11,000–8,500 ^{14}C yr BP (13,000–9,500 cal yr BP). In view of the information now available, West's estimation should be considered essentially correct.

At the present time, there are no other sites in this region that contain the same degree of internal complexity as Dry Creek. It should be noted that Component I at the Carlo Creek site might date roughly to the time of Component II at Dry Creek, although available radiocarbon determinations obscure the situation (Bowers 1978). That site does contain evidence of both small (*Citellus* sp.) and large (*Rangifer* sp., *Ovis dalli*) game procurement. In addition, the lithic remains are thought to represent butchering activity. Technologically, it is difficult to determine any relationship with Dry Creek, although one might very well be possible. Carlo Creek is especially important, as it lies well back in the Alaska Range along the Nenana River and thus affords a view of both local montane and riverine activity. Two dates, 8,400 ± 200 (WSU-1700) and 8,690 ± 330 (GX-5132) ^{14}C yr BP (9,360 and 9,760 cal yr BP, respectively) probably best represent the temporal interval of this site. Another date of 10,040 ± 435 ^{14}C yr BP (GX-5131; 11,630 cal yr BP) has a large counting error and might be anomalous, although it does bring this occupation closer in line with Component II at Dry Creek.[4]

Small test pits at Little Panguingue Creek just north of Dry Creek in 1976, 1977, and 1979 revealed dense and highly localized accumulations of microblade technology in the sod layer that lay at the top of a 2-m loess section of the Healy terrace. To date, no radiocarbon dates are available; however, the assemblage is very close to Component II (microblade clusters) at Dry Creek.[5] Microblade technology, again very similar to Dry Creek, was discovered in test pits at Panguingue Creek in 1976, although no chronometric age determinations are possible.[6] Apart from the sites mentioned previously, no other remains have been dated to this time period, although the Teklanika River sites (West 1967) are typologically consistent with Component II at Dry Creek.[7]

The cultural materials from the different clusters at Dry Creek, on typological grounds, correspond perfectly with the established categories of artifacts in the Denali Complex (West 1967). However, the spatial separation of projectile points from artifact clusters typical of the Denali Complex creates a special problem of interpretation.

Bifacial projectile points have not been considered a diagnostic feature of the Denali Complex or related assemblages in Alaska. Those industries most closely related to the Denali

Complex are the Akmak Complex (Anderson 1970a) in northwestern Alaska (9,857 ± 155 ^{14}C cal yr BP [11,330 cal yr BP]) and the Ugashik Narrows Phase on the Alaska Peninsula (ca. 9,000 ^{14}C yr BP [10,200 cal yr BP]; Dumond 1977; Henn 1975). Like Component II at Dry Creek, these industries combine wedge-shaped core and microblade and bifacial (knife) technology. Likewise, other early Alaskan lithic assemblages such as the Gallagher Flint Station (Dixon 1975), dated to 10,540 ± 150 ^{14}C yr BP (12,430 cal yr BP), and Anangula in the Aleutians (Aigner 1970), dated to about 8,400 ^{14}C yr BP (9,450 cal yr BP), are both characterized as core/blade technologies.

Those industries characterized by the codominance of both wedge-shaped core and bifacial technology (Denali, Akmak, Ugashik) are considered to be derivatives of the Siberian late Paleolithic (Abramova 1973; West 1967, 1981). Some authors have pointed specifically to the Diuktai Culture, which was widespread in northeastern Siberia at the end of the Pleistocene (Dumond 1977; Haynes 1978; Mochanov 1973; Powers 1978; Powers and Hamilton 1978; West 1981). There is a great degree of similarity between Component II at Dry Creek and the Diuktai Culture (20,000–10,000 ^{14}C yr BP [24,000–11,500 cal yr BP]). The cultural remains from central Yakutia (Mochanov 1978), with dates ranging from at least 13,110 ± 110 ^{14}C yr BP (LE-908; 15,720 cal yr BP) to 12,100 ± 120 ^{14}C yr BP (LE-907; 13,970 cal yr BP), display considerable continuity with levels VI–V at Ushki Lake on the Kamchatka Peninsula with a date of 10,360 ± 350 ^{14}C yr BP (MO-345; 12,060 cal yr BP; Dikov 1978). Mochanov (1978) states that bifacial knives and spearpoints were present in Diuktai assemblages. It must be reemphasized, however, that these points coexist with a microblade technology. The complete Diuktai points are few in number and all characterized as triangular or bipointed. One specimen resembles a lanceolate point with a straight, slightly constricted base (Mochanov 1977: figure 2:1), but it is very small and more similar to later Neolithic (Siberian) point forms than to lanceolate points in North America. These generalized Diuktai bifacial forms could be seen as prototypical for the bifacial projectile points at Dry Creek. The notion that there might have been a continuity in bifacial projectile points from Siberia to Alaska seems reasonable.

Thus both the Diuktai and Dry Creek hunters employed the microblade inset technique and bifacial projectile points for game procurement.

It was mentioned in chapter 3 that relationships for the Dry Creek projectile points could be sought in North America. The attributes of these basal fragments compare best with point types such as the Casper site's Hell Gap points or Haskett points from eastern Idaho (Butler 1978; Frison 1974). This again raises the issue of direct relationships with the interior Great Plains of North America. The age of Hell Gap points at the Casper site is roughly concurrent with the age of Component II at Dry Creek, so similar point styles were being employed in both areas at about the same time associated with similar fauna and maybe used in a similar way. Does this represent a convergence of technology from a common technological base or cultural-historic connections resulting from the spread of new projectile-point techniques through a still-common environment?

Conclusion

Thus we are left with two alternative approaches to explain the spatial and temporal differences within the Dry Creek site. The first approach would be to portray the lithic variants as simply representations of different activities of basically the same stock of people. According to this view, people would likely use several different kinds of projectile points for several different purposes. The second approach would be to portray the earlier occupants at Dry Creek as representing a different culture and employing exclusively triangular stone points (or at least deemphasizing inset points). From this view, the occurrence of microblades postdated (or, at least, experienced a major shift in increased emphasis on) the arrival of the earlier stone projectile point. There are numerous permutations of possible scenarios between these extremes.

This question was never presented so clearly before the excavation of the Dry Creek site. At present, the answer does not appear to inherently lie within the data from Dry Creek, or other known sites, but can only come from future sites. Until we have better sites, carefully dug, a clear, unambiguous answer cannot be given. We can state, however, that the technological separations at Dry Creek do show that these early people did not always rely on microblade inset points and a core-blade technology. The sole characterization of their sites on that basis would be misleading.

At this time, our energies should be directed at discovering sites that are in well-stratified and datable contexts with good faunal preservation and that display spatial patterning of activity. These searches should be regional in scope—for example, a single riverine system or limestone ridges that have habitable caves or rock shelters. These study regions should transect various ecological zones. This should permit investigators to deal with such questions as settlement patterns and seasonal activity to define the total technological inventory of a past cultural system. Then slowly, valley by valley, we may begin to understand just how the New World was populated and when. For this approach to be successful, it will have to be applied on a broad front, including our Soviet colleagues working in Siberia. Far better cooperation and communication are necessary between the Old and the New Worlds before the archaeology of these regions can become one.

Notes

1. More recent work at Moose Creek confirmed the presence of an early "non-microblade" component similar in age to Component I at Dry Creek, as well as a later "microblade" component coeval with Component II at Dry Creek (Pearson 1999).
2. An early component at the Usibelli site was never confirmed despite additional testing (Powers and Hoffecker 1989).
3. The slotted points from Trail Creek Cave No. 2 have been directly dated to no earlier than c. 11,300 cal yr BP (Lee and Goebel 2016), so none of the cultural remains there have been demonstrated to be directly associated with the bones of extinct fauna.
4. More recent AMS ^{14}C dating of the Carlo Creek site indicates this Denali Complex component dates to c. 11,270 cal yr BP (Bowers and Reuther 2008).
5. Later testing by Powers and Hoffecker (1989) suggested a late Holocene age for the Denali Complex component at Little Panguingue Creek; however, new excavations in 2015 by K. Graf and the editor (Goebel) suggest an early-middle Holocene age.
6. Excavations in the late 1980s and early 1990s indicated the presence of two Denali Complex components, one roughly coeval with Dry Creek's Component II and the other dating to the early-middle Holocene (Goebel and Bigelow 1992).
7. Coffman and Potter (2011) confirmed the presence of an early Denali Complex cultural occupation at Teklanika West.

APPENDIX A

Component IV at the Dry Creek Site

W. ROGER POWERS

Introduction

Component IV is the only cultural horizon for the later part of the Holocene epoch preserved at the Dry Creek site. Artifacts assigned to this component were recovered from Loess 6/Paleosol 4a, which averages 100 mm in thickness (figures 2.9, 3.1–3.6). This buried soil unit appears to have developed during the establishment of the taiga over the Alaskan interior during the mid-Holocene (Thorson and Hamilton 1977).

Paleosol 4 is thick and continuous with well-developed oxidized horizons. The presence of a discontinuous B horizon indicates that different parts of the site were better drained than others. Near the bluff edge, the paleosol is well oxidized, indicating good drainage. Farther into the site, it becomes similar to low humic gleyed soils typical of poor drainage conditions. In general, Paleosol 4a is a subarctic brown forest soil similar to those found currently developing beneath the interior forests (Thorson and Hamilton 1977).

Charcoal occurs either as scattered flecks or as burned roots scattered throughout this soil horizon but is not encountered in high concentrations in areas of intense cultural activity. Charcoal used for radiocarbon dating probably represents the residuum of several forest fires that swept the area during the development of the soil. The samples were collected from the wall of the 1974 test trench and cannot be related directly to the activity areas discussed in the following sections. These dates, therefore, represent only upper and lower limiting dates for Component IV. There are three radiocarbon dates for Paleosol 4a. Two dates apply to the top of this soil: 3,430 ± 75 and 3,655 ± 60 ^{14}C yr BP (3,690 and 3,980 cal yr BP, respectively). The remaining date of 4,670 ± 95 ^{14}C yr BP (5,400 cal yr BP) is derived from the base of the paleosol (see table 3.1).

Unfortunately, the faunal remains from Component IV are unidentifiable. All specimens were badly broken and smashed, probably to extract bone grease. However, the spatial distribution of many of these fragments coincides with areas of intense flaking activity.

The cultural assemblage from this occupation is composed of 2,372 cataloged specimens, which include tools, flakes, bone fragments, pebbles, and rocks. For purposes of analysis, a sample of 2,131 pieces (or 90 percent of the total assemblage) was used. This sample is composed of 16 tools and 2,115 flakes. The flakes occur in either of two clusters, A and B (figure A.1), while, with few exceptions, the tools are scattered beyond the perimeters of the clusters.

Raw materials for tool manufacture during this occupation were rhyolites, quartzites, degraded quartzites, obsidian, cherts, and siltstones.

Component IV: Artifacts

The tools in Component IV can be subdivided into the following categories: bifaces (4), biface base (1), end scrapers (8), retouched flakes (2), and boulder spall knife (1).

Bifaces (4)

Four complete bifaces were found in this component and can best be characterized as weakly stemmed or slightly side-notched. The pattern of edge treatment just above the base really falls into neither category. These specimens probably underwent several episodes of reworking before they were discarded, which might account for these ambiguous hafting techniques. No freshly manufactured specimens were found with which to compare these pieces.

The tips are basically asymmetric triangles with excurvate edges. This form is the result of repair or resharpening episodes. The cross sections are lenticular. The bases of the bifaces were formed by chipping in broad concavities. This technique left slight ears at the corners of the basal extremities. The bases themselves are concave on three specimens and straight on one. There are varying degrees of grinding or polishing on some portion of all specimens.

All these pieces were manufactured by bifacial reduction on flakes of rhyolite (2), degraded quartzite (1), and obsidian (1).

Of the rhyolite bifaces, the largest (figure A.2A) displays marginal retouch on the ventral surface and facial flattening on the dorsal surface. In addition to very minor edge polish near the tip, there is polish on the facial facets near the base. This is probably the result of hafting. Retouching along one edge miscarried during a resharpening episode, leaving a marked concavity. This might have been the reason it was discarded. This piece measures 67 × 34 × 5 mm.

The second rhyolite biface has an extremely sharp tip (figure A.2B). It is more stemmed than notched. It shows minor polish along both edges below the shoulders and across the base. There might be some minor hafting wear. There were several resharpening episodes that resulted in a highly asymmetric tip. Reworking of this piece appears to have been interrupted by the inability to remove a knob at the center of one face. This biface measures 45 × 30 × 6 mm.

The degraded quartzite biface is likewise asymmetric (figure A.2C). It also has a very sharp tip and retains heavy polish across the base. Its edges are fresh as a result of a resharpening episode. There is also wear on the facial facets near the base. The piece measures 33 × 27 × 6 mm.

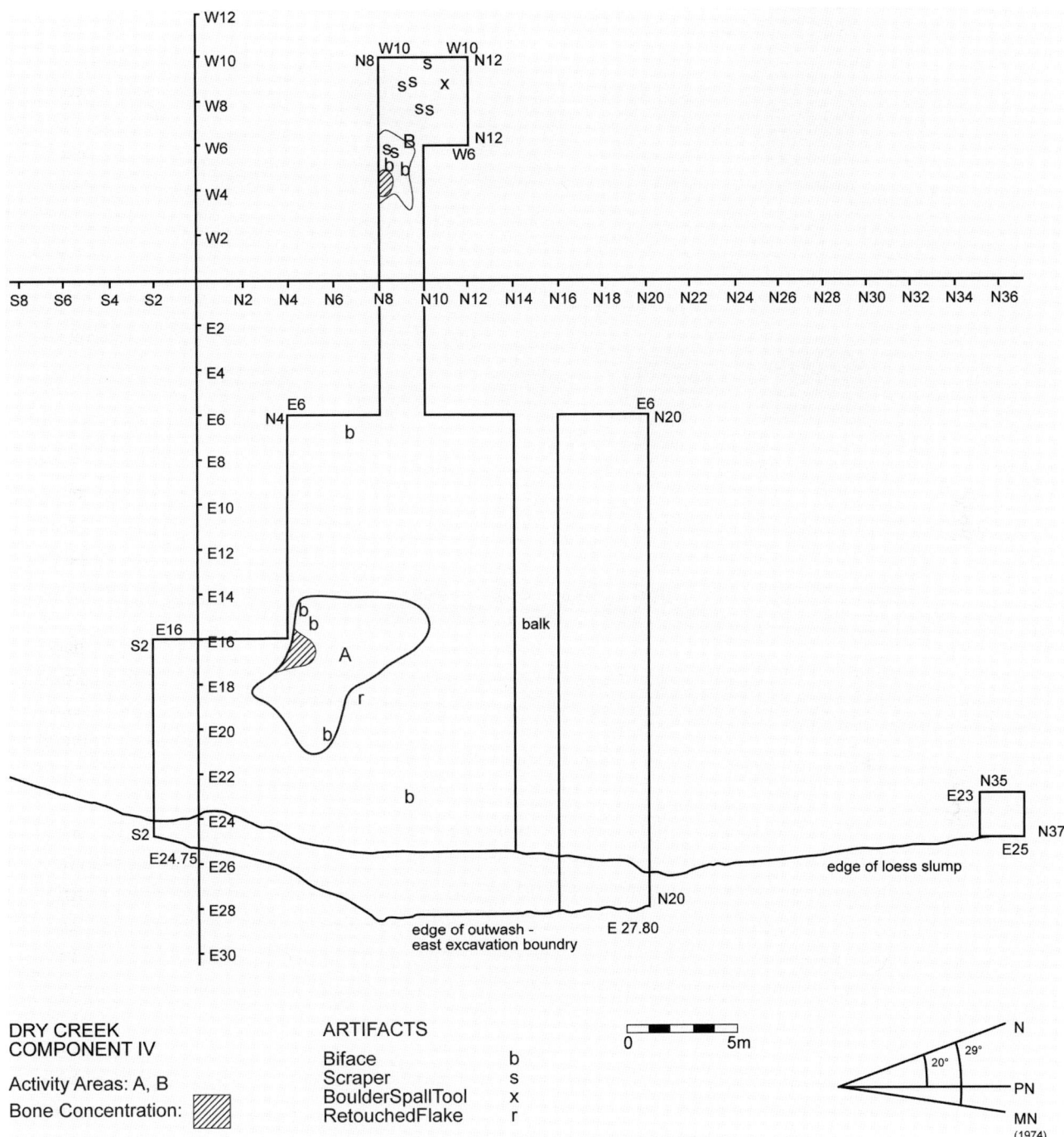

FIGURE A.1. Map of Component IV at the Dry Creek site showing activity areas A and B.

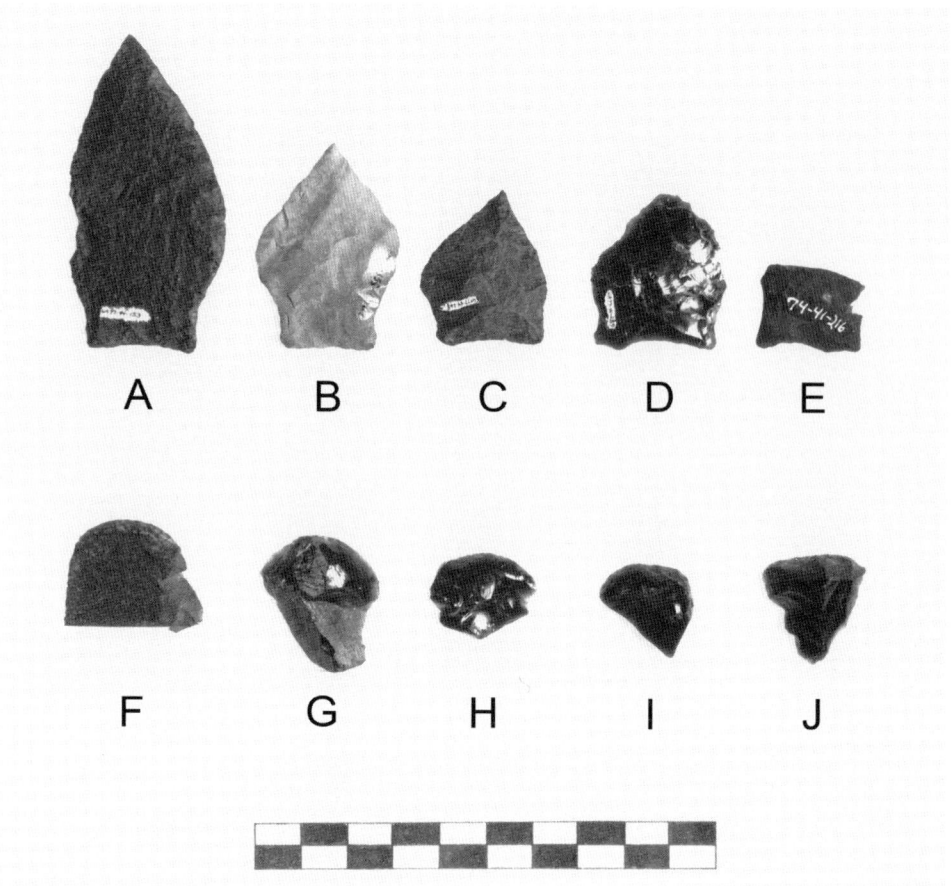

FIGURE A.2. Artifacts from Component IV at the Dry Creek site (A–D, bifaces; E, biface base; F, flat end scraper; G–J, steep end scrapers; scale = 10 cm).

The obsidian biface (figure A.2D) is the most heavily used tool in this category. The artifact was discarded because either repair failed or the tool was used so heavily that resharpening reduced it to an unusable size. The tip is very blunt, and the edges are dull, thick, and heavily crushed. Likewise, the facial facets are heavily worn and polished from extensive use. The edges along the lower part of the piece display heavy grinding, which is also evident across the base. The artifact measures 36 × 29 × 6 mm.

Biface Base (1)

This specimen is the basal remnant of a tool manufactured on a rhyolite flake by bifacial reduction (figure A.2E). The edges and base are polished. No polish on the facial facets could be detected. A break, which removed the tip, is a simple hinge fracture. Along one edge is a series of longitudinal fractures (facets) beginning at the snap and extending about one-third of the way to the base. They appear to be purposeful burin blows. The intersection of the hinge fracture and the longitudinal facets shows fine crushing, indicating use as a scraper on a hard material. The artifact measures 19+ × 23+ × 5+ mm.

The Component IV bifaces might have been intended for use as projectile points when first made, but judging from their asymmetry and edge wear, they were probably functioning as cutting tools when they were lost or discarded.

End Scrapers (8)

Of the end scrapers, seven were in situ and one was recovered from a block of Paleosol 4a that had slumped into the 1974 trench.

All these specimens are on flakes and were manufactured by unifacial retouching. They display at least one well-formed, convex working edge.

As in Component I, the end scrapers can be subdivided into two categories: steep end scrapers and flat end scrapers.

Steep End Scrapers (5)

These have at least one working edge that lies at a 60° to 90° angle to the ventral surface of the flake. Three were made on subcortical flakes of obsidian, and two were made on cortical flakes of the same material. Four of the steep end scrapers have single, steeply flaked, convex working edges (figure A.2G–J). One specimen is a double end scraper with working edges of the same morphology that lie on the end and one side of the piece (figure A.2J). This scraper also displays crushing along the edge opposite the lateral working edge.

All the specimens show crushing along the working edges and lateral margins. One (figure A.2G) displays polishing on the dorsal facets at the end opposite the working edge, which possibly indicates hafting.

Flat End Scrapers (3)

These were made on flakes of obsidian (2) and degraded quartzite (1). The working edges on these scrapers lie at a 30° to 60° angle to the ventral surface of the flake. Two have symmetrical, convex working edges (figure A.2F). One is asymmetric and has a narrow retouched working edge at one corner of the flake. There is also crushing along one side on an unretouched facet, indicating that this edge likewise functioned as a scraper. The facets on the dorsal face show polishing. Length measurements range from 15 to 38 mm with a mean of 26 mm. Widths range from 20 to 27 mm with a mean of 23 mm. Thicknesses range from 6 to 13 mm with a mean of 9 mm.

Retouched Flakes (2)

One flake of degraded quartzite and one flake of rhyolite were retouched slightly along two edges to form small scrapers or knives.

Boulder Spall Tool (1)

This piece is the only example of a large implement from Component IV. Bifacial edge flaking was used to develop both a symmetrical outline and a cross section on a slab of degraded quartzite. Flaking was carried out along the entire perimeter, although it was more concentrated on one face. The opposite face had approximately one-third of its surface sheared away. Edge flaking on this face was much less intense. The tool resulting from this activity is roughly ovate in outline and has a rhomboidal cross section.

The edges of this tool display considerable wear and crushing, possibly as a result of use as a butchering tool. The tool measures 207 × 118 × 32 mm.

Activity Areas

Analysis of flaking events and raw material distribution for Component IV has been conducted by Tim Smith (1981). This analysis is still in progress as part of his doctoral dissertation research. However, the data at hand are more than sufficient to demonstrate the main activities conducted at the site during this occupation.

The definition of activity areas was accomplished by mapping the distributions of 13 raw-material types that had been reduced to differing degrees in this component. The combined perimeters of the individual raw-material clusters revealed two areas of intense flaking activity. The first, and largest, of these (Cluster A) lies near the southern margin of the excavation (figure A.1), and the other (Cluster B) is located near the western extremity of the excavated area.

In order to associate the assorted flakes with specific flaking episodes, it was necessary to further refine the types of raw materials. Of the five rock types mentioned previously, the rhyolites and degraded quartzites were further subdivided into five and four subgroups, respectively. Of the rhyolites, five different types can be distinguished: tan-greenish (64 flakes), reddish (162 flakes), heat-affected (153 flakes), greenish-phenocrysted (92 flakes), and gray-grainy (14 flakes). There are four types of degraded quartzite: rugose (283 flakes), "matte" (238 flakes), smooth (231 flakes), and phenocrysted (89 flakes). The remaining categories contain the following number of flakes each: obsidian (104), chert (93), siltstone (6), and granite (12). Oddly enough, a piece of granite was smashed in Cluster A, and many of the "flakes" could be fit back together. The purpose of this activity is unknown.

Of the sample of 2,130 specimens, 1,974 flakes and 2 tools fall within Cluster A. Cluster B is composed of 141 flakes and 10 tools. In addition, 2 tools, both bifaces of smooth degraded quartzite, fall outside the clusters. One is situated about midway between the clusters, and the other lies to the east of Cluster A (figure A.1).

Three tools were found within the cluster for which there is no flaking debris within the site. The first is a biface of a light-tan rhyolite that is in Cluster A, the second is a boulder spall tool of quartzite that is located in Cluster B, and the third is a retouched flake of light-tan rhyolite that also falls in Cluster B.

A summary of the raw materials, tools, and flakes found in the clusters and tools that fall outside the clusters can be found in tables A.1 and A.2.

Several conclusions can be drawn from these observations and data:

1. The clusters are composed almost entirely of small bifacial reduction flakes, and their greatest concentration is in Cluster A.
2. The clusters share only three raw-material subtypes: tan-greenish and heat-affected rhyolite and obsidian. Tools of either type of rhyolite were not located inside or outside of the clusters. The obsidian flakes (3) in Cluster B are unrelated to the obsidian tools in that cluster. Almost all the obsidian flakes occurred in Cluster A, although no tools of that material were discovered in association.

TABLE A.1. Summary of raw materials, tools, and flakes for Cluster A of Component IV at the Dry Creek site

Raw Materials	Tools (N = 2)	Flakes (N = 1,794)
Rhyolites		
Tan-greenish and heat-affected	No tools found	669
Reddish	No tools found	162
Greenish-phenocrysted	No tools found	92
Degraded quartzites		
Rugose	One retouched flake	283
Matte	No tools found	283
Smooth	No tools in cluster	231
Phenocrysted	One biface	89
Obsidian	No tools found	104
Cherts	No tools found	93
Siltstone	No tools found	13

TABLE A.2. Summary of raw materials, tools, and flakes for Cluster B of Component IV at the Dry Creek site

Raw Materials	Tools (N = 10)	Flakes (N = 141)
Rhyolites		
Tan-greenish and heat-affected	No tools found	124
Gray grainy	No tools found	14
Tan	One retouched flake	0
Obsidian	One biface; 7 side scrapers	3
Quartzite	One boulder spall tool	0

3. The flake distributions of the remaining material types are mutually exclusive.
4. The bifaces are scattered within the site; two fall in Cluster A, one is in Cluster B, and two are situated outside of the clusters. Of the five bifaces, three were made at the site, and one of these was found in Cluster A. The obsidian biface in Cluster B might have been manufactured in Cluster A, although it could have been made elsewhere.
5. The scrapers are all situated in Cluster B. The obsidian end scrapers might have been made in Cluster A; however, this is uncertain.
6. There are tools in the site that apparently were manufactured elsewhere: one biface in Cluster A and both the boulder spall tool and the retouched flakes in Cluster B.

7. The majority of the flakes in the site (1,405) resulted from bifacial reduction activity for which no tools were found.

It is difficult to establish the contemporaneity of the clusters, since the flakes of one cluster could not be conjoined to tools in the other cluster. However, the distribution of the obsidian flakes and tools might indicate tool production in one cluster and tool use in the other cluster. Also, the co-occurrence of the tan-greenish and heat-affected rhyolites could indicate utilization of these specific rock types on the site at the same time. Unfortunately, the activities represented by the tool distributions in Component IV could just as well have resulted from separate and/or temporally different occupations.

It should be emphasized that the scrapers constitute the only clear concentration of tools in association with a flaking cluster (B), and this strongly suggests that this was a specialized activity area where the use of those tools was required.

Paleoecology and Regional Relationships

We still have an incomplete picture of the vegetation shifts and realignments in the Nenana Valley during the Holocene. At the present time, there is a clear steppe quality to the vegetation communities of the floodplains and the lower slopes of the valley and one that has probably been preserved throughout the Holocene. In addition, the taiga occurs in several forms. It is established as gallery forests along terrace edges, and water courses and forms continuous cover on most moderate south-facing slopes. It becomes continuous only below 300 m. In the immediate area of Dry Creek, there are broad expanses of scrub, heath, and bog communities that today form important feeding areas for moose and black bears. The mountain slopes are dryer and support an herbaceous cover important for both sheep and caribou.

We know from the Carlo Creek site (Bowers 1980) that caribou were present by 8,500 ^{14}C yr BP (9,500 cal yr BP), and sheep have utilized the area (Dry Creek and Carlo Creek) since Pleistocene times. The modern, large ungulates (moose, caribou, and sheep) have probably existed in the area throughout the Holocene, although we lack direct fossil evidence in this particular area.

The association of Component IV with Paleosol 4a and the radiocarbon dates for this horizon (ca. 5,000–3,000 ^{14}C yr BP [5,700–3,200 cal yr BP]) indicate that this set of tools and waste were left at the site during the development of the Holocene forest in the Nenana Valley, or at least during the development of the gallery forest along the northwest side of Dry Creek. Based on such contextual reasoning, it can be argued that the tool kit in Component IV was employed in the procurement of essentially modern species of large or small game.

Technologically, the single distinctive tool set in Component IV at Dry Creek is the side-notched points. They occur at other ecologically different localities in the nearby Alaska Range and Tanana Valley.

In the area closest to Dry Creek, asymmetric notched points were found at the Ratekin site deep in the Alaska Range (Skarland and Keim 1958). This material is undated and has undergone mixing with older and younger artifacts. While there are no faunal remains at the site, topography and altitude could be used to argue that seasonal caribou hunting was at least one of the major activities conducted at this locality.

A better sample of notched points, again mostly asymmetric, was found at the XMH-35 site near Tangle lakes, which dates somewhere in the range of 4,500 ^{14}C yr BP (5,160 cal yr BP; Mobley 1982). Faunal remains are unfortunately absent from this site also, but the regional setting is also conducive to caribou hunting.

Notched points are an important tool category in several phases at the Healy Lake site (Cook and McKennan 1970). They occur in the Tuktu Phase and continue through the Denali and Historic Athapaskan Phases along with lanceolate bifaces, microblades, and burins. The position of Healy Lake, adjacent to the Tanana Hills, certainly made this locality ideal for exploiting both upland and lowland resources.

Fortunately, a complex comparable to Component IV at Dry Creek and in association with faunal remains has recently been discovered and studied at the XBD-106 site on Rainbow Lake near the Tanana River between the Little Delta River and Delta Creek (Bacon and Holmes 1980).

Here a typical taiga fauna composed of caribou, moose, black bear, and beaver occur with side-notched points, possible microblades, and end scrapers. Although there is no date for this site, the association of side-notched points with typical modern fauna is significant and places a more concrete basis for our understanding of the cultural ecology of the Tanana Valley.

Perhaps one of the most important discoveries of recent years for the late Holocene archaeology of the Alaskan interior was made by Holmes and Bacon (1982) at the XMH-297 site on the Delta River north of the Alaska Range. A small collection of artifacts (Component 5) has been recovered from Block B, Loess 6. This component is composed of flakes, one chi-tho, and two hammerstones. In addition, Loess 6 yielded the proximal end of a bison tibia that, unfortunately, was not directly associated with archaeological materials. However, the bison specimen is bracketed by an overlying date of 2,280 ± 145 ^{14}C yr BP (2,010 cal yr BP) and an underlying date of 3,980 ± 150 ^{14}C yr BP (4,450 cal yr BP). Based on this information, it appears that bison were in the Tanana Valley during the later Holocene. Bison remains have been radiocarbon dated to 500 ^{14}C yr BP (530 cal yr BP) near Anchorage (see chapter 6).

On the basis of this evidence, we should modify our view of the Holocene northern hunter as an exploiter of only a modern northern fauna. This new evidence, while not categorically placing cultural remains in association with bison during the late Holocene, does establish that bison existed in certain localities in Alaska where limited and probably relict steppe environments existed—for example, the Delta River valley. The Nenana River valley in the vicinity of Dry Creek is another area where these relict xeric plant communities occur today.

In the broader view, it is probably more correct to view the later Holocene hunters of the north Alaska Range foothills and the Tanana Valley as possessing a broad-spectrum subsistence base geared to the mosaic of plant-animal patterns still typical for the area today. The subsistence base included—besides the possibility for fishing and bird resources coupled with small mammals—the procurement of large mammals of both taiga and tundra. In addition, it appears that the steppe grazing niche remained in existence and was occupied by herds of bison and possibly wapiti long after Component IV times at Dry Creek.

The Northern Archaic Tradition

Beyond the Tanana Valley and the central Alaska Range, the broader typological affinities of Component IV at Dry Creek relate to the widespread occurrence of notched points in Alaska that date as early as 11,000 ^{14}C yr BP (13,000 cal yr BP; Gal and Hall 1982) and as late as 1,250 ^{14}C yr BP (1,220 cal yr BP; Anderson 1978). These notched-point complexes are usually linked to the Northern Archaic Tradition (Anderson 1978; Dumond 1978), although the temporal boundaries and cultural content of this tradition can vary considerably.

If one attempts to develop a composite understanding of the Northern Archaic by incorporating its many attributes from the differing formulations in the literature, a very complicated cultural-historical entity emerges that displays differing ecological adaptations, artifact variability, and a considerable temporal and spatial spread. Its sites are found in the taiga (Anderson 1968b; Dumond 1978) and in the tundra (Davis et al. 1981; Gal and Hall 1982). While notched points are important throughout, stemmed or even lanceolate points can also form a significant part of the assemblages (Anderson 1968). Any one of these combinations of point styles can co-occur with microblades, or microblades may be completely absent (Anderson 1968a, 1968b). The microblades might have been detached from tabular Tuktu cores (Anderson 1978; Campbell 1961) or from wedge-shaped Denali cores (Cook and McKennan 1970). This particular situation depends on whether the Tuktu Complex and the later phases at Healy Lake are included in the Northern Archaic Tradition.

Anderson (1968a, 1968b) first used the term *Northern Archaic Tradition* with regard to a set of cultural strata at Onion Portage that were positioned between and contrasted with the preceding Paleoarctic and succeeding Arctic Small Tool assemblages. The earlier part of this tradition, the Palisades II Complex, was characterized by notched points, coarse stones, crude workmanship, and a lack of microblades. However, the close affinity of Palisades II projectile-point styles with those of the Tuktu Complex (Campbell 1961) was recognized, although the lanceolate points and microblades of the latter set it apart. The later phases of the Northern Archaic at Onion Portage witnessed the appearance of more projectile-point variability, including corner notching and lanceolate forms.

More recently, Anderson (1978) has refined his thinking on the problem of the Northern Archaic, and we see added to the earlier Palisades II Complex an implied relationship with the Tuktu Complex. This Palisades/Tuktu Phase dates between 6,500 and 6,000 ^{14}C yr BP (7,420 and 6,840 cal yr BP). The succeeding Portage Complex displays a shift from side- or corner-notched points to lanceolate or pentagonal forms. Following a Denbigh Flint occupation, we see both elements of the Northern Archaic and Arctic Small Tool Traditions in the Choris levels that date to between roughly 3,000 and 2,000 ^{14}C yr BP (3,200 and 1,950 cal yr BP). Again, after an interval of various occupations, the Northern Archaic reappears as the Itkillik Complex, which is dated to roughly 1250 ^{14}C yr BP (1,220 cal yr BP).

As such, the Northern Archaic has a great time depth and internal typological complexity and is ethnogenetically linked to the Athapaskans.

Ecologically, the Northern Archaic is viewed as an adaptation to a forest way of life with an emphasis on caribou hunting primarily and fishing secondarily. The fishing aspect of the economy is based on the presence of notched pebble sinkers in the Northern Archaic levels at Onion Portage (Anderson 1968).

Dumond (1977:47) related a number of Alaskan sites to the Northern Archaic that are typified by "somewhat asymmetrical projectile points with deep, wide, side-notches, and bases that are commonly rather convex; large unifacially chipped knives; and chipped endscrapers." Again, in later times, we see more projectile-point variability with corner notching and lanceolate forms. Other assemblages such as the Tuktu Complex and the Denali Complex at Healy Lake, which contain microblades, are also included in this discussion. However, Dumond (1977) notes the absence of notched pebble sinkers in the archaeological record of the Northern Archaic for Ugashik and Upper Naknek drainages.

With respect to the origins of the Northern Archaic, Dumond (1977) traces its affinities to peoples moving from the south into the northern regions as forests were spreading around 7,000–6,000 ^{14}C yr BP (7,840–6,840 cal yr BP). The occurrence of notched and lanceolate points with microblade technologies is explained by the mixing of different peoples in contact situations.

On the other hand, Cook (1969) and Henn (1978) have both argued that the continuity of microblade technologies into later Holocene cultures, especially in the eastern interior and on the Alaska Peninsula, respectively, is sound evidence for continuity in populations themselves.

The position of Dry Creek Component IV in the later Holocene prehistory of the interior is fairly straightforward. The small complex of tools in Component IV is fully consistent with the definition of the Northern Archaic proposed by Dumond (1977). It likewise compares well with the early part of the Northern Archaic as manifested in Anderson's (1968a, 1968b) Palisades II Complex, although the radiocarbon date at Dry Creek overlaps only with the later part of Palisades II as it is dated at Onion Portage. The notable features of Component IV shared with both these formulations are side-notched points; larger, more crudely worked implements; and an absence of microblades. While these similarities are important, we should also note the striking contrast among Component IV at Dry Creek and those complexes that contain, in addition to side-notched points, lanceolate forms and microblade industries. As already pointed out, none of these typological complexes are present in Component IV at Dry Creek. Also, when Component IV is compared with Component II, we are struck by the fact that they share not a single artifact type except retouched flakes. Our view of the situation from the Nenana Valley at the present time is that the early microblade and biface technologies of Component II and the artifacts from Component IV are as different as night and day. There are simply no grounds for developing the idea of continuity between Components II and IV.

Based on the evidence from the Dry Creek site, the proposition that an entirely different population occupied the Nenana Valley during the mid-Holocene would appear completely justified. However, evidence from other areas of Alaska summarized previously points to the fact that Dry Creek is more the exception rather than the rule. Since side-notched points similar to those in Component IV do occur with microblades elsewhere, their absence at Dry Creek should be viewed cautiously. The Component IV occupation does not appear to have been as intense as the occupation in Component II. However, it is possible that the major area of settlement was in a different, unexcavated part of the site. Likewise, a late microblade technology could have been part of the tool kit used by the later Holocene inhabitants of the Nenana Valley, and they simply did not bring it to the part of the site that we excavated, if they brought it to the site at all.

Summary

It can safely be said that the Northern Archaic Tradition is represented at the Dry Creek site by Component IV and that it occurred somewhere between 5,000 and 3,000 ^{14}C yr BP (5,720 and 3,190 cal yr BP).

We see two definable activity areas where tool manufacture and repair were conducted and where bone was smashed to bits. Side-notched points were located both within and outside these areas. Only one of these pieces appears functional as a weapon tip. The remainder were probably repaired or reworked into knives.

We cannot establish that the two areas were used at the same time. While it is possible that they were, it is just as likely that they resulted from two separate occupations. In either case, the occupations appear to have been brief.

In the absence of identifiable faunal remains, no specific aspect of the subsistence economy can be clarified. We can only suggest that large or small mammal procurement was conducted from the site. We have also suggested that besides modern large ungulates, bison might still have been available in the Nenana Valley during the late occupation of the Dry Creek site.

Finally, the artifacts from Component IV stand in striking contrast to those of Component II. There is no apparent continuity between the two components. However, evidence from elsewhere in Alaska suggests that we might be missing part of the picture at Dry Creek with respect to the absence of microblades as well as other types of bifaces commonly found with notched-point assemblages.

Clearly, we have only scratched the surface of Northern Archaic archaeology in the Nenana Valley, and until the pattern we see at Dry Creek is confirmed, it would be wise to not press the data too far.

PART 2
Dry Creek Update

larger inventory were available from Level 1" (Schweger et al. 1982:440). Dumond (1980) and Colinvaux and West (1984) also expressed this concern. In reality, though, this argument was never tenable because at 347 m², the Dry Creek excavation surpassed all other horizontal areas excavated at early period sites in Alaska (except perhaps Broken Mammoth [see Potter et al. 2014]). Moreover, the Component I assemblage of 4,524 artifacts (cores, tools, and debitage) surpasses the reported sizes of nearly all other late Pleistocene assemblages predating 13,000 cal yr BP in Alaska, for which artifact numbers can be obtained from the published literature (Moose Creek Component I yielded 2,259 [Pearson 1999]; Walker Road Component I, 4,980 [Goebel et al. 1991, 1996]; Owl Ridge Component I, 1,038 [Gore and Graf, in press]; and Broken Mammoth Cultural Zone 4, 1,319 [Krasinski and Yesner 2008]). Further, most early assemblages post-dating 13,000 cal yr BP in Alaska contain far fewer artifacts (Donnelly Ridge, 1,513 [West 1996b]; Chugwater, 1,223 [Lively 1996]; Phipps, 1,628 [West et al. 1996a]; Sparks Point, 586 [West et al. 1996b]; Hidden Falls, 612 [Davis 1996]; Panguingue Creek, 72 [Goebel and Bigelow 1996]; Owl Ridge Component II, 1,386 [Gore and Graf, in press]; Broken Mammoth Cultural Zone 3, 4,065 [Krasinski and Yesner 2008]). Given the central Alaskan record that has emerged over the past thirty years, it is very difficult to argue that Dry Creek Component I is too small to make determinations about its technological character. Even if it was, the existence of similar Nenana Complex assemblages at other sites—especially Walker Road, Moose Creek, and Owl Ridge—make the sample-size argument moot.

The more enduring problem raised about Dry Creek Component I is related to its geological formation. Dumond (2001) pointed out that Component I's artifact concentrations underlaid Component II clusters (figure 8.1; see also chapter 5), a pattern Dumond (2001) argued might have resulted from the postdepositional movement of Component II artifacts downward to form an artificial lower component. Thorson (2006) made three additional observations in support of Dumond's interpretation. First, Sand 1, which separated the two components, was thin and discontinuous across the site so that in some places, Loess 3 was found directly overlying Loess 2. In other words, sediment containing Component II was in places found to be directly superimposed on sediment containing Component I. Second, when existing radiocarbon dates for Loesses 2 and 3 were calibrated, they overlapped at two sigma (σ), making their ages statistically contemporaneous. Third, because Component I artifact clusters were found only near the terrace edge, where deposits annually freeze and thaw and during summer months become desiccated, they were highly susceptible to postdepositional deformation, including crack formation, faunal burrowing, and solifluction.

For Dumond and Thorson to be correct, one would expect there to be technological similarities between the artifacts of Components I and II, and Odess and Shirar (2007) presented evidence they thought was supportive of this. While examining the 1970s Dry Creek lithic assemblage housed at the University of Alaska Museum of the North, they identified a Component I artifact as a microblade core tablet (figure 8.2A), suggesting it had been misidentified as undiagnostic debitage by previous investigators. They described the piece as having been "burinated from the distal end, a practice often seen in Denali-Complex material from other sites" (Odess and Shirar 2007:130). If this was indeed the case, then the Component I assemblage could no longer be called "nonmicroblade."

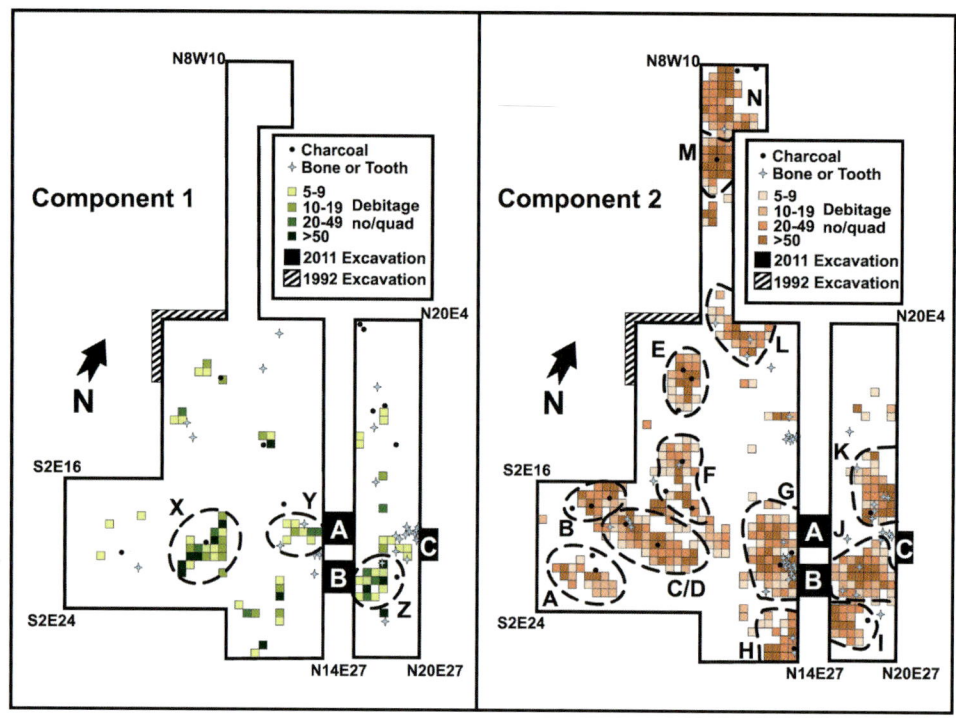

FIGURE 8.1. Excavation maps of Components I and II (left and right, respectively) from the original excavations at Dry Creek (adapted from chapter 5). Note locations of Blocks A, B, and C of the 2011 excavation.

FIGURE 8.2. Comparison of the Component I shatter (artifact UA76-155-1443) (A) with a Component II microblade core tablet (B) found during the 2011 excavation. The artifacts are oriented so that attributes of the pieces can be compared with each piece's platform in the center, lateral margins to either side of the platform, and dorsal surface overlying the platform. White arrows highlight directions of previous removals.

Given these concerns, the Dry Creek site deserved a second look using modern techniques to assess its geological and stratigraphic integrity and the character of its presumed distinctive assemblages. In 2011 we reopened Dry Creek to document site-formation processes, testing the hypothesis that Component I resulted from the postdepositional movement and mixing of artifacts originally from Component II. Our excavations also gave us an opportunity to test whether Component I's lithic assemblage was indeed distinct from Component II, not just in tool types and technology, but also in lithic raw materials. In the following sections, we present our findings, focusing mostly on Components I and II, but first we review the methods applied.

Excavation Methods and Analytical Procedures

The 2011 excavations at Dry Creek covered an area of 10 m². Two 2 × 2 m blocks (A and B) were placed along a 2-m-wide balk preserved during the 1970s excavation, and a 1 × 2 m block (C) was placed along the northern margin of the 1970s excavation (Figure 8.1). These three areas were chosen because in each case, we expected to find remnants of Component II artifact clusters (i.e., G and J) superimposed stratigraphically over Component I artifact clusters (i.e., Y and Z). In 1978 when the site was backfilled, the bulldozer operator inadvertently scraped away the top 50–80 cm of intact deposits of the preserved balk in this area, so we were unable to experience the entire profile in blocks A and B. However, we were able to quickly access the lower portion of the profile, our immediate objective. Only the deposits from the base of Loess 5 (Paleosol 3) were preserved. Block C was unscathed by the 1970s bulldozer, so it provided a view of the entire stratigraphic profile. In this area we also retrieved materials from Component IV.

Excavation and Sample Extraction Techniques

Excavations followed standard procedures. Site sediments were removed by hand troweling. All excavated sediment was dry-screened through 1/8-in mesh. Three-point provenience of artifacts, bones, and charcoal samples recovered in situ was recorded with a Sokkia EDM total station. Provenience of materials recovered from the screen was assigned according to 50-cm² horizontal quadrant and 5-cm vertical level (within each recognizable stratigraphic unit). The top of each new stratum was exposed, mapped, and photographed across the entire block before its excavation commenced. Trend and plunge (i.e., dip direction and angle) of artifacts were measured using Silva Ranger clinometer compasses when the artifact's original aspect could be confidently assessed. All postdepositional disturbances were documented in plan views for each stratigraphic boundary and reflected in site profiles.

Sediment and stratigraphic descriptions and profiles were completed in the field. We collected micromorphological samples by driving plastic conduit boxes into exposed profile walls, labelling the boxes with provenience and orientation. We also collected a 2.5-cm-interval sediment column from the naturally exposed profile along the terrace edge for standard sedimentological analyses.

Geological Laboratory Analyses

Sediment samples were used for granulometric and geochemical analyses. Granulometric samples were pretreated with 10 mL of 10 percent solution of sodium hexametaphosphate and sonicated for three minutes to ensure complete dispersion. Particle size distributions were determined via laser granulometry using a Malvern Mastersizer2000 with a HydroMU dispersion unit and classified at one-phi intervals following Blott and Pye (2012). Granulometric intervals included very coarse sand (2,000–1,000 μm), coarse sand (1,000–500 μm), medium sand (500–250 μm), fine sand (250–125 μm), very fine sand (125–62.5 μm), very coarse silt (62.5–31.3 μm), coarse silt (31.3–15.6 μm), medium silt (15.6–7.8 μm), fine silt (7.8–3.9 μm), very fine silt (3.9–2 μm), and clay (< 2 μm) fractions. Standards were run periodically throughout the analysis to ensure accurate data.

To prepare soil samples for geochemical analysis, we first removed macroscopic organic materials (e.g., charcoal and rootlets) from samples and then dried and powdered them. The Dry Creek sediments are noncalcareous, so acidification to remove inorganic carbon was unnecessary. Carbon and nitrogen concentrations of soil organic matter were determined from bulk soil material using a Thermo Finnegan Flash 1112 Series elemental analyzer. Concentrations of other major elements were determined through a combination of inductively coupled plasma mass spectrometry and inductively coupled plasma atomic emission spectrometry (ICP-MS/ICP-AES) done at a commercial laboratory. Before submitting the bulk soil samples for analysis, we put them into solution using a four-acid near-total digestion process to remove all background material except basic chemical elements for identification. Ammonium oxalate extractable iron and aluminum concentrations were determined by the Texas A&M Soil Characterization Laboratory, following USDA-NRCS standard procedures using a 0.2 M ammonium oxalate solution for extractions and ICP-AES for analysis (R. Burt and Soil Survey Staff 2014).

Micromorphological samples were air dried, impregnated with Hillquist epoxy, and prepared into 2 × 3-in thin sections to facilitate micromorphological and mineralogical observations. We accomplished the micromorphological analyses with an Olympus BX-51 research microscope with both polarized light and UV fluorescence, and we photographed the thin sections using a Leica DFC 450 camera attachment.

Radiocarbon Analyses

Given that geochronology was an important component of this study, here we provide a detailed description of our analysis protocols. We accomplished taxonomic identifications of charcoal samples using plant reference collections and libraries at the Desert Research Institute (Reno) and Department of Anthropology at Texas A&M University. For radiocarbon dating, we only analyzed samples collected directly from hearth features during our excavations. Two samples were prepared, pretreated, and analyzed at Beta Analytic, Inc., and four samples were prepared and pretreated at the Human Paleoecology and Isotope Geochemistry Lab, Pennsylvania State University. These latter samples were then analyzed at the W. M. Keck Carbon Cycle Accelerator Mass Spectrometry facility, University of California, Irvine (UCIAMS).

Physical preparation and chemical pretreatment of samples at Penn State followed regular procedures. First, datable material was physically separated from rootlets and other modern contaminants, followed by standard acid/base/acid (ABA) pretreatment consisting of repeated baths in 1N HCl and NaOH at 70°C for thirty minutes on a heater block. The initial acid wash dissolved any potential carbonate contamination, while base washes extracted humic acids (signaled by brown discoloration in the NaOH solution). Base washes were repeated until the solution was clear, indicating the complete removal of potential contaminants. A final acid wash removed secondary carbonates formed during the base treatment. Samples were then returned to neutral pH with two one-minute baths in Nanopure water at 70°C to remove chlorides. Subsequently, they were dried on a heater block.

Sample CO_2 was produced by combustion at 900°C for three hours in evacuated sealed quartz tubes using a CuO oxygen source and Ag wire to remove sulfur and chloride compounds. Primary (OX-1) and secondary (FIRI-F and IAEA-C5) standards were selected to match the sample type and expected age; these underwent the same chemical steps for quality assurance. The CO_2 generated was reduced to graphite at 550°C using a modified-hydrogen-reduction method onto a Fe catalyst (Alfa Aesar mesh −325 lots JO2M27 and L16P22), with reaction water drawn off with $Mg(ClO_4)_2$. To reduce modern carbon contamination, the Fe catalyst was baked at 400°C in H_2 for 45 minutes prior to analysis, in addition to being baked monthly at 300°C for 3 hours in air.

The resulting solid graphite samples were pressed into Al targets and loaded on a target wheel for AMS analysis, along with OX-1 (oxalic acid), other known-age standards, and wood blanks. AMS ^{14}C measurements were made on a modified National Electronics Corporation compact spectrometer with a 0.5MV accelerator (NEC 1.5SDH-1). This accelerator is equipped with a spherical ionizer ion source operating at high cathode voltage (9kV) to generate intense $\bar{c}$ beams, as well as an enhanced injection beam line for better ion-optical matching. Injector modifications included use of a second einzel lens, an increased ion source voltage of 65.5 kV, and a redesigned large-gap injector magnet (DF01319). These alterations reduced analytical error to the 2–3‰ range for near-modern samples under currents of up to 225 μA of $^{12}\bar{c}$ (40 to 50 μA, higher than normal). This led to errors in the ± 30 to 40 ^{14}C yr range for the samples in this study.

All ^{14}C ages were $\delta^{13}C$ corrected for mass-dependent fractionation with measured $\delta^{13}C$ values following Stuiver and Polach (1977). Paired dates from hearth features were averaged using the method of Ward and Wilson (1978) after testing for contemporaneity (chi-square test). Calibrations and modeling were accomplished with OxCal 4.2.3 (Bronk Ramsey 2009, 2013) using the IntCal13 Northern Hemisphere atmospheric curve (Reimer et al. 2013).

Lithic Analyses

We cataloged all materials using a collection number supplied by the University of Alaska Museum of the North, UA2011-064. Lithic artifacts were visually inspected using a 10x-hand lens, and analyses followed an analytical protocol developed by the lead author of this chapter for early period archaeological assemblages of Alaska and Siberia (Graf 2008, 2010; Graf and Goebel 2009). Analyses focused on basic descriptive variables, including artifact class (core, debitage, tool), artifact type (flake core, bipolar core, microblade core, angular shatter, cortical spall, flake, bladelike flake, bipolar flake, resharpening chip, biface thinning flake, burin spall, microblade, microblade-core technical spall, blade-core technical spall, side scraper, end scraper, angle burin, retouched blade, retouched microblade, retouched flake, biface, chopper, and hammerstone), raw-material class (cryptocrystalline silicate [CCS], quartzite, rhyolite, basalt and other [quartz, granite, hematite]), raw-material type (further breakdown of raw-material class based on visible macroscopic characteristics), cortex presence, cortex type (primary [from outcrop] or secondary [from alluvial cobbles]), artifact condition (complete or broken), and artifact size value (< 1 cm, 1–3 cm, 3–5 cm, and > 5 cm in overall dimension).

To assess behavioral site-formation processes, additional variables were calculated: frequency of primary reduction activities (combining numbers of cores, cortical flakes, flakes, bladelike flakes, and bipolar flakes), frequency of secondary reduction activities (combining resharpening chips, biface thinning flakes, burin spalls, and tools), and raw-material selection (considering frequencies of raw materials by artifact types). Variables for assessment of natural site-formation processes included artifact size and artifact plunge. These permitted an assessment of the degree of artifact sorting and vertical movement.

Faunal Analyses

Archaeofaunal assemblages in the Alaska Interior are typically comminuted and small (Hoffecker et al. 1996; Holmes 2011; Potter 2007; Yesner 1994, 1996, 2001), and the Dry Creek faunal assemblage is no exception (see chapter 6). To nonetheless extract as much information as possible from the poorly preserved remains retrieved in 2011, fragments smaller than 1 cm that could be manipulated and analyzed were included. However, no refitting was attempted due to the fragile nature of the materials. Comparative collections at the Alaska Consortium of Zooarchaeology in Anchorage and the Mammalogy Department of the University of Alaska Museum of the North were used for elemental and taxonomic comparisons, with taxon, element, portion, age, and side being recorded when possible (see Reitz and Wing 1999). Number of identified specimens (NISP) and percent NISP were calculated for each stratigraphic unit to facilitate interstratigraphic comparisons. Other measures of abundance (e.g., minimum number of individuals and minimum number of elements) could not be calculated.

Additionally, we inspected each specimen for taphonomic traces of butchering, carnivore gnawing, percussion damage, and trampling through fracture attributes and cortical surface modifications (following Krasinski 2010). Observations were made with a 10x-hand lens using a strong oblique light source. Attributes scored included length, width, and thickness of specimen (mm); presence or absence of cortical bone and surface bone modifications (e.g., cutmarks, gnawmarks, root etching); and weathering stage and degree of burning (following Behrensmeyer 1978 and Stiner 1994).

2011 Results

Here we present data collected from the 2011 Dry Creek project, focusing on analyses facilitating assessment of the integrity of the site's terminal Pleistocene cultural deposits; however, we also present the few materials recovered from Component IV.

Geology, Stratigraphy, and Site Formation

Dry Creek's archaeology rests within a mantle of eolian silts and sands reaching up to 200 cm in thickness, sloping approximately 120° southeast of true north at a 10.5 percent, or 6°,

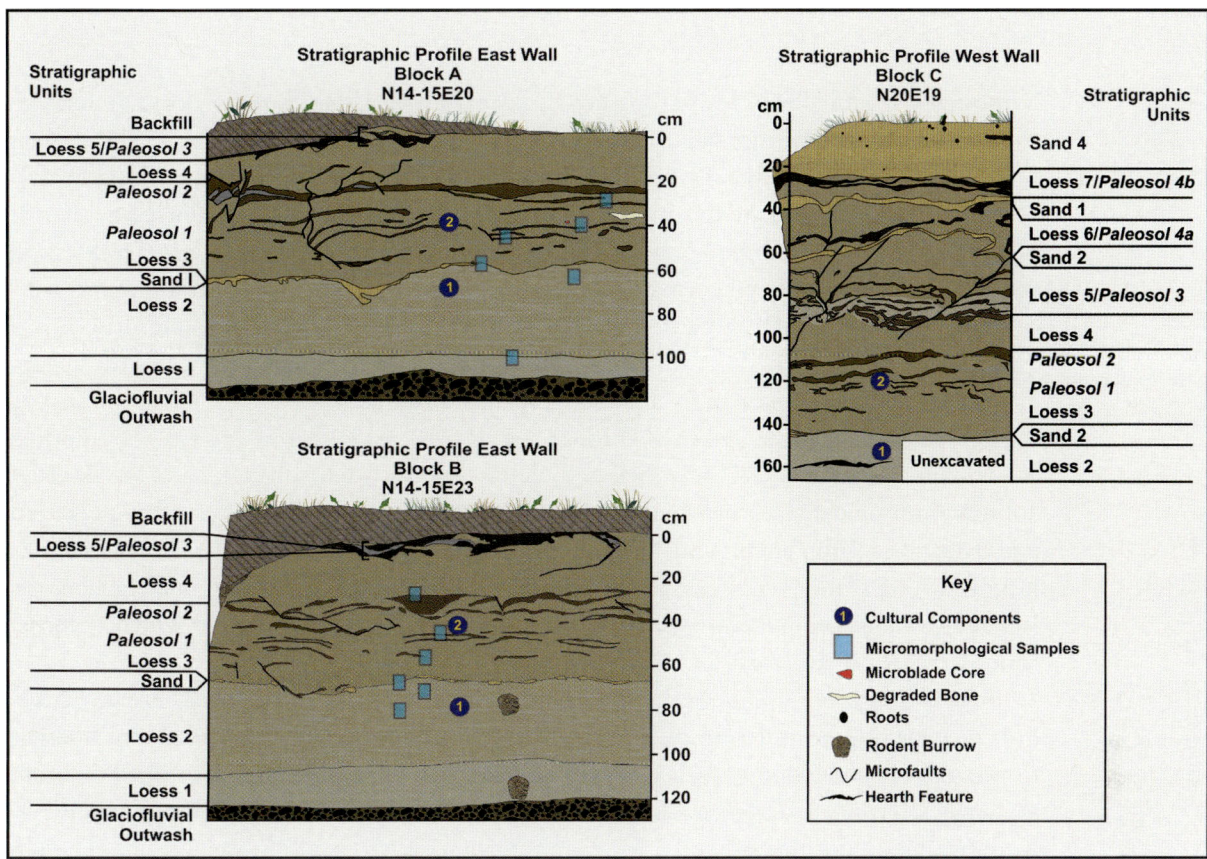

FIGURE 8.3. Stratigraphic profiles from the 2011 excavations. Profiles from blocks A and B are the eastern walls of those blocks, while the profile from block C is of the western wall.

gradient. This loess mantle lies unconformably above early Upper Pleistocene glacial outwash on a terrace remnant of the local Healy Glaciation (Dortch et al. 2010; Ritter 1982; Thorson 1986; Wahrhaftig 1958). We recognized 11 sedimentological units, all eolian in origin, the same as those defined by Thorson and Hamilton (1977) and presented in chapter 3 (figure 8.3). For consistency, we maintained the same stratigraphic designations for these units; however, our observations led us to interpret some of the stratigraphic profile differently than the earlier studies.

Stratigraphic descriptions that follow are based on all three of our excavated blocks. The lower portion of the profile, which contains Components I and II, is presented first. The description of the upper portion of the profile considers the depositional context of Component IV.

Loess 1 forms the base of the unconsolidated late Pleistocene-Holocene loess mantle. It is a compact grayish-brown (2.5Y 5/2) sandy loam measuring 10–20 cm thick, with unsorted subangular clasts and tight cohesiveness. The unit forms a straight, abrupt boundary with the underlying glacial outwash, but occasionally isolated lenses of pebble-sized clasts, likely frost heaved up from below, are seen near its lower boundary. The upper boundary of Loess 1 is diffuse, with a gradual coarsening in texture in its upper 5–8 cm.

Although it is difficult to clearly separate Loess 1 from overlying Loess 2 based on texture, there is a reasonably abrupt color change at this boundary that helps delineate the contact originally defined by Thorson and Hamilton (1977). However, we attribute the gray coloration of Loess 1 to postdepositional gleying from reduction, perhaps denoting a recent permafrost zone. While removing backfill to expose the northernmost original excavation wall (N20), we encountered permanently frozen sediment at this precise stratigraphic position and elevation, 8–10 m back from the present-day terrace edge (around grid points N20E15-16). This permafrost surface might explain the gleying in Loess 1 and the abrupt Loess 1/2 boundary recorded by Thorson and Hamilton (1977).

Loess 2 is a cohesive, yellowish-brown (10YR 5/8) to light-gray (10YR 7/1) mottled sandy loam reaching 25–40 cm thick and containing unsorted subangular clasts. Mottles are redoximorphic masses and filaments that resulted from periods of alternating reduction and oxidation of iron compounds in the sediment. A few ancient animal burrow casts (i.e., krotovinas) were observed. Loess 2's upper boundary with Sand 1 is abrupt but irregular. We isolated cultural Component I within a 5-to-10-cm-thick layer, the top of which occurred about 5 cm below the Loess 2–Sand 1 contact.

Sand 1 consists of well-sorted rounded to subrounded clasts, is yellowish-brown (10YR 5/4) in color, and measures up to 5 cm in thickness. It occasionally expresses minor downslope creep or solifluction folds 1–2 cm tall and 5–20 cm broad. Its upper boundary is abrupt and wavy. Contrary to Thorson (2006) but in agreement with Thorson and Hamilton (1977), we were able to document Sand 1 as a continuous unit across all our excavations. Though quite thin (< 1 cm) in some places, while excavating in these areas, we were able to consistently identify and map it as a horizontal concentration of medium-sized sand grains.

Loess 3 is a cohesive, strong-brown (7.5YR 5/8) to light-gray (10YR 7/1) mottled sandy loam reaching 25–35 cm thick. It contains unsorted subangular clasts as well as two separate paleosols. Loess 3 expresses weak microfaulting and forms an abrupt, wavy upper boundary with Loess 4. Paleosol 1 was described as a complex of three to four A horizons within Loess 3 (Thorson and Hamilton 1977). We observed this in the field, too, but we attribute it to a single A horizon with several minor involutions caused by minimal solifluction, the folds of which are 2–3 cm tall and 25–50 cm broad. Paleosol 1 is 7.5YR 4/3 brown to 5YR 5/4 reddish-brown in color. Cultural Component II was reported to be associated with Paleosol 1 (Thorson and Hamilton 1977; see also chapter 3), but we isolated its artifacts both in the paleosol as well as in the mottled loess around it. Paleosol 2 was laterally continuous in all our excavations. It consists of two sets of weakly developed A/Eg/B soils measuring about 8–10 cm in thickness. Like in the 1970s excavations, we did not find any artifacts associated with Paleosol 2 in upper Loess 3.

For consistency, we have maintained the scheme presented by Thorson and Hamilton (1977) and Powers (chapter 3) that has Paleosols 1 and 2 combined into a single stratum. However, the two paleosols in Loess 3 represent two periods of stability that interrupted the deposition of two silty sand units. Because these represent two separate depositional events, they should be distinguished as two separate units (i.e., Loess 3a and 3b). Paleosol 1 represents an Ab horizon without an associated clear Bb horizon. Perhaps it represents a soil that formed in a shrub-tundra setting without the soil acidifiers or leaching zone of boreal forest soils in spodic or podzol contexts (Lukac and Godbold 2011). Paleosol 2 represents a better-preserved soil with spodic characteristics, including a well-developed A horizon overlying a thin, intermittent Eg horizon representing the eluviation of iron and aluminum oxides in cold, humid conditions to an underlying B horizon. This soil likely formed in an incipient boreal-forest setting.

Loess 4 is a cohesive, strong-brown (7.5YR 5/8) to light-gray (10YR 7/1) mottled sandy loam measuring 15–20 cm thick, with unsorted subangular clasts. It generally forms a diffuse boundary with overlying Paleosol 3 at the bottom of Loess 5, except where it has been microfaulted and soliflucted. In these contexts, portions of overlying Loess 5 (reaching 15–20 cm long) have been thrust down into Loess 4.

Loess 5 is a cohesive, strong-brown (7.5YR 5/8) to light-gray (10YR 7/1) mottled sandy loam. It reaches 20–30 cm thick and contains unsorted, subangular clasts. Loess 5 forms a distinct, wavy boundary with Sand 2 above. Postdepositional features such as frost cracks, microfaults, and major solifluction lobes are well represented. Solifluction involutions are 8–12 cm tall and about 20 cm broad, and microfaulting has led to nearly 15 cm of vertical displacement in some areas. Five to eight alternating buried A/Eg/Bs soils, originally combined into Paleosol 3 (Thorson and Hamilton 1977; see also chapter 3), were observed, beginning at the Loess 4/5 contact and continuing up through Loess 5.

We contend that the numerous A/Eg/Bs soils documented for Loess 5 should not be combined into a single paleosol, as Thorson and Hamilton (1977) did. The lowest A/Eg/Bs soil is robust and 5–7 cm thick, and it appears to be genetically tied to Loess 4, not Loess 5, considering that in principle, soils form on stable surfaces or at the tops of sedimentary units. The onset of deposition of Loess 5 interrupted the formation of the lowest of Paleosol 3's buried soil horizons, and therefore it should be described as part of Loess 4. Similarly, even though the six-eight overlying soils of Paleosol 3 are dispersed throughout Loess 5 as mapped by Thorson and Hamilton (1977), each of these thin, weakly developed soils should be presented as a stable surface under which occurred a loess event and over which occurred a new stratigraphically separate loess event. Technically, Loess 5 should be subdivided into at least six separate loess units (i.e., Loess 5a–f). Obviously, this portion of the Dry Creek stratigraphic profile is much more complex than originally reported, and its continued study could provide greater understanding of local and regional environmental conditions during the time of its deposition.

Above Loess 5, Sand 2 is a yellowish-brown (10YR 5/6) medium-to-fine silty sand varying from 1 cm to 3 cm thick. It is poorly sorted and contains subangular to subrounded grains. Sand 5 shows signs of frost cracking, microfaulting, and minor solifluction. Its upper boundary with Loess 6 is gradual.

Loess 6 is a brown (10YR 5/3) sandy loam measuring 12–25 cm thick and contains Paleosol 4a, a buried A/E/Bs soil. Evidence of some frost cracking, microfaulting, and minor solifluction is present, and Loess 6's boundary with overlying Sand 3 is distinct and wavy. Cultural Component IV occurs in association with Paleosol 4a.

Sand 3 is a yellowish-brown (10YR 5/6) medium-to-fine silty sand measuring 2–5 cm thick. It contains poorly sorted subangular to subrounded clasts. Some solifluction is present, and the upper boundary of Sand 3 with Loess 7 is gradual.

Loess 7 is a strong brown (7.5YR 5/8) sandy loam with poorly sorted subangular clasts. It is 7–10 cm thick, and its upper boundary with Sand 4 is abrupt. Loess 7 is minimally disturbed and contains Paleosol 4b, a set of at least three alternating A/Eg/B soils that greatly resemble Paleosol 3. Similar to Loess 5, the Loess 7 soils indicate at least three episodes of stability and soil formation immediately following separate loess depositional events.

The uppermost stratum of the Dry Creek profile, Sand 4, consists of very poorly sorted medium-to-fine sand with some silt, representing recent deposition of 10–25 cm of cliff-head sands.

As noted previously, we disagree with some of the original stratigraphic designations and interpretations of some of the specific processes that formed the site. First, Loesses 1 and 2 should be reinterpreted as a single loess depositional unit with the distinctive gray color of Loess 1 resulting from postdepositional gleying. Second, we argue that Paleosol 3 formed on the surface of Loess 4 prior to deposition of Loess 5 and is therefore genetically tied to Loess 4, not Loess 5, and that the overlying soil horizons that were originally tied to Paleosol 3 actually represent at least six later stable surfaces in which soils formed on the site during the deposition of Loess 5. Similarly, the overlying soil horizons in Loess 5 and the multiple horizons of Paleosol 4b in Loess 7 reflect repeated climatically stable soil-forming episodes distinguished by periods of significant loess accumulation. Precise radiocarbon dating of these Holocene soils could provide an important proxy record of regional environmental and climate change. Finally, we disagree with Thorson and Hamilton's (1977) conclusion that boreal-forest soil formation began with Paleosol 4. The character of Paleosol 3 also appears to reflect soil development in boreal-forest settings, while Paleosol 2 might have developed in an incipient boreal-forest setting. Only Paleosol 1 appears to reflect formation of a soil under shrub-tundra conditions.

Our field observations suggest stratigraphic integrity of the Dry Creek site. Stratigraphic contacts are well-defined. Cryoturbation is present at the macroscopic level in 6 of the 11 depositional units (Sands 1, 2, and 3 and Loesses 3, 5, and 6). Frost cracking and solifluction folds are minimal and restricted to within-stratum disturbances. These cryogenic features are completely absent from Loess 2 (and Component I), and even though they have affected Loess 3 (and Component II), their effect is limited to *within* the deposit. Similarly, microfault displacement is minor in the lower section of the profile, where Components I and II occur. Bioturbation is present throughout the profile, but it is rare and easy to identify and isolate in both plan view and vertically during excavation. Simply put, despite the presence of postdepositional disturbances, we found all stratigraphic units intact and recognizable across the excavations. We witnessed archaeological integrity with cultural components well-circumscribed within their associated sedimentological units. There was no evidence for the movement of artifacts *across* stratigraphic boundaries.

Sedimentology, Micromorphology, and Soil Geochemistry
Sedimentology
Sedimentological samples indicated remarkably similar grain size distributions throughout the Dry Creek profile (figure 8.4), a result that confirms field observations. Mean particle sizes of samples taken from the strata identified in the field as "loess" units (Loesses 1, 2, 3, 4, 5, and 7) range from 33 to 41 μm (very coarse silt fraction). Total silt ranges from 59 to 76 percent. The loess fraction (15.7–62.5 μm) includes the categories of coarse (15–22 μm) and very coarse (23–31 μm) silts ranging from 41 to 59 percent (with coarse silts

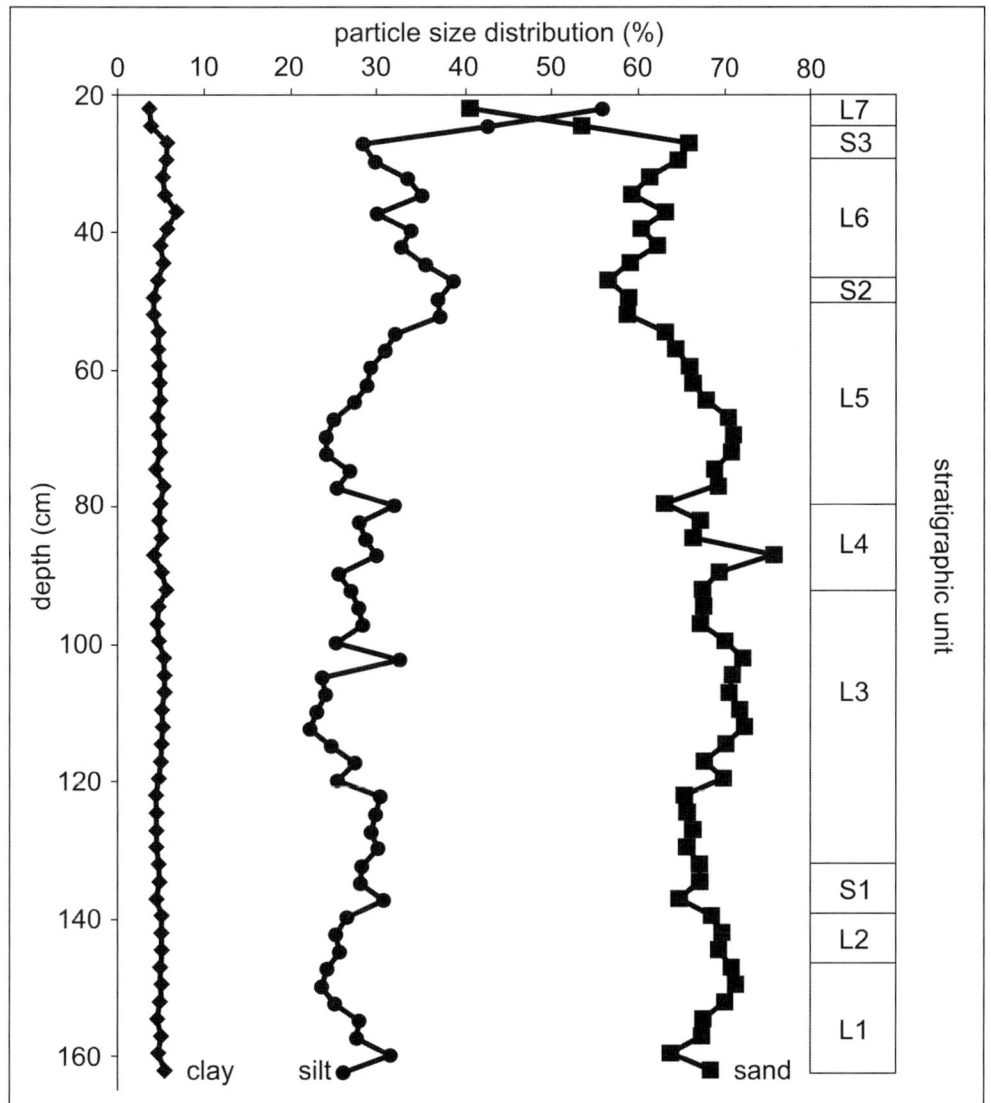

FIGURE 8.4. Particle size distributions showing relative percent of sand, silt, and clay by stratum.

ranging from 15 to 22 percent and very coarse silts from 23 to 31 percent). The medium to very fine silts (2–15.5 μm) are less abundant (16–26 percent), clays are nearly absent (2–3 percent), and sands range from 22 to 39 percent, a particle size distribution seemingly typical for Alaska (Muhs and Budahn 2006; Muhs et al. 2013). Interestingly, the majority of sand found in the Dry Creek loess units falls into the very fine sand (62.5–125 μm [18–26 percent]) fraction, with coarser fractions virtually absent (i.e., 125–2,000 μm fractions range from 4 to 13 percent). Thus strata designated as loess in the field are dominated by the "loess" fraction. The site's close proximity (~2.5 km) to the confluence of Dry Creek and the Nenana River, in conjunction with its relatively coarse loess fraction, indicates its immediate source of loess is the nearby Nenana River floodplain, an interpretation consistent with Muhs et al. (2013) for the Chitina section in the Copper River valley, which has a very similar particle size distribution to the Dry Creek section described here.

Traditionally, four sand lenses were recognized in the field (Sand 1 between Loesses 2 and 3, Sand 2 between Loesses 5 and 6, Sand 3 between Loesses 6 and 7, and Sand 4 overlying Loess 7). Particle size analysis confirms the presence of these four sand units as well as additional probable sand layers (e.g., a small peak of sand in the middle of Loess 4; figure 8.4). Despite collecting sediment samples every 2.5 cm, this sampling interval was too coarse to distinguish the individual sand lenses. Sands 1, 2, and 3, as well as the spike of sand in the middle of Loess 4, were recognized by minor increases (2–4 percent) in sand-sized particles and corresponding decreases in silt; however, changes in particle-size distribution of sand versus silt *and* clay content are only mildly distinguishable from the loess samples. It is important to reiterate that Sand 1 was notable both in the field and in thin sections as a conspicuous lens of coarse particles, with little fine material compared with adjacent loess, as figures 8.5 and 8.6 demonstrate. In fact, Sand 1 is a discrete unit with no evidence of having been infiltrated by sediment from above or below. Its contacts with Loesses 2 and 3 are sharp but wavy to irregular in form. A comparison of volume percent of Sand 1 compared with Loess 2 below and Loess 3 above indicates that Sand 1 is clearly different from the bracketing loesses and clearly made up of larger clasts that are not present in the loess units (figure 8.6).

The other sands have been described as more diffuse in nature, an observation confirmed by the particle-size analysis presented here, where fine material is present in proportions similar to the loess units, but coarse material is more prevalent (figure 8.4).

Particle-size data indicate a sediment profile at Dry Creek dominated by loess deposition of coarse silts to very fine sands from the terminal Pleistocene through the Holocene. However, this sequence is punctuated by periods of larger sand-clast deposition. As shown

FIGURE 8.5. Photomicrographs from Sand 1. (A) Sharp, unbroken contact between Loess 3 and Sand 1. (B) Sand 1 is coarser than the surrounding loess units and lacks fine matrix between the grains.

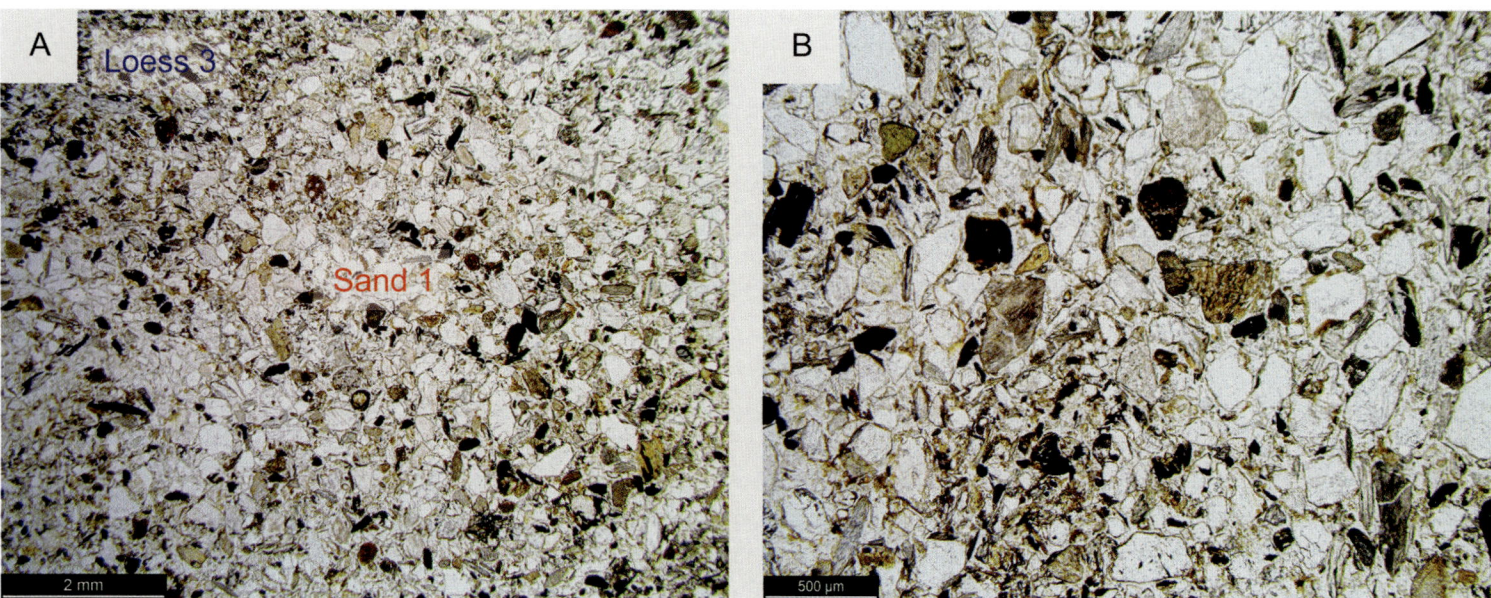

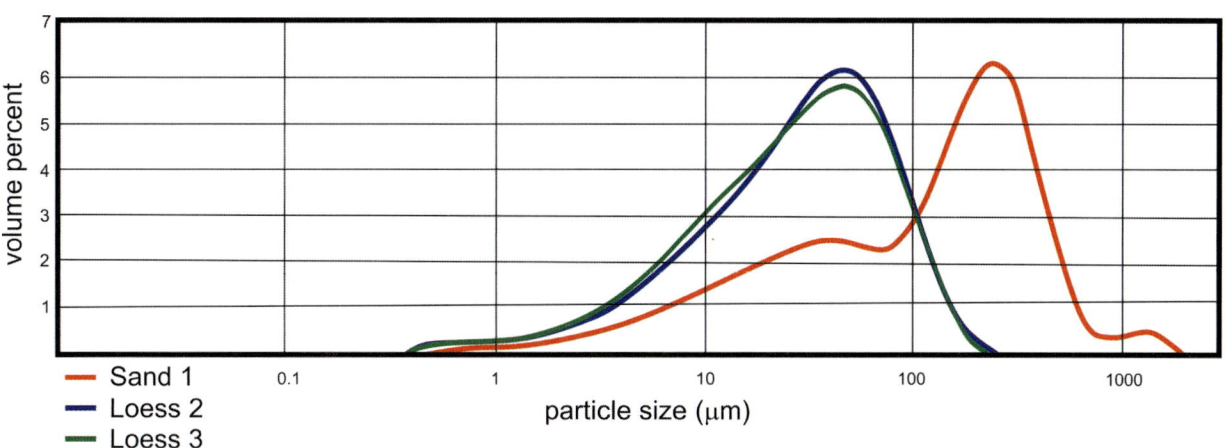

FIGURE 8.6. Comparison of volume percent of particle size for Sand 1, Loess 2, and Loess 3.

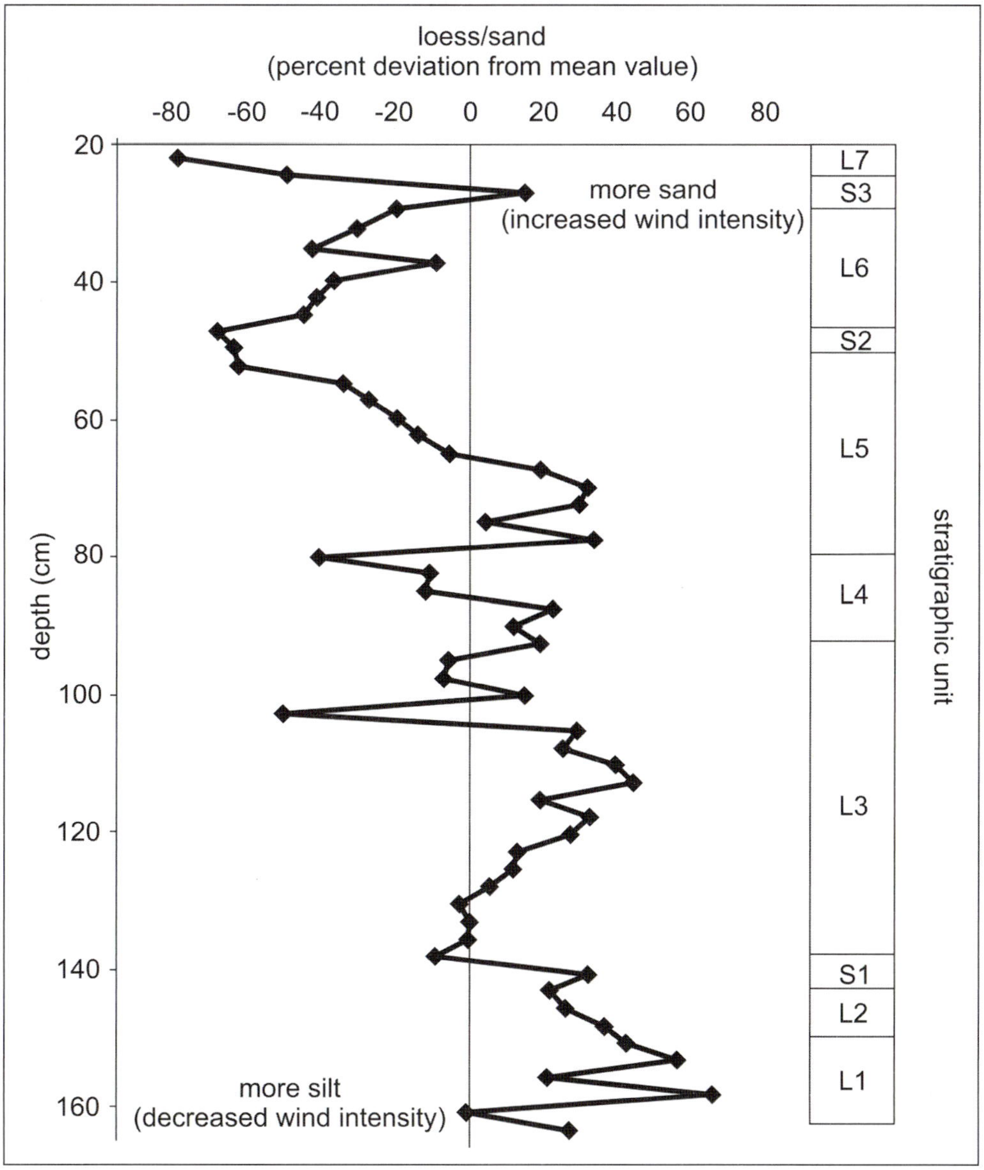

FIGURE 8.7. Wind intensity shown by comparing the deviation of volume percent very fine sand versus volume percent loess (medium and coarse silt) from the mean. Greater proportions of sand correspond to higher average wind velocities than do greater proportions of silt.

in figure 8.7, the deviation of percent of sand versus loess (i.e., medium and coarse silt) from the mean of loess and sand indicates conditions when more sand was being deposited than silt—in other words, periods when wind intensity was greater to entrain coarse sand clasts. The vicinity of the Dry Creek site experienced strong winds and perhaps significant storminess through the late Pleistocene and into the earliest Holocene (i.e., Loess 1 through the middle of Loess 3). After 10,000 cal yr BP (upper Loess 3), the pattern changed, with rapid, wide-swinging shifts in wind intensity (into Loess 5). Finally, during deposition of the upper half of the sediment profile in the late Holocene, there was a period of relative stability with lower wind intensity.

Micromorphology

Oriented micromorphological samples were taken from the lower strata (i.e., Loess 2, Sand 1, and Loess 3) containing and bracketing Components I and II. Thin sections of these samples provide details of soil micromorphology, microstratigraphic contacts, and mineralogy, together providing a characterization of the lower profile—not just soil development but also stratigraphic integrity (figures 8.5, 8.8–8.9). Examination of the thin sections indicates Loess 2 and Loess 3 are nearly identical in character. Their primary constituents are subangular to subrounded quartz silt and sand, platy muscovite that is generally coarser in size than the quartz, and highly weathered metamorphic rock fragments. Chlorite, biotite, and various heavy minerals are also present. Siliceous microdebitage is present in both units, associated with the cultural deposits. Paleosol 1 is primarily recognizable by accumulation of organic matter (mostly charcoal) and iron oxides. Mineral grains in the paleosol are no more or less chemically altered than in the parent material, and pedogenic fabrics are lacking. Sand 1 is composed of subangular to subrounded fine sand with little silt between the grains. Mineralogically, this stratum appears to have slightly more lithic fragments and heavy minerals than in the loess units, though quartz and muscovite are still the primary minerals in both types of deposits (figure 8.5).

Qualitative mineralogical analyses indicate the loess units at Dry Creek are extremely uniform. Quartz is the dominant mineral present, followed by muscovite (figure 8.9). Lithic fragments (generally mudrock and schist, as well as occasional granitic fragments) are also abundant. Visually identified trace minerals include rutile, chlorite, potassium and plagioclase feldspar, and biotite. Kaolinite is also present in minor abundance in the clay-size fraction.

FIGURE 8.8. Photomicrographs of Loess 3 and Paleosols 1 and 2, all taken in plane polarized light: (A) Paleosol 1 occurs as thin (mm-to-cm scale) discrete bands of accumulated organic matter and iron oxides in Loess 3; (B) accumulated charcoal (black) in Paleosol 1 and pellet-shaped iron oxides (red-brown), indicating the presence of soil microorganisms during pedogenesis; (C) charcoal in Paleosol 2 (note the largely unaltered nature of the framework grains of muscovite, quartz, and chlorite in the paleosol and general lack of weathering); (D) iron oxide nodules in Paleosol 1, indicating variable drainage during soil formation; (E) Paleosol 2, which occurs as a broad, less-discrete accumulation of organic matter at the top of Loess 3; (F) large, well-preserved charcoal fragments in Paleosol 2; (G–H) pellet-shaped iron oxide accumulations in Paleosol 2, indicating biological activity during soil formation (note the lack of nonpelletoid nodules as compared with Paleosol 1, suggesting that Paleosol 2 was better drained during its formation than Paleosol 1).

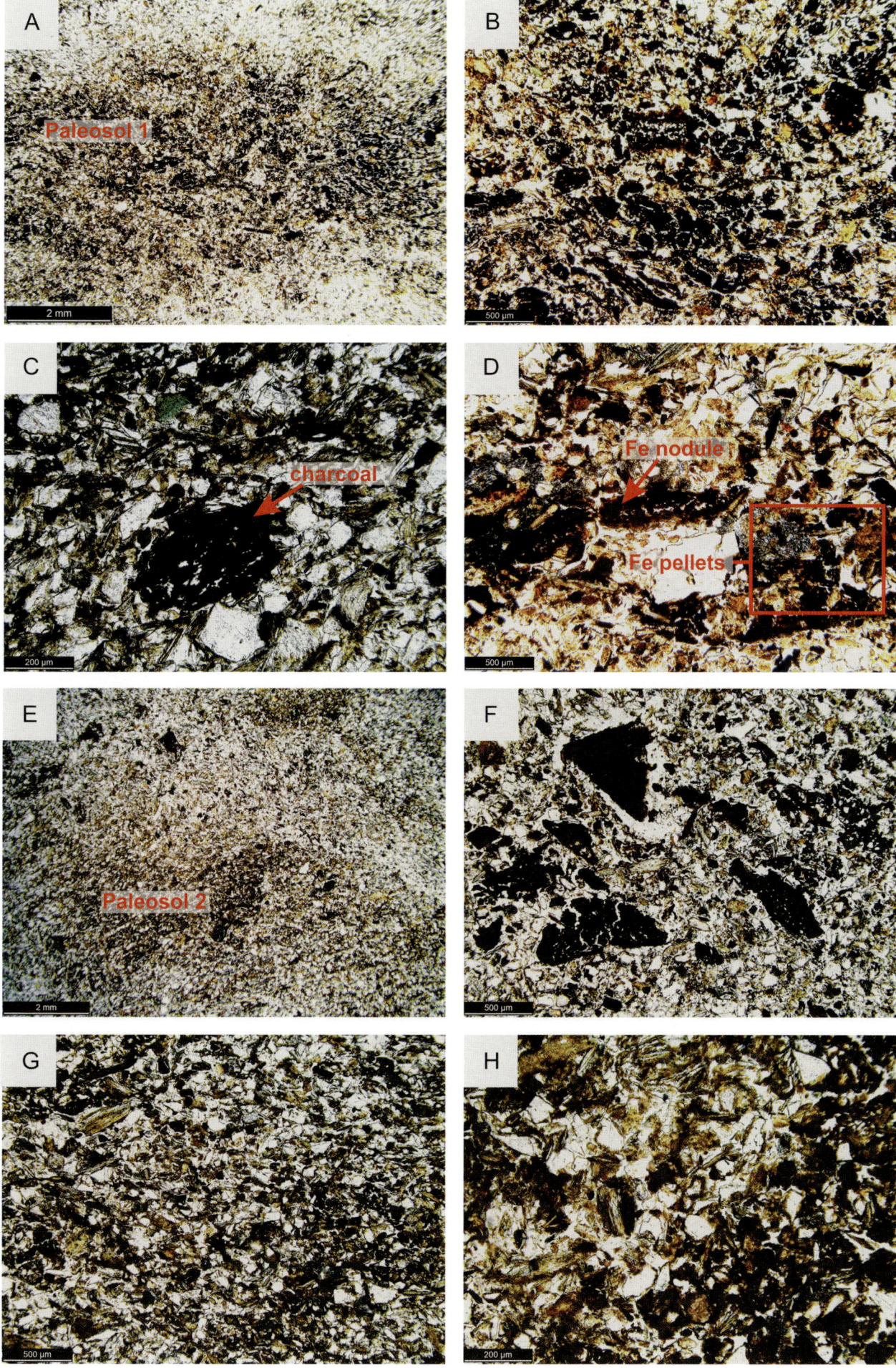

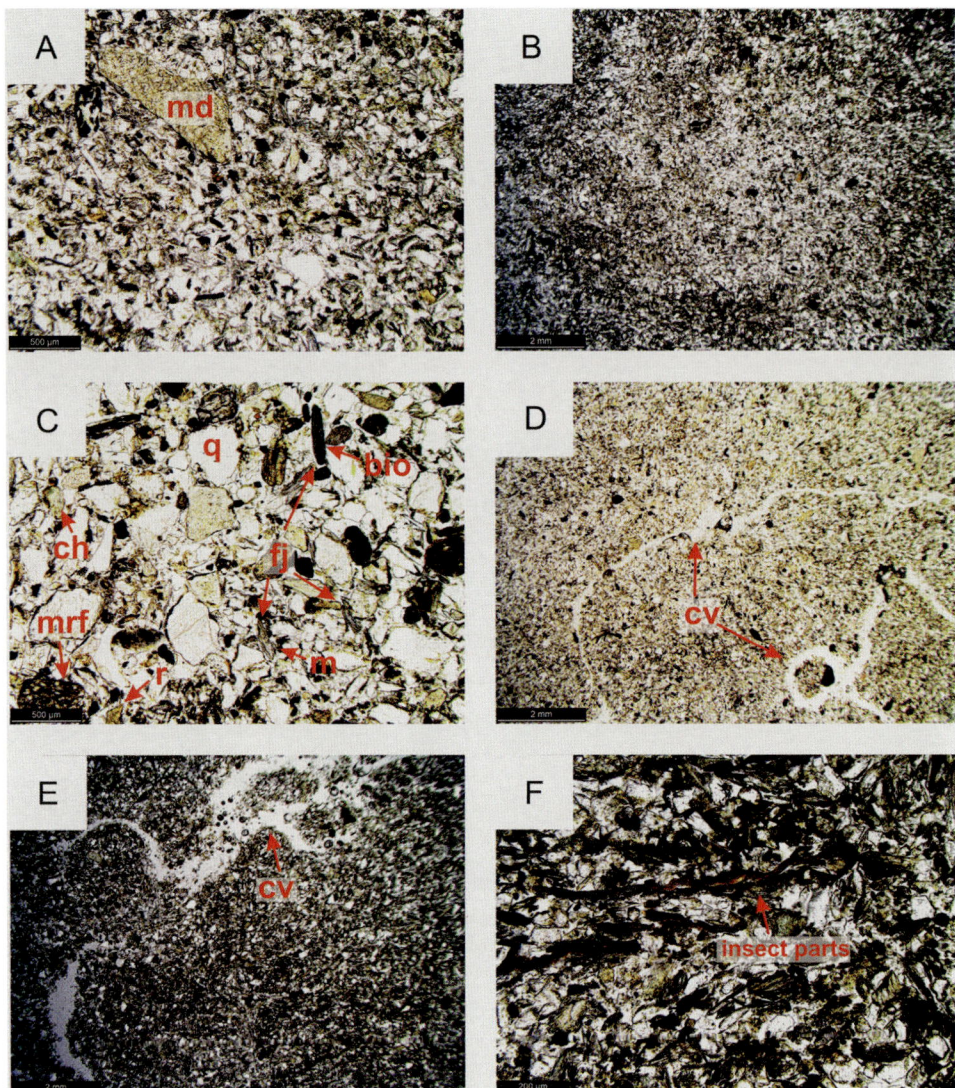

FIGURE 8.9. Photomicrographs of Loess 2, Sand 1, and Loess 3, all taken in plane polarized light: (A) nonvertical lithic microdebitage (md) surrounded by silty matrix that has not been cryogenically disturbed in Loess 2; (B) Loess 3 silt showing lack of cryogenic fabrics; (C) frost-jacked (fj) sand grains in Sand 1 as well as mineralogical identification of quartz (q), muscovite (m), chlorite (ch), metamorphic rock fragment (mrf), biotite (bio), and rutile (r); (D) channel voids (cv) caused by soil mesofauna in Loess 3; (E) channel voids caused by soil mesofauna in Loess 2; (F) preserved insect parts in Paleosol 2 (Loess 3).

These minerals reflect the metamorphic bedrock of the nearby Alaska Range and braided stream beds, like the Nenana River, flowing from glaciers in the range and are typical of interior Alaska (Muhs and Budahn 2006).

Soil Geochemistry

The major-element bulk chemistry of the loess packages shows little variation with depth (figure 8.10); however, the sand layers appear to have slightly different chemistry than the loess

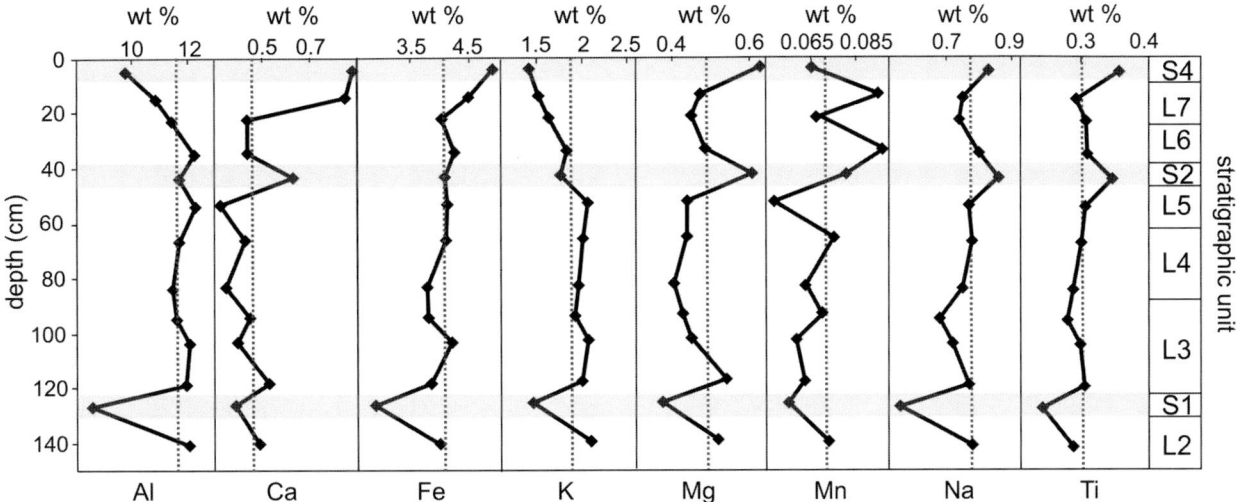

FIGURE 8.10. Weight percent diagram of bulk geochemistry, showing major elements by stratum.

units, suggesting they might have a slightly different source or that, to some extent, the mineralogical composition of the samples is controlled by particle size. Sand 1 is generally depleted in major element cations as compared to the loesses and other sands. Sands 2 and 4 are more similar to one another in composition than either is to Sand 1. The loess units are quite similar to one another compositionally. This trend is particularly highlighted when observing weight percent of Ti, a characteristically immobile trace element or proxy for sediment source. Cold mean annual temperature, relatively low mean annual precipitation, and frequent eolian deposition work in conjunction to ensure that the sediment has seen minimal chemical alteration.

The paleosols in the profile are mostly characterized by the accumulation of organic matter, rather than the alteration of primary minerals, and the bulk chemistry of the site reflects this pattern. Organic carbon is abundant, in excess of 2.5 percent by weight at the surface and dropping to approximately 0.5 percent at depth (figure 8.11). Organic carbon is both more abundant and more variable in concentration at the top of the profile, reflecting the presence of numerous thin organic horizons in Loesses 5, 6, and 7, a pattern observed in the field and supported by the decrease in wind intensity documented in other analyses (see the previous section). This serves as a contrast to the thicker single paleosols lower in the profile and might signify changes in landscape stability over time, with stable surfaces occurring infrequently low in the section but more often toward the top.

Although the character of the paleosols is difficult to determine from bulk chemistry, their colloidal chemistry is perhaps a little more telling (table 8.1). Extractable Fe and Al values represent the amount of mobilized Fe and Al in soil water, rather than the Fe and Al bound in minerals that ICP measures. Paleosols 4b and 4a have relatively high extractable Fe and Al values and an acidic pH, expected of modern forest soils, while Paleosols 3, 2, and 1 each have slightly decreasing values. All these values, however, are relatively low, indicating weak

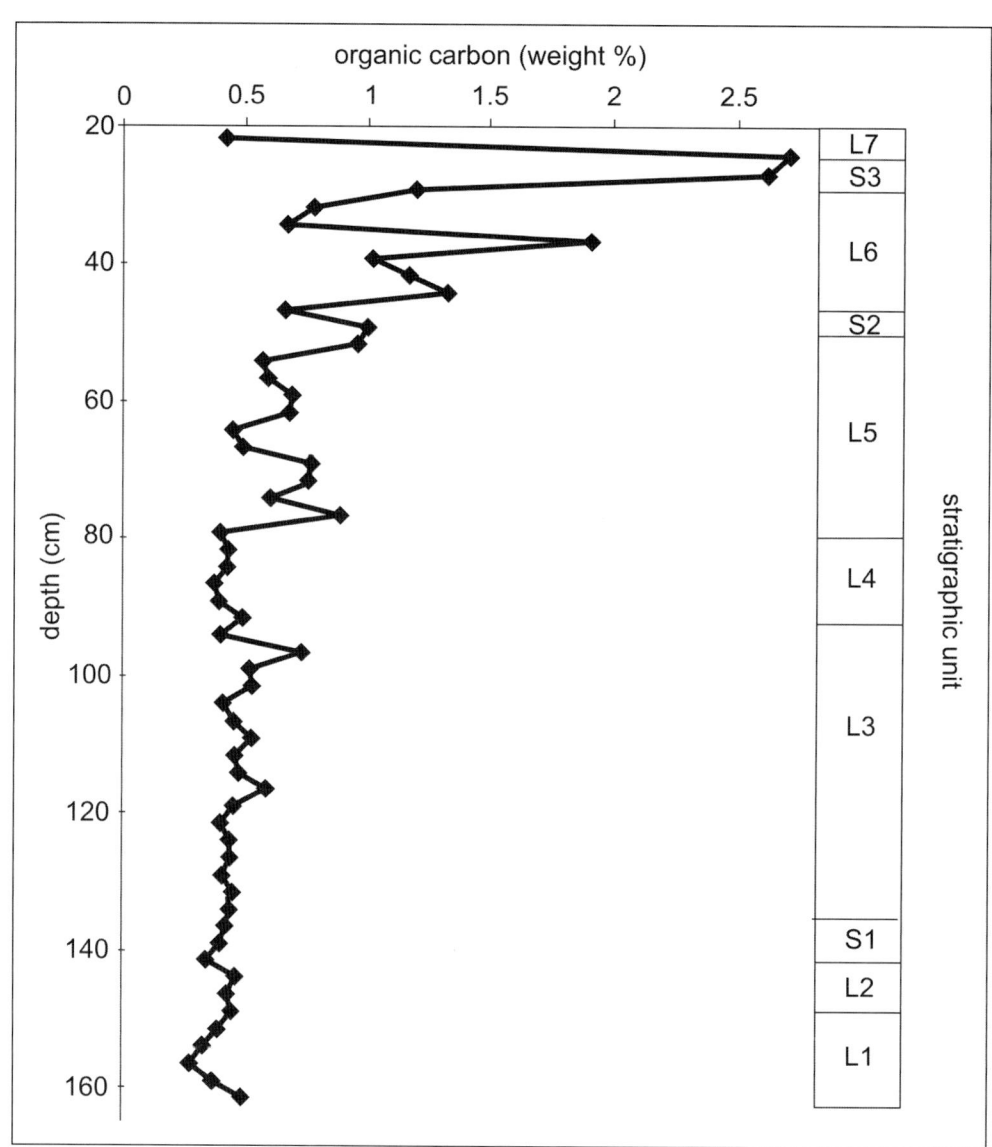

FIGURE 8.11. Weight percent diagram of organic carbon by stratum.

TABLE 8.1. Soil characterization data

Paleosol	pH	AmOx Fe (%)	AmOx Al (%)
4b	4.9	1.46	0.19
4a	4.9	1.41	0.20
3	5.1	0.92	0.12
2	5.7	0.77	0.10
1	6.0	0.53	0.08

soil development at this location. Given the soil-morphology patterns described in the field, as well as palynological records in the region that indicate poplar and spruce trees migrated into the region between about 11,000 and 10,000 cal yr BP (Ager 1983; Bigelow and Powers 2001; Hu et al. 1993; Schweger 1981; Tinner et al. 2006), we contend that Paleosol 2 formed during a transitional period as the boreal-forest biome was emerging in the Nenana Valley.

Soil Micromorphology and Formation of Loess 2, Sand 1, and Loess 3

Examination of microstratigraphic features preserved in micromorphological samples from the site's lower stratigraphy (Loess 2, Sand 1, and Loess 3) provides important information about the geological integrity of the site's older deposits. According to Thorson (1990), repeated freeze-thaw cycles and accumulation of ice lenses in soil voids lead to volumetric changes, causing shifts in both individual grains within a soil profile and soil material as a whole. Even more troublesome for northern sites like Dry Creek is the increased susceptibility of soils to the formation of ice lenses, especially when the soils are silty to sandy loams in texture (Van Vliet-Lanoe 2010). Our micromorphological observations, however, show that cryogenic activity in the lower portion of the profile was minimal (figure 8.9), despite the sediment package falling into this highly susceptible grain-size category. Microscale features typical of frost-affected and solifluctic soils are virtually absent from Loess 2 and Loess 3 (figure 8.9A, 8.9B). Moreover, a small number of elongated grains with nonhorizontal orientations and incipient silt capping were observed on a few coarser grains, and these were restricted mostly to Sand 1 (figure 8.9C). Patterns of cryoturbation are minor and do not constitute the degree of disturbance typifying translocation of materials from one stratum to another.

Thorson (2006) proposed bioturbation as the primary mechanism for artifact displacement at Dry Creek. In this scenario, subsurface voids created by burrowing fauna collapse in on themselves, translocating sediment, soil material, and associated artifacts down through the profile (Johnson and Watson-Stegner 1990). Relict voids measuring 0.1–0.2 mm in diameter and created by mesofauna were observed in most thin sections taken from Loess 3 (figure 8.9D) and in all thin sections from Loess 2 (figure 8.9E). These voids, however, are not collapsed or infilled. Small accumulations of iron oxides with fecal morphology are present in both Paleosols 1 and 2 but do not line void spaces, and preserved insect parts were observed in Paleosol 2 (figure 8.9F). Certainly there has been some biological activity within the lowermost loess units at Dry Creek; however, these features were not associated with microdebitage, and burrows with meniscate backfilling were not observed. Contrary to Thorson's (2006) suggestion, mesofaunal bioturbation within this portion of the soil profile did not cause the downward movement of lithic debitage.

Descriptions and Dating of Cultural Features

Three hearth features were observed in Components I and II and excavated during 2011 (figure 8.12). No features were identified in Component IV. In Block A, we encountered, described, and collected Feature 11.01, stratigraphically isolated in Loess 3 and associated with Paleosol 1 and Component II (figure 8.13). The feature's contents followed the same solifluction-fold pattern of the paleosol in this block and consisted of black-colored (10YR 2/1), charred, greasy sandy-loam sediment; charred and calcined bone fragments; and flakes

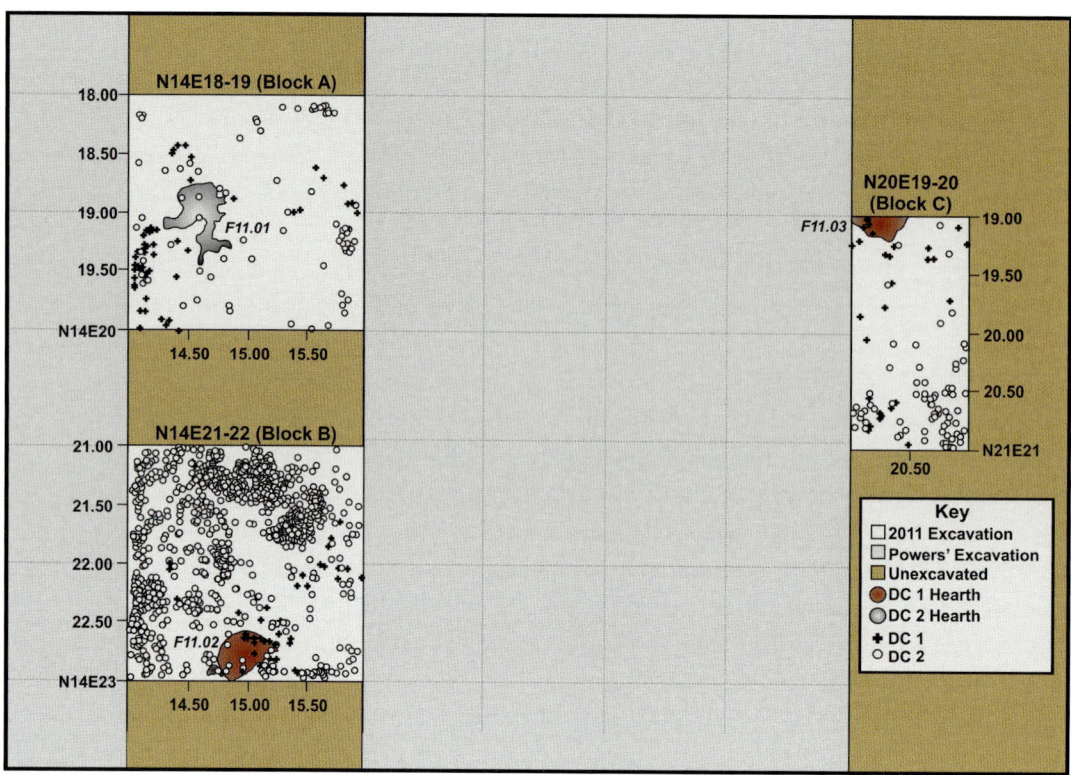

FIGURE 8.12. Map of 2011 excavation showing locations of blocks relative to Powers's 1970s excavation and horizontal distributions of cultural materials.

with evidence of heat alteration (i.e., discoloration, crazing). The feature measured 80 × 65 cm in plan view and 1–2 cm in thickness. Despite being disfigured from solifluction and no longer a circular basin, this feature was discrete in nature, isolated in a well-circumscribed area containing wood charcoal, charred bone, and a few fire-affected artifacts. Additionally, it was surrounded by numerous materials unaffected by burning, including lithic artifacts, faunal remains, and uncharred paleosol and loess. These parameters suggest the feature resulted from intentional, controlled burning by humans and represents a Component II hearth. Two AMS dates, obtained on two pieces of *Salix* sp. wood charcoal with three-point provenience and found within the feature, yielded radiocarbon dates of 9,480 ± 35 ^{14}C yr BP (PSU-5835/UCIAMS-135115) and 9,460 ± 50 ^{14}C yr BP (Beta-315410; table 8.2). Together, these dates produced a weighted mean of 9,473 ± 29 ^{14}C yr BP (χ^2 = 0.107; df = 1; T_{crit} = 3.841) and provided chronological control for Component II in our excavation of artifact Cluster G.

In Block B, we encountered a hearth feature isolated in Loess 2 and associated with Component I artifacts and bones (figures 8.12, 8.14). This hearth, F11.02, measured 50 × 35 cm in plan view, formed a 6-cm-thick basin after being excavated, and had a hemispheric shape. Its western and southern margins were distinctly circular, but its northern and eastern margins were truncated by an ancient krotovina. We are confident the burrow is ancient because it was

FIGURE 8.13. Multiple views of Feature 11.01 in Component II, interpreted to represent a hearth disturbed by solifluction. Top, illustration (left) and photograph (right) of feature in plan view; bottom, feature in profile after excavation of its northern half.

infilled with sediment that became weathered and thus mottled, similar to the surrounding intact loess. Hearth fill consisted of ashy, charcoal-rich silt ranging in color from very dark-brown (10YR 2/2) to dark reddish-brown (5YR 3/2) and reddish-brown (5YR 4/4) as well as several large (≥ 1 cm diameter) pieces of wood charcoal. It also contained degraded bone "powder" and a biface thinning flake. With charred pieces within and uncharred cultural

TABLE 8.2. New radiocarbon dates from Dry Creek

Stratum	Component	Square	Feature	Material	Lab Number	Age Estimate[a,b]	Age Calibrated[c]
Loess 3	II	N14E19	11.01	Hearth charcoal, *Salix* sp.	UCIAMS-135115	9,480 ± 35	10,580–11,070
Loess 3	II	N14E19	11.01	Hearth charcoal, *Salix* sp.	Beta-315410	9,460 ± 40	10,560–11,070
Loess 2	I	N14-15 E21-22	11.02	Hearth charcoal, *Salix* sp.	UCIAMS-135114	11,510 ± 40	13,270–13,450
Loess 2	I	N14-15 E21-22	11.02	Hearth charcoal, *Salix* sp.	Beta-315411	11,530 ± 50	13,280–13,460
Loess 2	I	N20E19	11.03	Hearth charcoal, *Salix* sp.	UCIAMS-135113	11,580 ± 40	13,300–13,490
Loess 2	I	N20E19	11.03	Hearth charcoal, *Salix* sp.	UCIAMS-135112	11,635 ± 40	13,380–13,570

[a]Age in radiocarbon years before present (^{14}C yr BP); reported with 1σ (standard deviation).
[b]Dates corrected for isotopic fractionation following Stuiver and Polach (1977).
[c]Calibrated using OxCal v4.2.3 (Bronk Ramsey 2013); r:5 IntCal13 atmospheric curve (Reimer et al. 2013); 2σ range.

materials—including bone and lithics—surrounding it, the feature appeared well-bounded. Despite being truncated by a rodent burrow, this basin-like feature was largely preserved. Two pieces of *Salix* sp. wood charcoal collected within the hearth yielded dates of 11,510 ± 40 ^{14}C yr BP (PSU-5834/UCIAMS-135114) and 11,530 ± 50 ^{14}C yr BP (Beta-315411; table 8.2), respectively, together producing a weighted mean of 11,520 ± 28 ^{14}C yr BP (χ^2 = 0.125; df = 1; T_{crit} = 3.841) and providing an age estimate for Component I in this area of the excavation associated with the southeastern portion of artifact Cluster Z.

In Block C, we found another hearth feature, F11.03, also within Loess 2 and associated with Component I materials (figures 8.12, 8.15). This circular feature is still partially preserved in Dry Creek's deposits and visible in cross section in the western profile. We excavated a portion of it near the intersection of square N20E19 and the feature's easternmost edge. The exposed portion of the hearth measured 50 × 25 cm. It consisted of a concentration of ashy, charcoal-smeared silt ranging in color from black (5YR 2.5/1) to dark reddish-brown (5YR 3/4) and reddish-brown (5YR 5/4). Charcoal pieces, 44 pieces of burnt and calcined bone and ungulate (*Ovis dalli*) teeth, and a few lithics were found in the feature, while 21 pieces of unburnt bone, teeth, and lithics were found surrounding the feature. Two wood charcoal pieces from the feature, identified as *Salix* sp., yielded dates of 11,580 ± 40 ^{14}C yr BP (PSU-5833/UCIAMS-135113) and 11,635 ± 40 ^{14}C yr BP (PSU-5832/UCIAMS-135112; table 8.2), respectively, together producing a weighted mean of 11,608 ± 28 ^{14}C yr BP (χ^2 = 0.945; df = 1; T_{crit} = 3.841). This provides an age estimate for Component I in this area of the excavation associated with a northwestern extension of artifact Cluster Z.

Dates from the three features were placed in a stratigraphic model in OxCal for calibration and to better estimate the difference in age between Components I and II (figure 8.16A). Results of unmodeled and modeled (prior and posterior) calibrated ranges are reported in table 8.3. For each hearth, the paired dates were averaged using the *combine* command. Component I was modeled as a *phase*, with Features 11.02 and 11.03 constrained by *boundaries* approximating the beginning and end of activities during Component I. Feature 11.01 in Component II was modeled outside of the sequence as a stand-alone combined date.

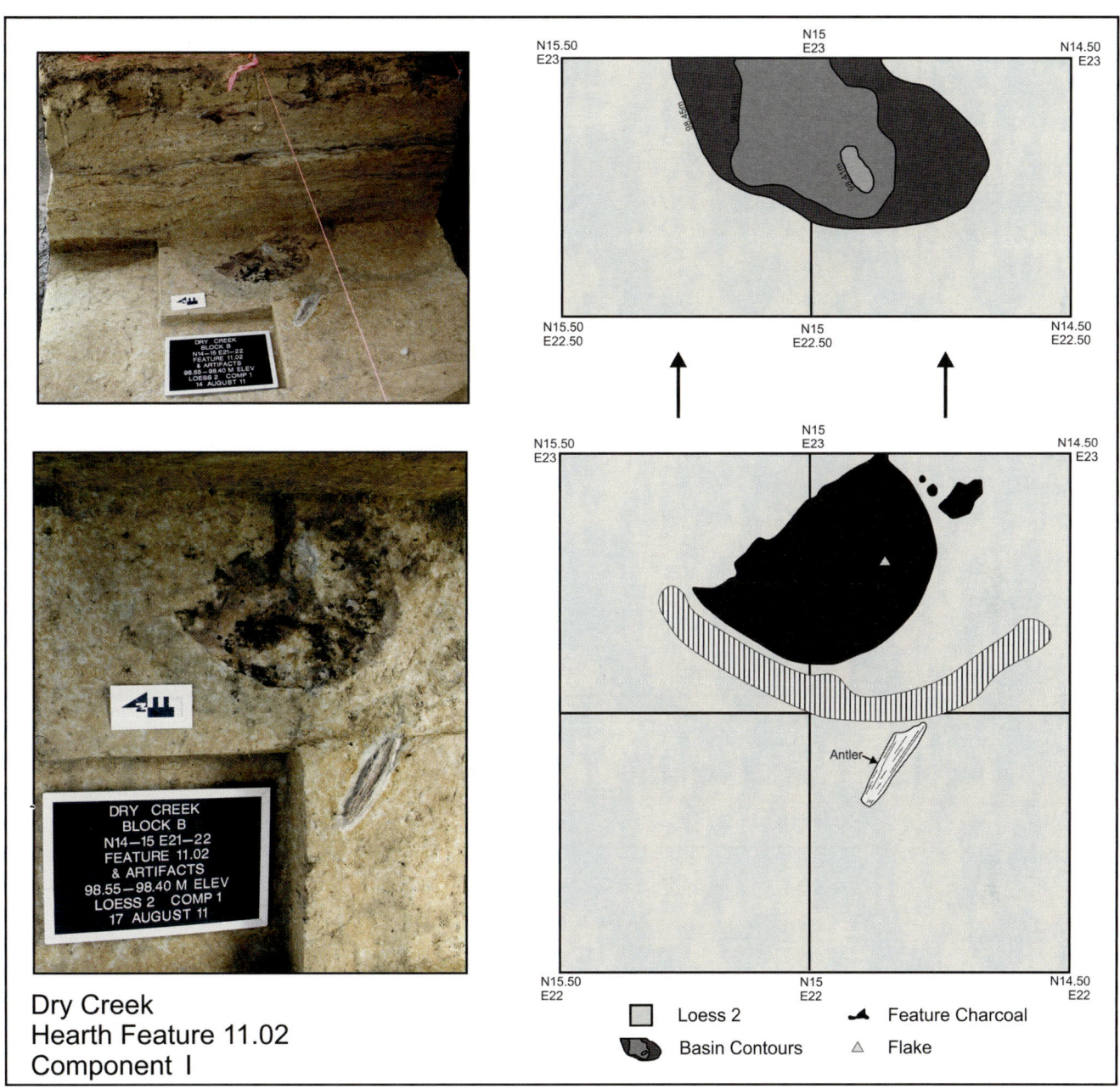

FIGURE 8.14. Feature 11.02 from Component I, interpreted to represent a hearth. Left, plan view photographs of feature; top right, contour map of feature basin upon complete excavation; bottom right, map of feature in plan view.

Modeled 2σ-calibrated ages for Features 11.03 and 11.02 are 13,365–13,485 cal yr BP and 13,307–13,441 cal yr BP, respectively. The boundary for the *end* of Component I activity is poorly constrained, ranging from 12,735 to 13,440 cal yr BP; this represents the whole range of a discontinuous distribution wherein 87.9 percent of the probability density covers

FIGURE 8.15. Feature 11.03 from Component I, interpreted to represent a hearth. Top left, photograph of feature in plan view; bottom left, map of feature in plan view; right, stratigraphic profile (west wall) with Feature 11.03 noted in the profile.

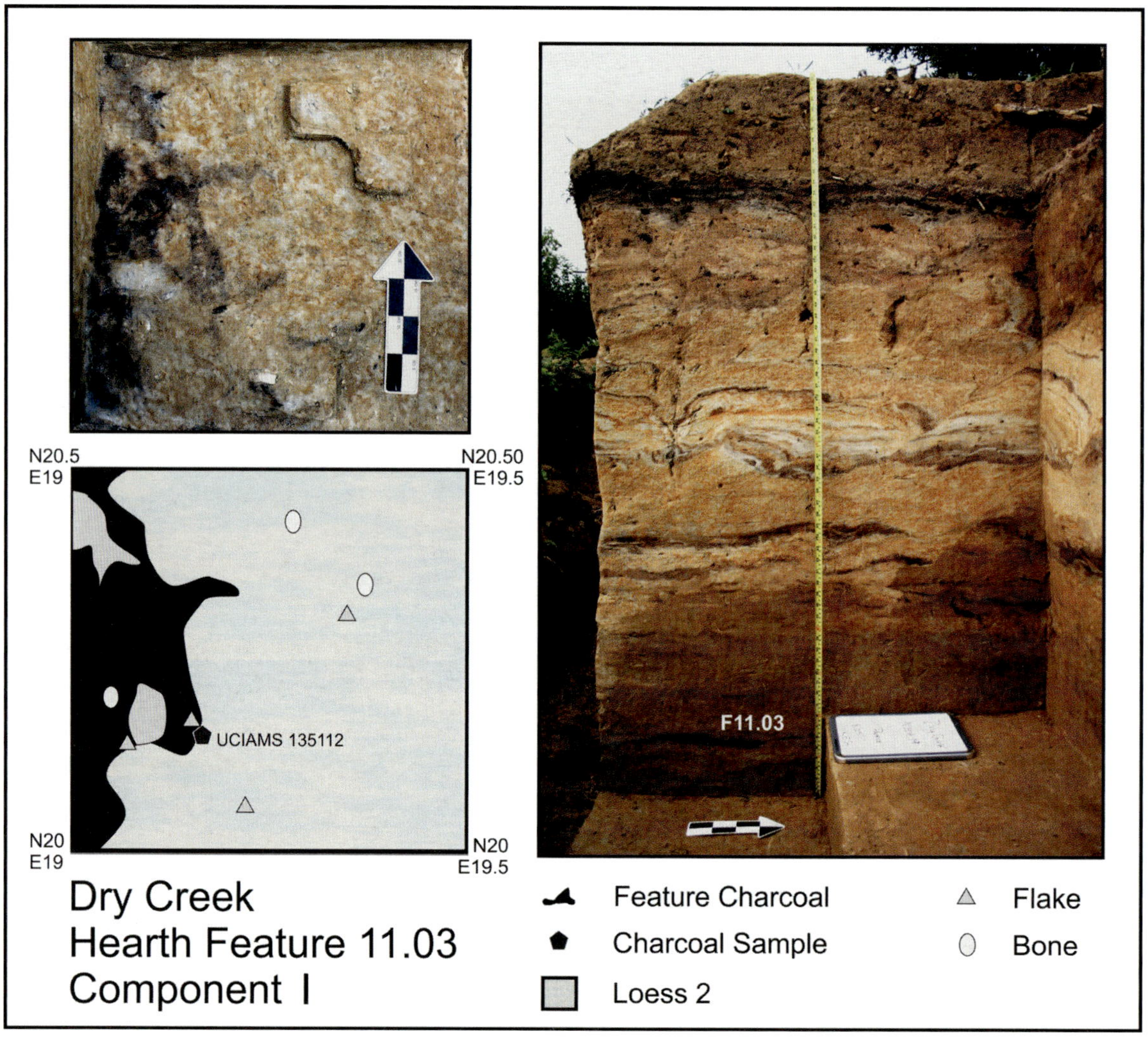

FIGURE 8.16. (A) Modeled calibrations of new radiocarbon dates from Dry Creek Components I and II in OxCal v.4.2.3 (Bronk Ramsey 2009, 2013) using the IntCal13 curve (Reimer et al. 2013). Paired dates are combined as weighted averages within a sequence for Component I and as a stand-alone weighted average for Component II. The shaded bar indicates the span of the Younger Dryas. (B) Estimated length of the hiatus between Components I and II, calculated using the *difference* command to compare the ages of Feature 11.02 and Feature 11.01, and the estimated end of Component I and Feature 11.01.

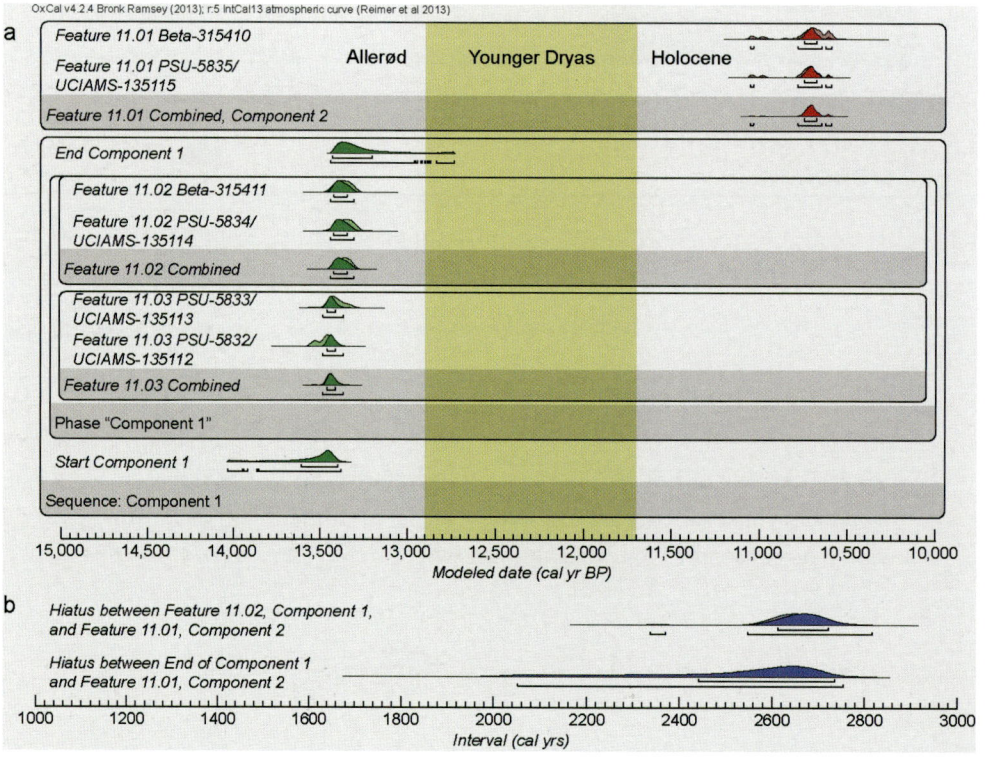

TABLE 8.3. Modeling calibrations of new radiocarbon dates from Dry Creek Components I and II

Lab Number	Square	Feature	¹⁴C Age yr BP	Unmodeled 2σ cal yr BP	%	Modeled 2σ cal yr BP	%
Loess 3, Component II							
PSU-5835/ UCIAMS-135115	N14E19	11.01	9,480 ± 35	10,960–11,070	13.5		
				10,850–10,860	0.4		
				10,580–10,800	81.5		
Beta-315410	N14E19	11.01	9,460 ± 40	10,950–11,070	12.1		
				10,840–10,870	0.9		
				10,560–10,810	82.4		
			Feature 11.01 Combined			11,030–11,060	1.6
						10,980–10,990	0.2
						10,640–10,790	88.1
						10,590–10,630	5.5
Loess 2, Component I							
			End of Component I			12,930–13,440	87.9
						12,730–12,880	7.5
PSU-5834/ UCIAMS-135114	N14-15 E21-22	11.02	11,510 ± 40	13,270–13,450	95.4		
Beta-315411	N14-15 E21-22	11.02	11,530 ± 50	13,280–13,460	95.4		
			Feature 11.02 Combined			13,300–13,450	95.4
PSU-5833/ UCIAMS-135113	N20E19	11.03	11,580 ± 40	13,300–13,490	95.4		
PSU-5832/ UCIAMS-135112	N20E19	11.03	11,635 ± 40	13,380–13,570	95.4		
			Feature 11.03 Combined			13,360–13,490	95.4
			Beginning of Component I			13,910–14,030	6.9
						13,880–13,900	0.5
						13,370–13,860	88.0
			Hiatus Estimates				
			Between Features 11.02 and 11.01			2,330–2,370 cal year	1.2
						2,540–2,820 cal year	94.2
			Between End of Component I and Feature 11.01			2,050–2,760 cal year	95.4

12,930–13,440 cal yr BP. Feature 11.01 in Component II has a modeled 2σ-calibrated range of 10,590–11,060 cal yr BP, also a discontinuous distribution, with 88.1 percent probability density from 10,645 to 10,785 cal yr BP. Agreement indices for individual and combined dates and the overall model are all above 100 percent, indicating good agreement between the model and the dates themselves.

Components I and II are clearly separate cultural occupations, and two methods were used to estimate the length of the hiatus between the occupations. Using the *difference* command in OxCal, we estimated the length of time elapsed between Feature 11.02 in Component I and Feature 11.01 in Component II as 2,335–2,820 calendar years at 2σ, with a weighted mean of 2,655 ± 85 calendar years. A more conservative estimate of the gap takes the difference of the boundary estimate for the end of Component I and Feature 11.01, which is 2,050–2,755 calendar years at 2σ, with a weighted mean of 2,510 ± 195 calendar years (figure 8.16B). The hiatus between Components I and II is therefore at least 2,000 calendar years and possibly as many as 2,800 calendar years in the areas of our excavation.

Lithic Artifact Assemblages

During our excavations, we found a total of 3,886 artifacts from three cultural occupations (Components I, II, and IV), confirming the site's known cultural stratigraphy (chapter 3). The 2011 assemblage constitutes a 9.8 percent increase in the number of artifacts now reported from Dry Creek. We found three artifacts in the 1970s backdirt in the vicinity of Blocks A and B, which we removed to expose intact deposits for excavation. The remainder of the 2011 assemblage, 3,883 pieces, came from the excavation of intact deposits. Of these, three-point provenience was recorded for 1,984 artifacts, or 51 percent of the assemblage (figure 8.12), compared to 26 percent during Powers' excavations. For Component I, 251 artifacts were recovered, and for Component II, 3,629 (table 8.4). Only three artifacts were found in Component IV, including one rhyolite flake, one chalcedony resharpening chip, and one basalt chopper. The very light expression of Component IV is consistent with original investigations. Due to specific site-formation goals of our project and the incredibly small artifact count from Component IV, we focus on Components I and II.

In addition to examining site stratigraphy and micromorphology, another means to test the geoarchaeological integrity of Components I and II was through examination of the lithic artifact assemblage collected during our controlled field investigations. If Component I represents down-drifted Component II artifacts, we expected to find no significant differences between components in behavioral site-formation variables (i.e., primary versus secondary technological activities and raw material procurement and selection behaviors) but significant differences between components in natural site-formation variables (i.e., artifact-size sorting and artifact plunge).

Lithic Technological Organization and Behavioral Site Formation

Core and tool numbers are quite low. The only core found in Component I is a bipolar core manufactured on a flake. Component II cores include three flake cores, three bipolar cores, and eight microblade cores (figure 8.17). Component I tools include one retouched blade, one retouched flake, and two end scrapers. Component II tools include one side scraper, one burin, one biface fragment, three hammerstones, and nine marginally retouched unifacial tools. In Component I, primary-reduction debitage (i.e., cortical spalls, flakes, bladelike flakes, and bipolar flakes) comprises nearly half (46.3 percent) of the debitage assemblage and likely resulted from reduction of simple flake cores. Secondary-reduction activities focused on core trimming and tool resharpening, with just over half (53.7 percent) of the debitage being resharpening chips or biface thinning flakes. No

TABLE 8.4. Artifact types by component from the 2011 excavations of Dry Creek

Artifact Typology	n	%	I	II	IV	Powers's Backdirt
Cores						
1. Flake cores	3	< 0.1		3 (0.1%)		
2. Bipolar cores	4	0.1	1 (0.4%)	3 (0.1%)		
3. Microblade cores	8	0.2		8 (0.2%)		
Subtotal	15	0.4	1 (0.4%)	14 (0.4%)		
Debitage						
1. Angular shatter	87	2.2	7 (2.8%)	80 (2.2%)		
2. Cortical spalls	71	1.8	36 (14.3%)	34 (0.9%)		1 (33.3%)
3. Flakes	797	20.5	66 (26.3%)	729 (20.1%)	1 (33.3%)	1 (33.3%)
4. Blade-like flakes	10	0.3	2 (0.8%)	8 (0.3%)		
5. Bipolar flakes	15	0.4	2 (0.8%)	13 (0.4%)		
6. Resharpening chips	2,167	55.8	118 (47.0%)	2,048 (56.4%)	1 (33.3%)	
7. Biface thinning flakes	158	4.1	14 (5.6%)	143 (3.9%)		1 (33.3%)
8. Burin spalls	30	0.8		30 (0.9%)		
9. Microblades	491	12.6		491 (13.5%)		
10. Microblade technical spalls	24	0.6		24 (0.6%)		
11. Blade technical spalls	1	< 0.1	1 (0.4%)			
Subtotal	3,851	99.1	246 (98.0%)	3,600 (99.2%)	2 (66.7%)	3 (100.0%)
Tools						
12. Side scrapers	1	< 0.1		1 (< 0.1%)		
13. End scrapers	1	< 0.1	2 (0.8%)			
14. Angle burins	1	< 0.1		1 (< 0.1%)		
15. Retouched blades	1	< 0.1	1 (0.4%)			
16. Retouched microblades	3	< 0.1		3 (0.1%)		
17. Retouched flakes	7	0.2	1 (0.4%)	6 (0.2%)		
18. Bifaces	1	< 0.1		1 (< 0.1%)		
19. Choppers	1	< 0.1			1 (33.3%)	
20. Hammerstones	3	< 0.1		3 (< 0.1%)		
Subtotal	20	0.5	4 (1.6%)	15 (0.4%)	1 (33.3%)	
Component totals			251 (100.0%)	3,629 (100.0%)	3 (100.0%)	3 (100.0%)
Assemblage totals	3,886	100.0%	251 (6.5%)	3,629 (93.4%)	3 (< 0.1%)	3 (< 0.1%)

microblades are present in the Component I assemblage. In contrast, Component II is dominated by secondary-reduction debitage (61.7 percent), including resharpening chips, biface thinning flakes, and burin spalls, with only 38.7 percent of debitage representing primary-reduction activities. Chi-square analysis, comparing primary- and secondary-reduction

FIGURE 8.17. Sample of artifacts from Components I (A–F) and II (G–N) of the 2011 excavations, including end scrapers (A, D), proximal blade fragment (B), retouched blade fragment (C), blade core tablet (E), retouched flake (F), wedge-shaped microblade cores (G–K), wedge-shaped microblade core tablet (L), wedge-shaped microblade core fragment (M), and wedge-shaped microblade core preform (N).

TABLE 8.5. Lithic variables reflecting behavioral site formation

Reduction Types[a]		Component I	Component II	Total
Primary	Count	114	1,377	1,491
	Expected Count	95.4	1,395.6	1,491.0
	% of Total	3.0	35.8	38.8
Secondary	Count	132	2,222	2,354
	Expected Count	150.6	2,203.4	2,354.0
	% of Total	3.4	57.8	61.2%
Total	Count	246	3,599	3,845
	Expected Count	246.0	3,599.0	3,845.0
	% of Total	6.4	93.6	100.0
Presence of Cortex[b]				
Present	Count	215	3,581	3,796
	Expected Count	245.6	3,550.4	3,796.0
	% of Total	5.5	92.3	97.8
Absent	Count	36	48	84
	Expected Count	5.4	78.6	84.0
	% of Total	0.9	1.2	2.2
Total	Count	251	3,629	3,880
	Expected Count	251.0	3,629.0	3,880.0
	% of Total	6.5	93.5	100.0
Raw Materials[c]				
CCS	Count	51	3,001	3,052
	Expected Count	197.5	2,854.5	3,052.0
	% of Total	1.3	77.7	79.0
Quartzite	Count	91	347	438
	Expected Count	28.3	409.7	438.0
	% of Total	2.4	9.0	11.3
Basalt	Count	31	39	70
	Expected Count	4.5	65.5	70.0
	% of Total	0.8	1.0	1.8
Rhyolite	Count	77	227	304
	Expected Count	19.7	284.3	304.0
	% of Total	2.0	5.9	7.9
Total	Count	250	3,614	3,864
	Expected Count	250.0	3,614.0	3,864.0
	% of Total	6.5	93.5	100.0

[a] $\chi^2 = 6.333$; df = 1; $p < 0.02$. No cells have expected counts of less than 5. The minimum expected count is 95.39.

[b] $\chi^2 = 187.891$; df = 1; $p < 0.001$. No cells have expected counts of less than 5. The minimum expected count is 5.43.

[c] $\chi^2 = 608.383$; df = 3; $p < 0.001$. One (12.5 percent) cell has an expected count of less than 5. The minimum expected count is 4.53.

activities, indicates that the observed differences in technological activities between the components are significant ($\chi^2 = 6.333$; df = 1; $p < 0.02$; table 8.5).

Additionally, we examined presence and absence of cortex between component assemblages and found significantly more than expected Component I artifacts with cortex on their surfaces and more than expected pieces in Component II without cortex ($\chi^2 = 187.891$;

df = 1; p < 0.001; table 8.5). These results suggest a focus on primary-reduction activities in Component I and an alternative focus on secondary-reduction activities in Component II.

Seven different classes of lithic raw materials were identified in the 2011 assemblage. Of these, 23 individual raw-material types were observed and scored (table 8.6), including 4 types of rhyolite, 12 varieties of fine-grained cherts and chalcedonies collectively called cryptocrystalline silicates (CCS), 2 varieties of quartzite, 1 basalt, and 3 "other" raw materials (quartz, granite, and hematite) numbering so few they were combined together. Several raw-material types make up the Component I assemblage and include a mixture of quartzites (29 percent), rhyolites (30 percent), CCS (20 percent), and basalt (21 percent). In contrast, Component II artifacts are almost entirely manufactured on CCS (nearly 82 percent). Significantly, more quartzites, rhyolites, and basalts than expected occur in Component I, but more CCS than expected in Component II (χ^2 = 608.383; df = 3; p < 0.01; table 8.6). Nearly 70 percent of Component II was manufactured on a single CCS-type named caramel chalcedony, while in sharp contrast, no caramel chalcedony was found in Component I. If Component II artifacts moved down through the profile to form Component I or mixed with it, altering its character, then we would have found this raw material in Component I, yet this was not the case. Clear differences between the components in technological activities and raw-material selection support the original interpretation of two independent cultural components reflecting separate occupation events. Thorson's (2006) supposition that Component I represents down-drifted material from Component II is not supported.

Lithics and Natural Site Formation

If cryoturbation significantly affected the site, we should expect to see smaller clasts lower in the profile and larger clasts thrust up and therefore higher in the profile (Bowers et al. 1983; Johnson et al. 1977; Thorson 1990). If bioturbation significantly affected the profile, we should expect to find larger clasts near the bottom of the artifact-rich zone, as burrowing animals would have brought smaller clasts up and out of burrows and left larger clasts behind in lower positions (Hansen and Morris 1968; Johnson 1989). To test for these patterns, each artifact was assigned a ranked size value, with 1 being assigned to artifacts measuring < 1 cm in overall dimension, 2 to artifacts measuring 1–3 cm, 3 to artifacts measuring 3–5 cm, and 4 to artifacts measuring > 5 cm. Both assemblages contained mostly small-sized artifacts falling into size values 1 and 2. Less than 4 percent of artifacts from each component were larger than size value 2. No significant relationship (Mann-Whitney U = 444,020.0, P = 0.403) between components in size distributions indicates no size sorting between or within the components. Instead, both components appear to be in place in their primary positions.

Artifact plunge was scored on 1,114 in situ artifacts to better understand their vertical orientations. Because artifacts with plunges > 45° are more likely to have moved in the profile (Goebel et al. 2000; Johnson and Hansen 1974; Schweger 1985; Wood and Johnson 1978), we expected that if the majority of artifact plunges are > 45°, then there was significant cryoturbation and displacement of artifacts. We found the majority of artifact plunges for both assemblages to be < 45°, with data being skewed positively with means < 23° (figure 8.18). Most artifacts in both components were lying closer to horizontal than vertical with no significant difference between the assemblages. Thus both components were only minimally affected by postdepositional processes. Lithic data from Components I and II indicate little

TABLE 8.6. Raw material types by component from the 2011 excavations of Dry Creek

Raw Materials	n	%	Component I	Component II	Component IV	Powers's Backdirt
Rhyolite						
1. Light gray with black specks	115	3.0	17 (6.8%)	97 (2.7%)	1 (33.3%)	
2. Medium gray	134	3.4	32 (12.7%)	101 (2.8%)		1 (33.3%)
3. Tan/gray banded	66	1.7	25 (10.0%)	41 (1.2%)		
4. Grayish green	1	< 0.1		1 (< 0.1%)		
Subtotal	316	8.1	74 (29.5%)	240 (6.7%)	1 (33.3%)	1 (33.3%)
CCS						
5. Medium gray	59	1.5	1 (0.4%)	58 (1.6%)		
6. Dark gray	32	0.8	3 (1.2%)	29 (0.8%)		
7. Black	8	0.2		8 (0.2%)		
8. Gray/black banded	19	0.5		19 (0.5%)		
9. Grayish purple	1	< 0.1		1 (< 0.1%)		
10. Gray/tan/brown banded	4	0.1		4 (0.1%)		
11. Light grayish green	1	< 0.1		1 (< 0.1%)		
12. Olive gray	1	< 0.1		1 (< 0.1%)		
13. Caramel/clear chalcedony[a]	2,518	64.9		2517 (69.3%)		1 (33.3%)
14. Gray chalcedony[a]	256	6.6	46 (18.3%)	209 (5.8%)	1 (33.3%)	
15. Brown/black chalcedony[a]	109	2.8	1 (0.4%)	108 (3.0%)		
16. Black/clear chalcedony[a]	21	0.5		21 (0.6%)		
Subtotal	3,029	78.0	51 (20.3%)	2,976 (81.9%)	1 (33.3%)	1 (33.3%)
Quartzite						
17. Medium gray	345	8.9	38 (15.1%)	306 (8.4%)		1 (33.3%)
18. Tan/gray banded	46	1.2	35 (13.9%)	11 (0.3%)		
Subtotal	391	10.1	73 (29.0%)	317 (8.7%)		1 (33.3%)
Basalt						
19. Dark gray	134	3.4	52 (20.7%)	81 (2.3%)	1 (33.3%)	
Other						
20. Gray quartz	2	< 0.1		2 (< 0.1%)		
21. White quartz	8	0.2	1 (0.4%)	7 (0.2%)		
22. Granite	5	0.1		5 (0.1%)		
23. Hematite	1	< 0.1		1 (< 0.1%)		
Subtotal	16	0.4	1 (0.4%)	15 (0.4%)		
Component totals			251 (100.0%)	3,629 (100.0%)	3 (100.0%)	3 (100.0%)
Assemblage totals	3,886	100.0%	251 (6.5%)	3,629 (93.4%)	3 (< 0.1%)	3 (< 0.1%)

[a]These are banded chalcedonies. Colors listed reflect solid or opaque portions of the raw material that typically grade to milky-colored translucent stone (or "clear" or transparent as noted).

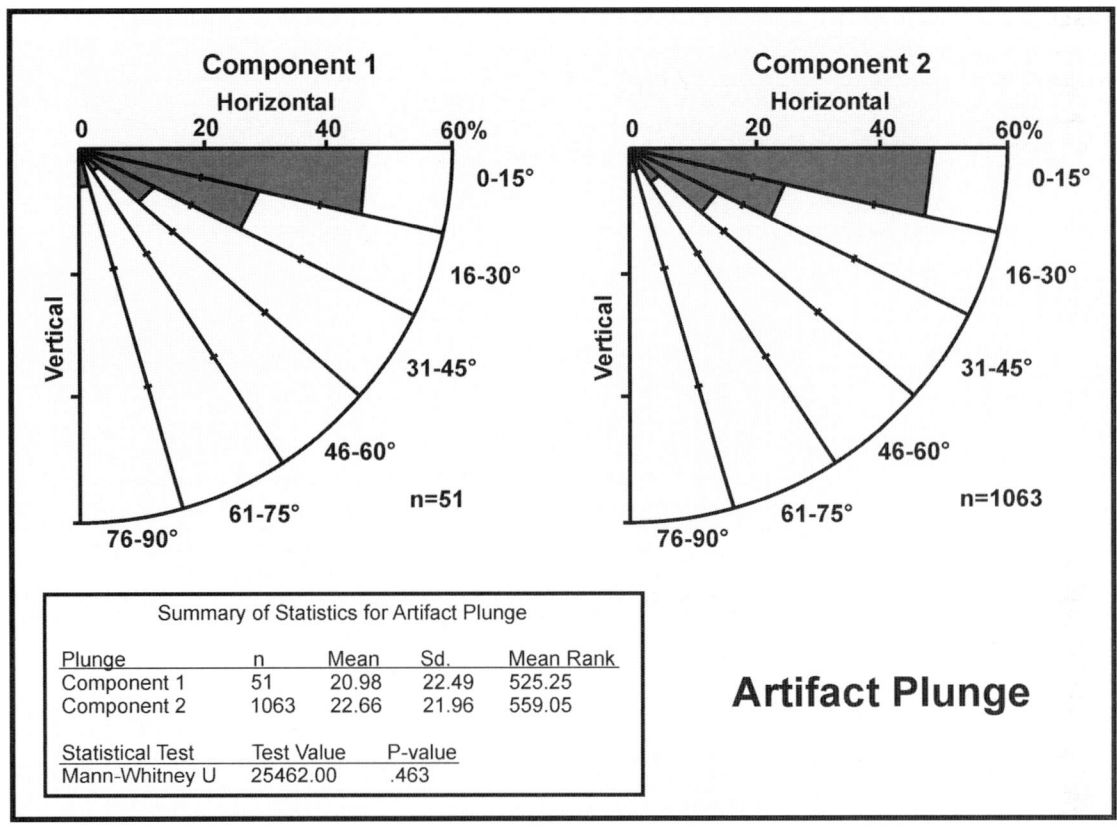

FIGURE 8.18. Frequencies of artifact plunge by component and statistical results from analysis of this variable.

postdepositional movement and represent independent assemblages left behind by at least two site visits.

Faunal Analyses

The complete faunal assemblage from the 2011 excavations includes 1,078 specimens, of which 832 were large enough (> 1 mm in maximum dimension) to confidently be analyzed. Of the 832 analyzed specimens, 137 (16.5 percent NISP) are from Component I, 54 (6.5 percent NISP) are from Component II, and 641 (77 percent NISP) are from Component IV. Each assemblage is composed primarily of comminuted specimens generally lacking cortical surfaces, limiting taxonomic identification and precluding identification of typical surface-bone modifications (e.g., cut marks, tooth marks, percussion damage).

The presence of 57 (41.6 percent NISP) maxillary molar and molar-enamel fragments in Component I led to the identification of Dall sheep, *Ovis dalli* (compared to UAMN 15703; table 8.7; figure 8.19). The remaining Component I specimens were primarily unidentifiable to element but were assigned to medium-to-large mammal (73, 53.3 percent NISP) and indeterminate mammal (3, 2.2 percent NISP) based on specimen metrics, especially thickness. None of the Component II assemblage was identifiable to taxon, but

TABLE 8.7. Measures of abundance and description of the Dry Creek fauna from Components I and II

Stratum	Taxon	NISP	Element	NISP
Component I	*Ovis dalli*	57 (41.6%)	Molar-enamel fragments	62 (45.3%)
	Indeterminate	80 (58.4%)	Indeterminate	75 (54.7%)
	Component totals	137 (100.0%)	Component total	137 (100.0%)
Component II	Medium-to-large mammal	17 (31.4%)	Molar-enamel fragments	4 (7.4%)
			Indeterminate	13 (24.1%)
	Indeterminate	37 (68.5%)	Indeterminate	37 (68.5%)
	Component total	54 (100.0%)	Component total	54 (100.0%)

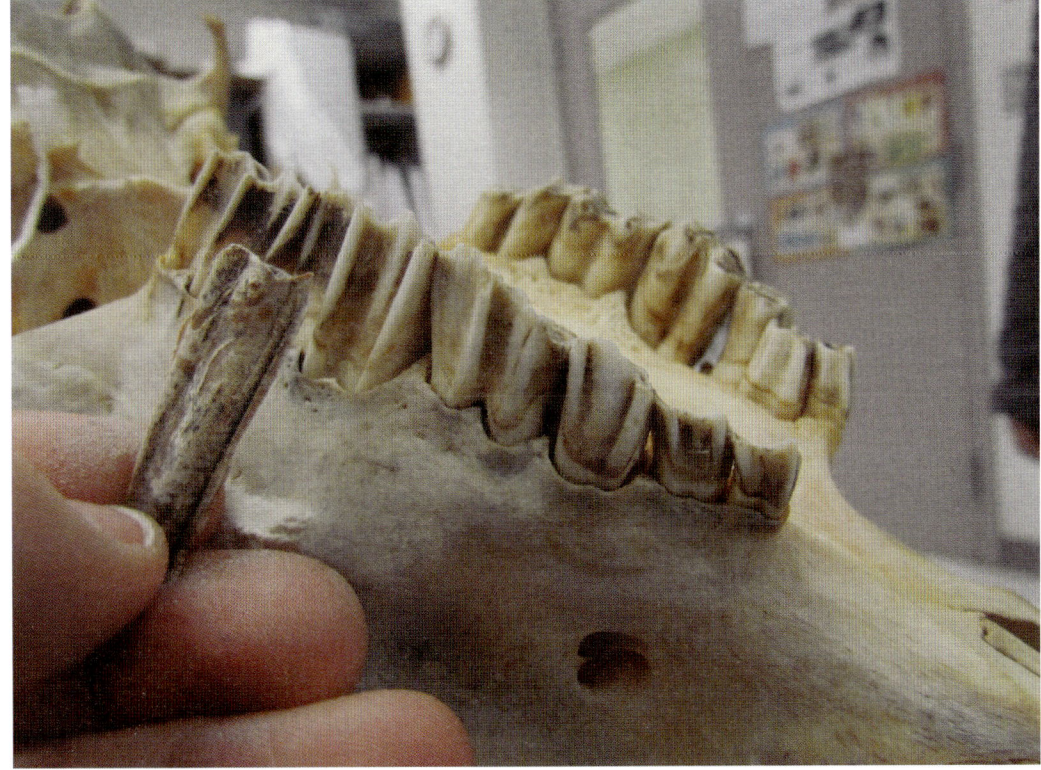

FIGURE 8.19. Faunal specimen UA 2011-064-1751 compared with a Dall sheep maxillary comparative specimen (UAMN 15703).

four specimens (7.4 percent NISP) were molar-enamel fragments identifiable as medium-to-large mammal. While the remaining specimens in Component II were unidentifiable to element, 17 of these (31.4 percent NISP) were attributed to medium-to-large mammal. Component IV specimens were characterized by 482 (75.2 percent NISP) unidentifiable,

11 (1.7 percent NISP) medium-to-large mammal, 1 (0.2 percent NISP) small mammal, and 147 (22.9 percent NISP) small mammal/bird specimens. Small mammal/bird specimens included 4 (0.6 percent NISP) ribs, 1 (0.2 percent NISP) scapula, 4 (0.6 percent NISP) vertebrae, 7 (1.1 percent NISP) long bones, and 2 (0.3 percent NISP) cranial fragments, as well as 129 (20.1 percent NISP) unidentifiable pieces.

When comparing mean length, width, and thickness of faunal specimens, we found some differences among components. Component I means are 17.7 mm for length, 5.78 mm for width, and 2.67 mm for thickness. Component II means are 5.12 mm for length, 6.15 mm for width, and 4.79 mm for thickness. Component IV means are 5.53 mm for length, 4.17 mm for width, and 2.30 mm for thickness. These small measurements reflect the highly fragmentary nature of the Dry Creek faunal assemblage on the whole. Comparing means among the components, the Component I material is statistically longer than the other components' specimens (F Statistic = 322.010; df = 2; $p < 0.01$), but there is no significant difference in the means of width and thickness across the components. These data generally corroborate the lithic data—namely, that there are recognizable differences in behavioral site-formation processes between Components I and II, suggesting they represent two different occupations.

Zooarchaeological and taphonomic attributes of assemblages are valuable for evaluating site formation and integrity of components because they demonstrate unique trends in bone preservation and human behavior. Because our study focused primarily on behavioral and natural site formation with regards to whether Components I and II were separate, the remainder of our results concentrates on comparisons between Components I and II. Due to the preservation level of faunal materials from these components, we report the presence/absence of three variables that inform us about behavioral practices and bone preservation, including burning, cortical bone, and root etching. We expected faunal data from Components I and II to document two different sets of behaviors and levels of preservation if they

TABLE 8.8. Burning modifications on bones from Dry Creek Components I and II

Modifications		Component I	Component II	Total
0% burning	Count	81	6	87
	Expected Count	62.4	24.6	87.0
	% of Total	42.4	3.1	45.5
≤ 50% burning	Count	5	4	9
	Expected Count	6.5	2.5	9.0
	% of Total	2.6	2.1	4.7
100% charring	Count	0	44	44
	Expected Count	31.6	21.4	44.0
	% of Total	0.0	23.0	23.0
100% calcining	Count	51	0	51
	Expected Count	36.6	14.4	51.0
	% of Total	26.7	0.0	26.7
Total NISP	Count	137	54	191
	Expected Count	137.0	54.0	191.0
	% of Total	71.7	28.3	100.0

$\chi^2 = 152.495$; df = 3; $p < 0.001$. One (12.5 percent) cell has an expected count of less than 5. The minimum expected count is 2.54.

represent two different cultural occupations and thus two different depositional events. If, however, Components I and II resulted from the same occupation and depositional event, we expected to find overlapping trends in faunal data. In Component I, there are significantly more than expected specimens expressing no burning, but in Component II, there are significantly more than expected bones with 100 percent charring (χ^2 = 152.495; df = 3; p < 0.001; table 8.8). These patterns indicate different levels in charring/burning of bones between the components, thereby reflecting different ways people were treating fauna.

With regards to level of preservation, there are a couple of interesting patterns between the components. When present, cortical bone was patchy and discontinuous; however, all Component I material preserved cortical bone, while only 9.5 percent NISP of the 54 Component II specimens preserved cortical bone. With regards to root etching, all Component II specimens were root etched; however, not all Component I specimens (92.7 percent NISP) showed root etching. Contingency-table analyses comparing numbers between components indicate that Component I specimens preserved more cortical bone and less root etching than expected, whereas Component II preserved less cortical bone and more root etching than expected. Expected counts of these analyses are too low for reliable chi-square analysis (table 8.9); however, the patterns are interesting and indicate different levels of preservation in the two components and, therefore, taphonomic patterns unique to each component. Component I bones appear to be better preserved compared to the bones from Component II. These faunal data also suggest there was little to no postdepositional mixing between the components, because the components do not seem to have shared the same postdepositional environments.

TABLE 8.9. Preservation of bones from Dry Creek Components I and II

Cortical Bone		Component I	Component II	Total
Present	Count	137	49	186
	Expected Count	133.4	52.6	186.0
	% of Total	71.7	25.0	97.4
Absent	Count	0	5	5
	Expected Count	3.6	1.4	5.0
	% of Total	0.0	2.6	2.6
Total NISP	Count	137	54	191
	Expected Count	137.0	54.0	191.0
	% of Total	71.7	28.3	100.0
Root etching				
Present	Count	127	54	181
	Expected Count	129.8	51.2	181.0
	% of Total	66.5	28.3	94.8
Absent	Count	10	0	10
	Expected Count	7.2	2.8	10.0
	% of Total	5.2	0.0	5.2
Total NISP	Count	137	54	191
	Expected Count	137.0	54.0	191.0
	% of Total	71.7	28.3	100.0

Expected counts are too low for reliable chi-square analyses.

Reanalysis of Artifact UA76-155-1443

In 2010, while visiting the University of Alaska Museum of the North, Graf studied the putative Component I microblade core tablet (artifact number UA1976-155-1443), concluding that Odess and Shirar (2007) were mistaken and that it is not a microblade core tablet. To rejuvenate a microblade core, a core tablet is removed from the top of the core to create a new platform. It originates from a transverse blow directed at the core front (i.e., the fluted face), terminating toward the back of the core (i.e., the wedge element in the case of wedge-shaped core technology; Anderson 1970b; Morlan 1970). In wedge-shaped-core platform rejuvenation, the force of removal often takes some of the wedge element, so the tablet is overshot and J-shaped in profile (figure 8.2B). In other cases, the tablet hinges or even snaps on removal (Anderson 1970b), giving its end a bullnose or stepped shape. To be unequivocally called a microblade core tablet, such an artifact should possess the following characteristics: (1) a diagnostic pattern of parallel microfluting on its platform, demonstrating that it was struck from the front of a microblade core; (2) clearly defined lateral margins representing the core's narrow wedge shape, with signs of remnant bifacial trimming or core-platform trimming; and (3) either a plunging or a stepped termination. Core tablets also often have negative bulbs of percussion on their dorsal surfaces, marking previous core-platform-removal events (Inizan et al. 1999). Such features are clearly evident in the core tablet shown in figure 8.2B, as well as in those reported by Powers (chapter 4) from Component II, but they are not obvious on Odess and Shirar's artifact UA76-155-1443 (figure 8.2A).

Artifact UA76-155-1443's morphology instead suggests that it is not a microblade core tablet. First, of the two facets on the platform of the artifact, the largest—covering more than half of its surface—was removed from the side and not the top, 90° from the direction one would expect if it represented a remnant microblade flute. Second, the lateral margins of the artifact do not have characteristic trimming scars (figure 8.2A), suggesting that they are not remnants of a narrow microblade core. Third, the artifact does not have a dorsal surface with signs of previous core tablet removals (i.e., a negative bulb of percussion). Instead, the artifact is probably a flake fragment resulting from the bipolar reduction of a larger chunk of brown chert. It displays bidirectional crushing and stepping from direct simultaneous impact originating from two opposing directions (see Odell 2004).

Thus Component I still lacks microblades as well as the cores and core-rejuvenation debitage related to their production. Technologically and typologically, the Component I assemblage fits the definition of the Nenana Complex as defined by Powers and Hoffecker (1989) and Goebel et al. (1991).

Discussion

Geoarchaeological data collected and presented here indicate that the Dry Creek site location experienced more than 13,500 years of eolian deposition of mainly loess but also coarser sands during periods of increased wind intensity. Wind intensity (storminess) was relatively high through the terminal Pleistocene and into the early Holocene. During the middle Holocene, winds were variable, and finally during the late Holocene, wind intensity was much lower than before. Increased winds reflect drier conditions, so the overall pattern at Dry Creek is a trend toward increased humidity throughout the Holocene. Both increased soil production and increased concentrations of colloidal Fe and Al observed from the bottom to top of the stratigraphic profile support this trend of increased humidity through time, leading to the development of a boreal-forest biome in the vicinity of the site after the early Holocene. This is also demonstrated by an apparent change in soil type from a shrub-tundra soil (Paleosol 1) to a forest soil (Paleosol 2), a transition occurring just after 10,500 cal yr BP. Certainly, regional pollen records support this interpretation (Ager 1983; Bigelow and Powers 2001; Hu et al. 1993; Schweger et al. 1982; Tinner et al. 2006).

Humans visited the Dry Creek site during three different time periods, as reflected by new AMS ^{14}C ages for Component I of 13,305–13,485 cal yr BP, Component II at 10,590–11,060 cal yr BP, and Component IV at 3,480–5,600 cal yr BP. Component I contains a flake-based and blade-based tool kit. Component II has a tool kit characterized by wedge-shaped microblade cores and lanceolate bifaces. From our study, the Component IV artifacts were too few to characterize the tool kit used; however, the 1970s lithic assemblage indicates a notched-biface-based and flake-based tool kit (Appendix A). Faunal remains were few, but humans were procuring Dall sheep and possibly other medium-to-large mammals during the Component I occupation, medium-to-large mammals during the Component II occupation, and both medium-to-large and small animals during the Component IV occupation. The presence of hearth features in Components I and II indicates that the site served as a camp for at least long enough to leave these types of features behind. They are ephemeral, unprepared hearths, so their presence in both components indicates very short-term use of the site—at least where our team excavated. A lack of more substantial features reported from Powers's excavations also supports this interpretation.

The 2011 excavations and analyses were ultimately aimed at testing the hypothesis that the Nenana Complex occupation (Component I) at Dry Creek cannot be confidently separated from the overlying Denali Complex occupation (Component II) stratigraphically, chronologically, or technologically (Dumond 2001; Odess and Shirar 2007; Thorson 2006). Our results clearly indicate this hypothesis should be rejected.

Stratigraphic integrity is maintained at Dry Creek. We found clear stratigraphic boundaries separating Loess 2 from Sand 1 as well as Sand 1 from Loess 3. Though minor signs of cryoturbation are present in these lower strata, such postdepositional features only minimally affected the units and were confined to disturbances within individual strata. Macrofaunal bioturbation was rare, isolated, and mappable. Mesofaunal bioturbation was present in Loess 2 and Loess 3, but there were no significant signs of soil and sediment displacement at the microscopic level and no signs of microdebitage displacement. Across our newly excavated blocks, we found complete stratigraphic separation between components, with Component I being separated from Component II by at least 15–20 cm of vertical space, which encompassed culturally sterile Sand 1 between the lower limits of Loess 3 and the upper limits of Loess 2. Artifacts, therefore, were found in well-circumscribed, separated components. In short, we found no signs in the field or laboratory indicating significant disturbance or mixing of Loess 2 and Loess 3—containing Components I and II—countering propositions by Thorson (2006) and others (Dumond 2001; Odess and Shirar 2007).

Until now, site chronology at Dry Creek was based on chronometric dates from naturally occurring materials, thus dating geological contexts only. The six new AMS dates reported here were on charcoal from features interpreted to represent hearths. Together they indicate that the ages of Components I and II are 13,305–13,485 cal yr BP and 10,590–11,060 cal yr BP (i.e., the 2σ ranges of modeled hearth ages), respectively. In other words, in the areas where our excavations were conducted, where both components occur in stratigraphic succession, they are separated by 2,000–2,800 years of time. Additionally, the presence of well-circumscribed features and associated cultural materials clearly indicates that two distinct cultural components are preserved, again belying suggestions that cultural materials have moved postdepositionally.

Our analyses of the lithic assemblages established clear technological and typological differences between Components I and II. Component I materials indicate reduction activities focused on production and maintenance of flake-based technologies, but Component II artifacts indicate mostly secondary-reduction activities directed at maintenance of unifacial, bifacial, and microblade-osseous-composite tool technologies. Moreover, we observed clear differences in raw-material selection, with the Component I assemblage composed of nearly equal amounts of quartzite, rhyolite, basalt, and CCS, whereas the Component II assemblage indicates a focus on CCS and chalcedony. No size-sorting between the components occurred after deposition, and orientations of artifacts were chiefly horizontal, indicating minimal vertical movement of artifacts either up or down and certainly not between components.

Faunal analyses highlight differences between components. Component I bones were less burned than those from Component II, and Component I specimens were generally better preserved than bones from Component II. In other words, Component I fauna are associated with site-formation processes unique to Loess 2, while Component II fauna were subjected to postdepositional processes particular to Loess 3.

We found little evidence to support critiques of the Dry Creek stratigraphy and stratigraphic separation of Components I and II. Hypothesized postdepositional movement of Component II materials down through the profile to the position of Component I (Dumond 2001; Odess and Shirar 2007; Thorson 2006) has found absolutely no support from our results. To the contrary, our findings strongly support Powers's original separation of the components into two distinctive complexes. The two components have different lithic assemblages, different faunal assemblages, different preservation problems, and their own respective hearth features. They are consistently and clearly separated stratigraphically and chronologically. The subtle postdepositional deformation processes evident in the two components also reflect different depositional regimes.

Conclusions

With the newly collected data from our 2011 reinvestigation of the important Dry Creek site in central Alaska, we tested the hypothesis that the deepest stratigraphic layers of the site, specifically Loess 2, Sand 1, and Loess 3, were affected by postdepositional processes so strong that the geoarchaeological separation of the site's late Pleistocene archaeological record into two components was arbitrary (Thorson 2006). None of our results support this revision of the site's early record. Instead, we found clear stratigraphic and chronological separation between components as well as clear technological and faunal differences. Dry Creek's Component I represents a unique cultural occupation of the site, one significantly different from overlying Component II, at least in the area where we excavated. On this basis, therefore, we unequivocally reject the hypothesis that the two cannot be separated. The Dry Creek site was visited during the terminal Pleistocene during two different times: once between 13,305 and 13,485 cal yr BP and again between 10,590 and 11,060 cal yr BP, just prior to the spread of the boreal forest in the region. The site appears to have been abandoned during a 2,000–2,800 cal yr period that overlaps with the global Younger Dryas chronozone (11,700–12,900 cal yr BP) and might reflect the driest and perhaps coldest period following the initial arrival of humans in the valley. Our studies provide chronological control over 1 of 3 of the original Component I activity areas and 1 of the 14 original Component II activity areas. Coupled with Bigelow and Powers's (1994) dates, the age of formation of Paleosol 1 is 10,590–12,730 cal yr BP. During Component I times, visitors to the site manufactured and maintained flake- and biface-based technologies around hearth features, where they cooked medium-to-large-bodied mammals, such as Dall sheep. After the hiatus, site visitors repaired tools—including scrapers, burins, and lanceolate bifaces—and manufactured microblades to maintain microblade-osseous composite tools also around hearths. Data presented here support the notion of a separate and early nonmicroblade Nenana Complex in the Nenana Valley region of central Alaska.

Before this investigation, we were uncertain of exactly when people first visited the Dry Creek site. Given new congruent dates on hearth charcoal from two hearth features, we now know that Dry Creek Component I predates other cultural occupations in the Nenana Valley by several centuries (see Goebel et al. 1996; Graf and Bigelow 2011; Holmes et al. 2010; Pearson 2000). In comparison with the nearby upper Tanana basin, only the lowest component at Swan Point predates Dry Creek Component I by nearly nine hundred years (Holmes 2011). The lowest occupations at FAI-2043, Broken Mammoth, and Mead are contemporaneous with Dry Creek Component I, albeit age ranges on hearth charcoal from Broken Mammoth and Mead make them slightly younger (Gaines et al. 2011; Potter et al. 2014). Now more than ten nonmicroblade assemblages from sites in the Nenana and Tanana Valleys have been found in well-stratified contexts dating to the Allerød. At about 13,000–13,500 cal yr BP, people in central Alaska were making biface-based technologies, not selecting microblade-osseous-composite technologies (Gaines et al. 2011; Graf and Bigelow 2011; Holmes 2011; Potter et al. 2014; Sattler et al. 2011). They were employing an industry Powers and colleagues (Hoffecker et al. 1993; Powers and Hoffecker 1989) defined as the Nenana Complex. A similar pattern exists at the Ushki Lake sites in Kamchatka, Russia, where nonmicroblade assemblages containing small bifacial projectile points also date to this timeframe (Dikov 1977; Goebel et al. 2003, 2010). Why millennial shifts in lithic industries—from microblade to nonmicroblade back to microblade technologies—occurred is still unknown. Either different human groups with different ways of making a living visited the region at different times, or the shifts reflect adaptive switches from pre-Allerød through post-Allerød times.

Some have argued that successful human occupation of the region came with the spread of shrub tundra (Hoffecker and Elias 2007). Pollen records indicate the presence of shrub-tundra vegetation across Beringia after 14,000 cal yr BP (Anderson et al. 2004; Bigelow and Powers 2001). New dates at Dry Creek and the appearance of the many coeval and subsequent cultural occupations reflect a sustained population following emergence of shrub-tundra resources. Perhaps eastern Beringian climatic conditions prior to the Allerød were too harsh to sustain human population beyond intermittent exploration of the region.

Based on a short chronology for Clovis (Waters and Stafford 2007), new mid-Allerød-aged dates from Dry Creek now make it a pre-Clovis site. Combined with data emerging from numerous other pre-Clovis-aged sites in central Alaska as well as northeast Asia, such as Yana and Berelekh (Pitulko et al. 2013, 2014), Beringia now has more well-dated pre-Clovis sites than any other region in the Americas. Given this, we need to pay close attention to its rich archaeological record. We think answers to important archaeological and ecological questions regarding late Pleistocene human dispersals to the New World lie in detailed geoarchaeological, chronological, technological, and subsistence studies of Beringia's rich archaeological record, including sites such as Dry Creek. Now more than ever, the Dry Creek site provides an important datum for Beringian archaeology and peopling of the Americas research.

Note

1. Parts of this chapter were originally published in Graf et al. (2015) but are presented again here along with expanded information about the 2011 field project at Dry Creek. The authors acknowledge the Elfrieda Frank Foundation, the Center for the Study of the First Americans, and the Department of Anthropology at Texas A&M University for financial and logistical support of this project. Thanks to the 2011 field crew and volunteers: Angela Younie, Josh Lynch, Marion Coe, Caroline Ketron, Jacque Kocer, Stephanie Rivera, Ashley Smallwood, Ted Goebel, Ashley Ahman, Hannah Atkinson, Ray Burnham, Grace Mills, James Zerbe, Julie Crisifulli, David McMahon, Jeremy Karchut, Brian Wygal, Sam Coffman, Nancy Bigelow, Rémi Méreuze, Angélique Neffe, Julie Esdale and her 2011 field crew, and Michael Grooms. Thanks also to Jim Whitney, Scott Shirar, and Sam Coffman for providing access to the original Dry Creek collection at the University of Alaska Museum of the North.

CHAPTER NINE

A Dry Creek Retrospective

TED GOEBEL AND JOHN F. HOFFECKER

Introduction

The Dry Creek report was originally drafted in 1983, but despite undergoing peer review and revisions later that decade, it was never published. In the years that have passed since then, major developments have occurred in Beringian Quaternary studies and archaeology as well as in the study of the peopling of the Americas. Nevertheless, the Dry Creek site is as important now as it was then. The purpose of this retrospective chapter is to consider the place of Dry Creek in the altered paleoecological and archaeological contexts of 2017, forty years after the site's groundbreaking excavation.

Assessing Dry Creek's place in the wider contexts of Beringian paleoecology and prehistory first requires consideration of research undertaken at the site or with its collections since completion of the original excavation (table 9.1). Since 1977, fieldwork at Dry Creek has been directed at clarifying the site's geochronology and the geological contexts of its early cultural components (Bigelow and Powers 1994; Bigelow et al. 1990; Graf et al. 2015; Hoffecker 1988; Thorson 2006), while collections research has focused on interpreting the technological and typological variability that Powers observed in the site's early cultural components (Goebel 1990; Goebel et al. 1991; Graf and Goebel 2009; Odess and Shirar 2007; Potter 2011; Reuther et al. 2011).

Also during the past four decades, paleoecological and archaeological research has occurred at other localities in the Nenana Valley and neighboring Teklanika Valley to the west (table 9.2; figure 9.1). Many of these activities were a direct outgrowth of Powers's original research program at Dry Creek, but not all of them. After the conclusion of the Dry Creek excavation in 1977, Powers's "North Alaska Range Project" continued through regional geomorphic studies (Ritter 1982; Ritter and Ten Brink 1986; Ten Brink and Waythomas 1985; Thorson 1986; Thorson and Bender 1985) as well as archaeological surveys that led to the discovery of additional terminal Pleistocene archaeological sites in geomorphic and stratigraphic contexts analogous to Dry Creek (Hoffecker 1988; Powers and

TABLE 9.1. Research related to the Dry Creek site since completion of the 1983 Dry Creek report

Year	Activity	Reference
1987	Thermoluminescence dating of the site's basal loess	Hoffecker 1988
1989	Analysis of the paleoclimatic significance of Sand 1	Bigelow et al. 1990
1988–89	Typological analysis of debitage and retouched artifacts from Component I and Component II	Goebel 1990; Goebel et al. 1991
1992	Geochronological sampling of the site's northwest profile (at grid point N4E6) and AMS radiocarbon dating of Paleosols 1 and 2	Bigelow and Powers 1994; Hoffecker et al. 1996a
1999–2000	Cluster analysis of tool kits represented in Components I and II	Potter 2005, 2011
2004	Analysis of tool-stone procurement and selection at Dry Creek	Graf and Goebel 2009
2006	Critique of Dry Creek geoarchaeology published	Thorson 2006
2006–7	X-ray fluorescence analysis of obsidian artifacts from Dry Creek	Reuther et al. 2011
2007	Description of a possible microblade core tablet from Component I	Odess and Shirar 2007
2011	Geoarchaeological excavations at Dry Creek in response to critique of Thorson (2006)	Graf et al. 2015

Hoffecker 1989). Powers and an evolving cadre of graduate students from the University of Alaska Fairbanks eventually excavated large blocks at several of these sites: John Hoffecker and subsequently Georges Pearson at Moose Creek (Hoffecker 1985, 1996; Pearson 1999); Peter Phippen at Owl Ridge (Hoffecker et al. 1996; Phippen 1988); Howard Maxwell and subsequently Ted Goebel and Nancy Bigelow at Panguingue Creek (Goebel and Bigelow 1992; Maxwell 1987; Pontti 1997; Powers and Maxwell 1986); and Tom Gillispie, Ted Goebel, and Nancy Bigelow at Walker Road (Bigelow 1992; Flannigan 2002; Goebel et al. 1996; Higgs 1992, 1994; Powers et al. 1990). Also during those years, Goebel (1996) conducted a limited geoarchaeological project at the Teklanika West site in Denali National Park and Preserve.

The "Powers Era" of field-based archaeological research in the Nenana Valley region effectively came to a close with the completion of the Moose Creek excavations in 1996, but into the 2000s, Powers's former students and others continued to carry out field and laboratory studies. Paleoecologically, Bigelow and Edwards (2001) produced the first detailed vegetation history for the valley with a pollen core drawn from the bottom of Windmill Lake, and Dortch et al. (2010) provided some of the first chronometric dates for the region's Pleistocene glaciations through Beryllium-10 (^{10}Be) dating of exposed rock surfaces on a series of moraines and outwash terraces. Archaeologically, a series of cultural-resource management projects following the Nenana River corridor led to the discovery and test excavation of several new sites ranging in age from the terminal Pleistocene to the early middle Holocene (i.e., Houdini Creek, Lucky Strike, and Eroadaway; Holmes et al. 2010; Potter et al. 2007).

Since 2007, there has been a resurgence of archaeological research activity in the Nenana and Teklanika Valleys, with the AMS ^{14}C dating of the oldest component at the Carlo Creek

TABLE 9.2. Published paleoecological and terminal Pleistocene–early Holocene archaeological research carried out in the Nenana Valley and neighboring Teklanika Valley since completion of the Dry Creek excavations in 1977

Year	Activity	Reference
1978–80	Glacial geology field studies in Nenana Valley	Ritter 1982; Ritter and Ten Brink 1986; Ten Brink and Waythomas 1985; Thorson 1986; Thorson and Bender 1985
1978–80	Geomorphologically based archaeological survey and discovery of the Moose Creek and Walker Road sites	Hoffecker 1988; Powers and Hoffecker 1989
1979, 1984	Test excavations at Moose Creek site	Hoffecker 1985, 1996
1980	Publication of Carlo Creek archaeological excavation report	Bowers 1980
1982–84	Excavations at Owl Ridge site	Phippen 1988; Hoffecker et al. 1996b
1984–89	Excavations at Walker Road site	Bigelow 1992; Flannigan 2002; Powers et al. 1990; Goebel et al. 1996; Higgs 1992, 1994
1985, 1991	Excavations at Panguingue Creek site	Goebel and Bigelow 1992; Maxwell 1987; Pontti 1997; Powers and Maxwell 1986
1987	Test excavations at Eroadaway site	Holmes et al. 2010
1992	Geoarchaeological testing and ^{14}C dating of the Teklanika West site	Goebel 1996
1995	Test excavations at Houdini Creek	Potter et al. 2007
1996	Excavations at Moose Creek site	Pearson 1999
1996	Coring of Windmill Lake for pollen analysis	Bigelow and Edwards 2001
2003	Test excavations at Lucky Strike site	Potter et al. 2007
2005	^{10}Be dating of Nenana Valley glacial moraines	Dortch 2006; Dortch et al. 2010
2007	AMS ^{14}C dating of Carlo Creek site hearths	Bowers and Reuther 2008
2007–10	Renewed excavations at Owl Ridge site	Graf et al. 2010; Graf and Bigelow 2011
2007	Test excavations at Bull River II site	Wygal 2010
2009	Resumed excavations at Teklanika West site	Coffman 2011; Potter and Coffman 2011

site (Bowers and Reuther 2008), renewed excavations at Owl Ridge (Graf and Bigelow 2011; Graf et al. 2010) and Teklanika West (Coffman 2011; Coffman and Potter 2011), and test excavations at Bull River II (Wygal 2010), the latter two sites being located within Denali National Park and Preserve. At the same time, previously excavated lithic materials have been analyzed using new methods (Goebel 2011; Goebel et al. 2008a; Gore and Graf, in press; Reuther et al. 2011). Obviously, the rich paleoecological and archaeological records of the central Alaska Range are not yet "played out," and the continuation of scientific research in the region makes the early work of Powers and his team still very relevant. Although center stage of terminal Pleistocene archaeological research shifted in the 1990s to the middle Tanana Valley with excavations of the Broken Mammoth and Swan Point sites (e.g., Holmes et al. 1996; Yesner 1996), a regional perspective incorporating information from all available sites, whether they be in the lowlands or uplands, is critical to achieving complete understanding

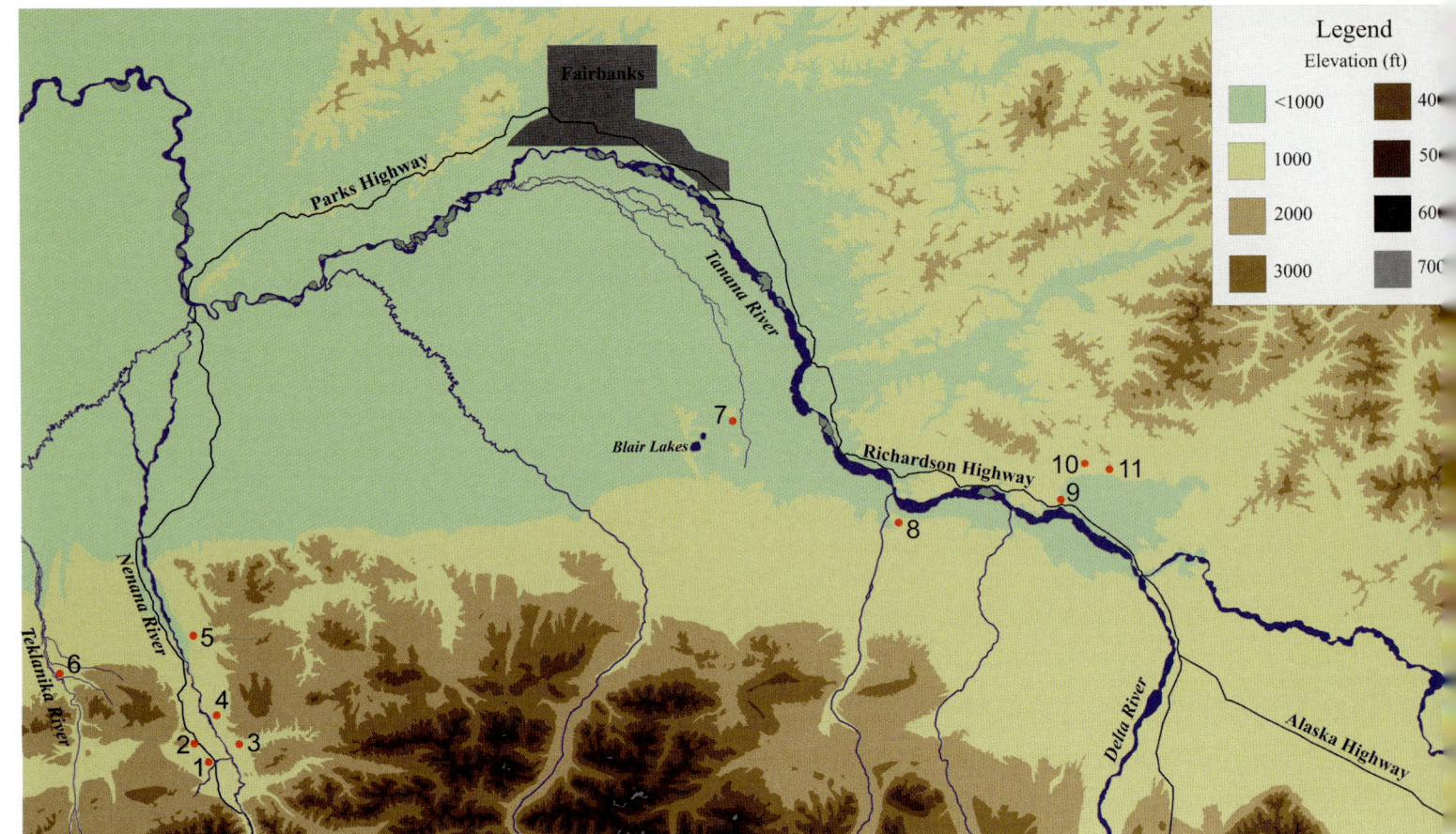

FIGURE 9.1. Map of central Alaska showing locations mentioned in the text (1, Dry Creek; 2, Panguingue Creek; 3, Houdini Creek; 4, Walker Road; 5, Moose Creek; 6, Owl Ridge; 7, McDonald Creek; 8, Upward Sun; 9, Broken Mammoth; 10, Mead; 11, Swan Point; 12, Healy Lake Village site/Linda's Point).

of the first Beringians—not just their environments and ecology but also their technology, subsistence, and settlement behavior. For this reason, in this chapter we address the important contexts in which the Dry Creek site is still central, including (1) Quaternary geomorphology and environments, (2) geoarchaeology and chronology, (3) early hunter-gatherer behavior, (4) human colonization of Beringia, and (5) the peopling of the Americas. For each, we begin with a summary of pertinent conclusions reached in the 1983 Dry Creek report and presented in part 1 of this book and then proceed with a review of post-1983 developments, culminating with a final assessment of the place of Dry Creek in today's study of Beringian paleoecology and prehistory.

Site Setting: Quaternary Geomorphology, Chronology, and Environments

The Dry Creek site is buried in eolian silts and sands deposited during the latest Pleistocene and Holocene. This loess mantle overlies a glacial outwash terrace that Wahrhaftig (1958) originally mapped and assigned to the Healy Glaciation, interpreting it to be Illinoian in age. In chapter 3, Powers refrained from assigning an age to the Healy terrace, perhaps reflecting an inability then to know it, but since the 1980s, geologists have tended to assign it to the early Wisconsinan (marine-isotope stage [MIS] 4; Ritter 1982; Thorson 1986; Thorson and Hamilton 1977). Confirmation of this recently has come from a cluster of seven ^{10}Be dates on exposed boulders associated with the Healy moraine, ranging from 60,000 to 53,000 cal yr BP; these serve as immediate minimum ages for the glaciation (Dortch et al. 2010).

The basal loesses and sands that contain the Dry Creek site's terminal Pleistocene cultural components are well understood given the pioneering stratigraphic and sedimentological work undertaken by Thorson and Hamilton (1977), as reviewed in chapter 3. They are primarily eolian in origin, with fluxes in grain size representing changes in wind intensity. Although originally thought to represent local fluctuations atop the Dry Creek bluff, more recent studies at other localities in the Nenana Valley (e.g., Walker Road, Panguingue Creek, and Moose Creek) have indicated that Loess 1, Loess 2, Sand 1, and Loess 3 are expressed valley-wide (Bigelow et al. 1990; Powers and Hoffecker 1989) and potentially represent regional climate patterns. During full-glacial conditions, high winds escaping the Nenana River gorge blew loess and sand far downriver into the Tanana flats and beyond (figure 9.1; Thorson and Bender 1985), but at some point during the late glacial, these winds diminished and loess began to accumulate on the valley's terrace surfaces. Just when this happened, however, is still unresolved, but based on radiocarbon evidence from the dating of Loess 2 at multiple localities, we now know that it happened prior to 13,500 cal yr BP (e.g., Goebel et al. 1996; Graf et al. 2015).

Do Dry Creek's Terminal Pleistocene and Holocene Sediments and Soils Represent Proxies of Regional Climate Change?

When the original studies of Dry Creek's geology were reported (Thorson and Hamilton 1977), too few late Quaternary loess sequences had been described for Powers and colleagues to interpret the site's sedimentological and pedological records as proxies for regional environmental or climatic conditions. However, with the excavation of additional sites in the region, geoarchaeologists began to recognize some regional patterning. Of the basal sedimentary units at Dry Creek, for example, Sand 1 took on special significance. After recognizing the same eolian unit at the Walker Road site in the late 1980s, Bigelow et al. (1990) proposed that it could represent increased wind intensity, a local proxy for the Younger Dryas, a period of widespread cold climate expressed in various paleoenvironmental records across the Northern Hemisphere (e.g., Alley 2000; Ballenger et al. 2011; Brauer et al. 2008; Broecker et al. 2010; de Vernal et al. 1996; Goebel et al. 2011; Hajdas et al. 1998; Lothrop et al. 2011). Since 1990, some proxy records also have indicated relatively cool, dry conditions for the Younger Dryas in central Alaska; however, others suggest that the effects of the Younger Dryas might have been minimal in the region (Abbott et al. 2000; Bigelow and Edwards 2001; Bigelow and Powers 2001; Edwards et al. 2001; Hu et al. 1993; Kokorowski et al. 2008). The current paleoclimatic model for Alaska attempts to explain these discrepancies by suggesting that during the Younger Dryas, sea-surface temperatures in the North Pacific sharply cooled, impacting southern Alaska but having a muted effect north of the Alaska Range, except in places like the Nenana Valley, where cool Pacific air was funneled through mountain passes into the interior (Bigelow and Edwards 2001; Graf and Bigelow 2011). Sand 1 at Dry Creek might indeed represent this phenomenon, and AMS radiocarbon ages reported since 1983 help pin down precisely when it occurred locally. Disregarding the conventional ages obtained in the 1970s and instead focusing on more recently obtained AMS ages (Bigelow and Powers 1994; Graf et al. 2015), the sand had to have been deposited *after* the youngest Component I hearth (< 13,250 cal yr BP) but *before* the most ancient Paleosol 1/Loess 3 charcoal sample (> 12,750 cal yr BP). Nowhere else in the Nenana Valley has Sand 1 been bracketed on both sides with AMS ^{14}C dates, but at Walker Road, lower-limiting dates indicate an age of < 12,900 cal yr BP for the same sand unit, and in the neighboring Teklanika Valley, charcoal samples from below and above the sand at Owl Ridge provide limiting ages of < 12,775 and > 12,560 cal yr BP, respectively (Graf and Bigelow 2011). Independent estimates place the Younger Dryas between 12,900 and 11,700 cal yr BP (Broeker et al. 2010; Straus and Goebel 2011), so if Sand 1 does represent this event in the central Alaska Range, it had to have occurred during the cold interval's first several centuries, while the later Younger Dryas saw a return to relatively warm conditions leading to the formation of the Nenana Valley's first postglacial soil (i.e., Paleosol 1 at Dry Creek). Future geoarchaeological research at the valley's sites should incorporate fine-scaled micromorphological analyses of Sand 1 to test whether it indeed represents regional eolian conditions and precisely when its deposition began and ended.

Dry Creek's early paleosol sequence has also been replicated at other localities in the Nenana Valley, further indicating the loess record's value as a terminal Pleistocene–early Holocene proxy for regional climate. Rarely in the valley, though, has the Holocene packet of sediments been found in such a deep profile as at Dry Creek, instead typically being only 50–80 cm thick. In these contexts, Paleosols 1 and 2 are not readily distinguishable from one another and have been typically referred to as Paleosol 1, or the "lower paleosol complex" (e.g., at Panguingue Creek and Moose Creek), and associated radiocarbon ages on natural charcoal span 12,250–8,750 cal yr BP (Hoffecker 1996; Pearson 1999). Dry Creek's Paleosol 3, however, is clearly evident at Panguingue Creek, Owl Ridge, and Teklanika West, while Paleosol 4 is evident at Panguingue Creek, Owl Ridge, Walker Road, and Moose Creek (Coffman and Potter 2011; Goebel 1996; Goebel and Bigelow 1996; Goebel et al. 1996; Graf et al. 2010; Pearson 1999). Formation of these repeatedly occurring pedocomplexes and intervening episodes of loess deposition could relate to region-wide climate oscillations—for example, the Holocene climatic optimum and so-called 8.2 ka event, or Younger Younger Dryas (e.g., Mason et al. 2001). These correlations should be explored through more refined dating of the sites' early and middle Holocene pedocomplexes.

What Is the Paleoenvironmental Context of Dry Creek's Late Pleistocene Components?

When the Dry Creek excavations were undertaken in the 1970s, the work was conducted largely in a "palynological vacuum." No detailed fossil-pollen studies had been accomplished in the Nenana Valley region, and at the time, the closest pollen cores spanning the late glacial were poorly dated, coarsely analyzed, and more than 125 km away in a lowland Alaska Interior setting (Ager 1975). As a result, Powers's and Guthrie's hypothesis that the area around Dry Creek represented a late-glacial refugium for the Pleistocene "mammoth steppe" (chapter 6) could not be independently corroborated. More recent palynological research in the region (e.g., Bigelow and Powers 2001), however, has provided a fuller (and much more precisely dated) portrayal of environmental change during the terminal Pleistocene and early Holocene, one that is at odds with the interpretation presented in the original Dry Creek report.

The most precise local pollen record comes from Windmill Lake, located near the confluence of the Nenana River and the Yanert Fork, about 30 km south of Dry Creek and 200 m higher in elevation. Cored by Bigelow and Edwards (2001), the Windmill pollen record begins about 13,000 ^{14}C yr BP (15,500 cal yr BP) and continues through the Holocene. The earliest part of the record, ~13,000–11,700 ^{14}C yr BP (15,500–13,500 cal yr BP), is characterized by relatively low pollen-accumulation rates but high frequencies of Poaceae (grasses), *Artemisia* (wormwood), Cyperaceae (sedges), and *Salix* (willows)—by all accounts a relatively unproductive tundra-steppe with few shrubs, what Bigelow and Edwards (2001) refer to chronologically as the "herb zone." An abrupt transition occurred about 11,700 ^{14}C yr BP (13,500 cal yr BP), with a sharp increase in pollen accumulation and very high frequencies of *Betula* (birch), chiefly at the expense of the grasses, wormwood, and willow. This new zone, which continues into the early Holocene, represents a shrub-tundra vegetation community, or the "birch zone" (Bigelow and Edwards 2001). The dramatic rise in shrub-birch pollen is evident in other cores in the region, too (e.g., at Eightmile Lake near the Nenana-Teklanika divide about 10 km west of the Dry Creek site), but the transition might have occurred at least a millennium earlier (Ager 1975, 1983)—that is, if the conventional radiocarbon dates for the pollen core are to be believed. The herb-to-birch-zone transition represented in these cores likely resulted from warmer summers and increasing effective moisture during the Allerød interstadial. It was under these conditions that humans first visited the Dry Creek site.

Within the birch zone of the Windmill Lake core, Bigelow and Edwards (2001) also noted a sharp fall in pollen-accumulation rates and birch-pollen frequencies as well as a concomitant rise in *Artemisia* pollen frequencies ca. 12,450–11,900 cal yr BP. This short-lived episode suggests cooler and/or drier conditions several centuries earlier than the presumed timing of the deposition of Sand 1 in the Nenana Valley's archaeological sites. The discrepancy could be a factor of the dating of this portion of the Windmill Lake core through sediment-accumulation rates, but taken together, the pollen and sedimentary records suggest a cooler and drier Younger Dryas than what came before and after.

Thus Powers's and Guthrie's notion of the Nenana Valley representing a terminal Pleistocene refugium for the tundra-steppe and its megafauna is not entirely accurate. The presence of wapiti remains in Component I fits neatly within a period of Allerød warming when shrub-tundra habitat—not open tundra-steppe habitat—blanketed the northern foothills of the Alaska Range, and the steppe-bison remains in Component II likely correspond to a return to cooler, drier conditions during the Younger Dryas and a local expansion of bison habitat (but see Potter et al. 2013). By 13,500 cal yr BP, the tundra-steppe was long gone from the vicinity of Dry Creek, and during the late-glacial period from 13,500 to 12,000 cal yr BP, climate and environment in the Nenana Valley fluctuated from interstadial to stadial conditions and back again. As discussed previously, given the Dry Creek site's location just north of the Alaska Range, the local environment would have been dynamically impacted by these climate oscillations, so we should expect potential modifications in human use of the site during the Component I and Component II chronozones.

The Temporal Context: Geoarchaeology and Chronology at Dry Creek

When Dry Creek was excavated in the mid-1970s, Alaskan archaeologists still were unaccustomed to considering geology as an important line of archaeological inquiry. Most sites excavated up to that time were in compressed settings—for example, Donnelly Ridge and the Tangle Lakes sites (West 1967, 1980)—and even the few sites with discernible stratigraphy were not excavated using the full benefit of stratigraphy. Frederick West's 1960s excavations at Teklanika West represent one such case: his reports characterized the site as "unstratified" (West 1967), but later studies quickly recognized a well-stratified site with multiple components (Coffman and Potter 2011; Goebel 1996). Similarly, at Healy Lake, John Cook (1969) recognized stratigraphy and described it, but he still dug the Village site using arbitrary levels, not using distinguishable strata to aid in defining clear cultural components.

Powers's early training in Idaho rockshelters and Paleolithic sites in Siberia instilled in him a strong sense for stratigraphy and Quaternary geology. Not surprisingly, therefore, when the Dry Creek excavation began, the team of researchers included Quaternary geologist Tom Hamilton, who had just completed his analysis of the geomorphology of the Onion Portage site (Hamilton 1970). The resulting study of Dry Creek's geology and geochronology (Thorson and Hamilton 1977; Powers and Hamilton 1978) became required reading for students as they learned how to approach the excavation of a multilayered Paleolithic site from a geological perspective; how to map, describe, and explain its sediments and soil units; and how to characterize and interpret the effects of postdepositional disturbances. No wonder Dry Creek has been regularly highlighted in geoarchaeological texts (e.g., Rapp and Hill 1998:100; Waters 1992), and no wonder all other stratigraphic excavations in the north Alaska Range have been accomplished from the basis of Dry Creek's "master" sequence. Nonetheless, the study of Dry Creek's geoarchaeology and geochronology continued long after the cessation of Powers's excavations in 1977 due to a pair of lingering questions: how old are the site's cultural components, and does Component I truly exist as an occupation separate from Component II?

How Old Are Dry Creek's Terminal Pleistocene Cultural Components?

Dating the Dry Creek site was not straightforward. In the excavations, charcoal was scarce, and bone practically nonexistent. The AMS "revolution" of the mid-1980s had not yet occurred when the site was excavated and the report prepared, so individual small charcoal samples needed to be combined, increasing the chance for contamination. Results were thus highly variable and unpredictable. Some samples yielded remarkably old dates with huge standard deviations, and it was then that fingers were pointed at the Tertiary lignite seams exposed across the Nenana Valley at the Usibelli coal mine (Powers and Hamilton 1978; Thorson and Hamilton 1977). Powers and his team reasoned that perhaps the material being dated was contaminated by ancient carbon from the lignite, blowing around the valley and being incorporated into Dry Creek's silt and sand matrix. Nonetheless, through trial and error, a rough radiocarbon chronology was built, and the terminal Pleistocene ages of the lower loesses and cultural components were provisionally established (Powers and Hamilton 1978; Thorson and Hamilton 1977; chapter 3). Through all this, though, only single ^{14}C dates were used to infer the chronometric ages of Components I and II, and these became two of the most often-cited dates for early Alaskan archaeology: 10,690 ± 250 ^{14}C yr BP (12,540 cal yr BP) for Component II and 11,020 ± 85

^{14}C yr BP (12,970 cal yr BP) for Component I. The 10,690 date established that the microblade-rich Denali Complex assemblage of Component II indeed dated to the late Pleistocene (like its Siberian analog, the Diuktai Culture), and the 11,020 date suggested for the first time that a biface industry lacking microblades existed in Alaska—an industry that looked right to represent a Beringian analog of Clovis. Unfortunately, no more radiocarbon dates were obtained that confirmed these two ages; suitable large charcoal samples did not exist.

After the introduction of AMS technology in the 1980s, however, Powers attempted to strengthen the Dry Creek chronology. In 1992, Bigelow and Powers (1994) reopened the original excavation block at the N4E6 corner to collect wood charcoal samples for AMS ^{14}C dating from the existing stratigraphic profile. Although no new dates were produced for Loess 2 and Component I, noncultural charcoal from Paleosols 1 and 2 confirmed that the loesses containing these ancient soils formed during the late Younger Dryas and earliest Holocene (12,700–10,250 cal yr BP), in agreement with Bigelow et al.'s (1990) hypothesis that lower-lying Sand 1 accumulated during the early Younger Dryas (starting about 12,900 cal yr BP). No new excavations occurred during this time, however—only the cleaning of the old profile—and none of the dated charcoal came from cultural features. The new AMS ages nonetheless provided a maximum age for Component II: its contents likely did not predate 12,700 cal yr BP.

At about the same time, Hoffecker (1988), along with geologist C. Waythomas, attempted to thermoluminescence (TL) date Dry Creek's basal loess (Loess 1), but they were unsuccessful. This work was aborted after the realization that other luminescence techniques (e.g., optically stimulated luminescence) might be more appropriate for dating central Alaska's loess deposits.

There the matter rested until 2011, when Graf et al. (2015; chapter 8) returned to Dry Creek to reopen the 1970s excavation and reevaluate the site's geoarchaeology. Luckily, Powers had left a 2-m-wide balk that extended perpendicular to the bluff edge through the entire excavation block. By judiciously selecting three areas to excavate, where Powers found high densities of both Component I and Component II artifacts in stratigraphic succession, Graf and her team found three charcoal features that they interpreted as hearths, two in Component I and one in Component II. In all three cases, these were clearly associated with artifacts and faunal remains. AMS dates on charcoal samples from these hearths provided the first unequivocal ages of the site's archaeological concentrations—Z in Component I and G in Component II (figures 5.1, 5.2). For Component I, the pair of hearths dated to 11,580 ± 40 and 11,635 ± 40 ^{14}C yr BP, significantly older than the initial conventional assessment of 11,120 ± 85 ^{14}C yr BP. Whether the other two Component I concentrations (X and Y) are synchronous with Z must await further research—either refitting the artifacts between the dated and undated concentrations or renewing the excavation of the balk adjacent to Cluster Y. We are now inclined to reject the 11,120 ^{14}C yr BP age and agree with Graf et al. (2015, chapter 7) that the real age of Component I is close to 11,600 ^{14}C yr BP (~13,440 cal yr BP).

AMS ^{14}C dating of Loess 3/Paleosol 1 and Component II was also successful in Graf's study. Two charcoal samples from a single hearth feature in Cluster G yielded dates of 9,480 ± 35 and 9,460 ± 50 ^{14}C yr BP. These dates are much younger than the original conventional age of 10,690 ± 250 ^{14}C yr BP, as well as Bigelow and Powers's (1994) pair of ages for Paleosol 1 (10,060 ± 75 and 10,615 ± 100 ^{14}C yr BP), but they overlap with the younger suite of ages obtained for Paleosol 2 (9,690 ± 75 and 9,340 ± 95 ^{14}C yr BP). Although we think it critical to point out that Graf et al.'s (2015; chapter 7) new results are the only dates so far obtained that actually provide chronological control on the human activity represented by Component II, the remaining AMS and conventional dates on natural charcoal for Paleosols 1 and 2 suggest that the component's numerous artifact concentrations could date to as much as a millennium or two earlier than Graf et al.'s (2015) dates suggest. In fact, when calibrated, the age span for the full suite of dates reported by Bigelow and Powers (1994) increases to 12,700–10,250 cal yr BP. Unfortunately, Graf et al.'s (2015) work does not help resolve whether Component II's biface clusters and microblade clusters, described by Hoffecker in chapter 5, are contemporaneous or date to different times. Obviously, to solve this problem, another round of excavations will be needed to collect additional charcoal samples from as many Component II concentrations as possible. After all, parts of 8 of 13 undated concentrations (i.e., excluding Cluster G) remain preserved in the site.

Is Component I a Stratigraphically Distinct Occupation from Component II?

Another geoarchaeological issue relates to site-formation processes and separation of Components I and II into two distinct entities. What to make of the distinctive bifacial-point assemblage occurring in Loess 2 was a controversy from the start. Early critiques of Powers's reports suggested it was a sampling anomaly (e.g., Dumond 1980; Hopkins et al. 1982)—that as more area was excavated, microblades would eventually be found. Colinvaux and West (1984) argued further that Component I–like bifaces were known to come from Denali Complex components with microblades at sites like Teklanika West and Healy Lake, suggesting the lack of Component I microblades was because humans were participating in activities that did not require microblades. Excavations at Walker Road, Moose Creek, and Owl Ridge partially "quieted" these critiques, given the repeated occurrences of Allerød-aged assemblages with small bifacial points but not microblades (Goebel et al. 1991; Hoffecker et al. 1993; Powers and Hoffecker 1989), a point we return to later in this chapter.

The controversy was refueled, however, in 2006 when Thorson published a paper severely critiquing Powers's excavation and claiming that Component I represented nothing more than down-drifted Component II artifacts, their translocation resulting largely from burrowing animals near the bluff edge. This was the impetus behind Graf's return to the site in 2011. As she unequivocally demonstrates in chapter 8, Thorson's (2006) claims can be dismissed in favor of Powers's original interpretation in the 1970s. Component I is clearly sealed below Sand 1, separated stratigraphically from Component II by as much as 35 cm of sterile, windblown sediment. In Graf's excavations, Component I's lithic tool stones, tool forms, and debitage classes were significantly different from immediately above-lying Component II artifacts. Hearth features from the two components are about 2,000 years apart chronologically. Graf et al. (2015; chapter 8) also noted minimal evidence of deformation from burrowing—nowhere near the degree of disturbance implied by Thorson (2006). Without question, Powers and Hamilton's (1978) original assessment of the site's terminal Pleistocene cultural stratigraphy, summarized in chapter 3, is correct. Dry Creek stands as the type-site for central Alaska's terminal Pleistocene archaeological record—the first place in Alaska where multiple Paleolithic cultural layers were encountered in a well-stratified setting.

Terminal Pleistocene Hunter-Gatherer Behavior: Interpreting the Archaeological Components at Dry Creek

Certainly, the theoretical perspective of the 1970s Dry Creek project was functional (e.g., Clark 1954). First and foremost, it was an investigation of how the recovered archaeological materials informed us about human ecology during the terminal Pleistocene, when the mammoth-steppe ecosystem of Beringia was rapidly vanishing. In this respect, the primary question driving research at Dry Creek was an exploration of the site's role in the adaptive system of its occupants, and nearly all the major findings of the 1983 report are presented in this light. Powers's presentation of the site's physical location promotes its likely use as a hunting overlook (chapter 2), and his review of the stratigraphy and dating of the site focuses on explaining the environmental setting of the cultural components as "an open herbaceous landscape" (chapter 3). Similarly, Guthrie's exploration of the faunal assemblage was environmental (using the remains to infer an ecological setting conducive to grazers as well as cervids) and behavioral (inferring human occupations during the autumn-winter seasons and opportunistic hunting from a "hunting spike camp and processing station"; chapter 6). Hoffecker's spatial analysis further suggested that these occupations were repeated and of short durations but perhaps reflecting a broader set of activities during the time of Component II (chapter 5). Even Powers's analysis of the lithic artifact assemblages recovered from the site's two terminal Pleistocene cultural components focused on cores and tools that provide information about hunter-gatherer activities, specifically the "production and maintenance of hunting equipment" and "the procurement and processing of game" (chapter 4). Moreover, the lithic analysis highlighted the presence of two tool kits in both components: a set of large and expedient tools prepared from locally available "low-grade" tool stones and a set of smaller, more formally produced and highly curated tools made on "high-quality" tool stones transported significant distances to the site. This organizational strategy, Powers concluded, was conducive to high mobility. Without question, the functional approach to archaeology is pervasive through the 1983 report. Many years later, it behooves us to review this aspect of the original Dry Creek project, to provide an updated archaeological perspective on the site's function.

Technology

Powers's analysis of the 1970s lithic assemblage from Dry Creek was unique in that he considered three dimensions of variability—technology, morphology, and function—to document and explain the similarities and differences between Components I and II. The stress was on the identification of tool kits with a functional basis to infer prehistoric human behavior, not on the construction of culture histories based on techno-typological comparisons with other sites and assemblages. As a result, the lithic analysis focused not only on the cores and core-reduction debitage recovered to explore aspects of technology but also on the retouched artifacts, or tools, recovered to explore aspects of form and function.

The analysis began with a description of the lithic raw materials used by the Dry Creek site's early inhabitants, with Powers pointing out the probable provenance of most of the more locally procured tool stones. Among these were a dark-gray "degraded" quartzite locally ubiquitous in Dry Creek's alluvium, variably colored rhyolite that likely originated in the hills on the other side of the Nenana River from Dry Creek, and locally available gray cherts and tan chalcedonies occasionally encountered in nearby exposures of the Nenana Valley's Tertiary-aged Nenana gravels. At the same time, Powers identified some tool stones that did not seem to have a local lithology—for example, obsidian and green, black, and "ferruginous" cherts.

Since 1983, the development of portable X-ray fluorescence (pXRF) technologies has led to some modifications and fine-tuning of these raw-material procurement patterns. On the one hand, we now know that the degraded quartzite found in both terminal Pleistocene components is actually a basalt that was easily obtained from Dry Creek's bedload; its ultimate point of origin is hypothesized to be a series of eroded dikes on the north face of nearby Mount Healy (Gore 2015, personal communication). The rhyolites, too, have discernible geochemical signatures (Coffman 2015, personal communication), and with continued exploration, soon we should know their precise origins as well. Not surprisingly, basalt and rhyolite were important components of both terminal Pleistocene occupations. Obsidian artifacts, however, occurred only in Component II, and their geochemical compositions match the Batza Téna source located 330 km northwest of the site (n = 10), as well as a "Group K" obsidian of unknown origin (n = 37; Reuther et al. 2011). (Earlier, Cook [1995] suggested the presence of obsidian artifacts in the assemblage from Wiki Peak near the Canadian border, but Reuther et al. [2011] pointed out that these artifacts came from the site's middle Holocene occupation, Component IV.) The provenance of Group K obsidian is still not known, but Reuther et al. (2011) suggest that given the relatively large sizes of the artifacts made on this obsidian, as well as the presence of cortex on some of them, the source might be somewhere nearby in the central Alaska Range. Similarly, the black chert that Powers originally interpreted to represent a nonlocal tool stone is now hypothesized to have originated from the alluvium of Panguingue Creek, a neighboring tributary of the Nenana River about 5 km north of Dry Creek (Graf and Goebel 2009), but this needs to be confirmed through geochemical and mineralogical comparisons. A continued search for the local rhyolite and obsidian sources, as well as careful geochemical analyses of the other tool stones represented in the Dry Creek assemblage, will ultimately lead to a precise understanding of raw-material procurement and technological organization at the site, a primary objective of Powers's original study.

Powers's technological analysis focused on cores and core-related debitage. For Component I, not much could be said—no cores occurred in the assemblage—but for Component II, Powers laid out in some detail the technology utilized in the creation and rejuvenation of wedge-shaped cores and the removal of microblades from them. For the time, this was one of the most detailed studies of microblade technology in Alaska, and it was among the first to recognize that early Alaskans, like their counterparts in northeast Asia, followed multiple recipes to produce wedge-shaped cores and microblades, not just the classic northeast Asian "Yubetsu" technique of preparing a wedge-shaped core's platform by removing a single longitudinal "ski spall," but also the more homegrown "Campus" method of side-striking a series of platform-rejuvenation flakes to fashion a platform (e.g., Hayashi 1968; Morlan 1970; see also Mobley 1991). This observation has been echoed strongly in the recent analyses by Gómez Coutouly (2012), who demonstrates that in most early Alaskan sites with evidence of microblade technology—including Component II at Dry Creek—the Campus method is most common and potentially represents an early Holocene variant of the Diuktai Tradition, one with roots in the more ancient Yubetsu method, which is only prevalent in Alaska's oldest known microblade industry, Cultural Zone 4 at the Swan Point site (Gómez Coutouly 2012; Holmes 2011).

Powers's technological analyses of the lithic assemblages from Components I and II focused on microblades, cores, and core-related debitage (e.g., core tablets and other platform-preparation or platform-rejuvenation spalls); he did not present a study of the more mundane parts of the debitage assemblage—the cortical spalls, core-reduction flakes, biface-thinning flakes, and so on—useful in defining technological activities carried out at hunter-gatherer sites. Graf and Goebel (2009) attempted to rectify this by presenting a debitage analysis that Goebel initiated as a graduate student at the University of Alaska in the late 1980s (see Goebel et al. 1991). Their analysis included a complete sample of the Component I assemblage (n = 4,491) and a partial sample of the Component II assemblage (n = 14,754). Contrary to Powers's original interpretation, the study of the debitage, when compared to cores and tools, suggested that both Component I and Component II were dominated by local tool stones, and very few actual cores or tools could be shown to have been transported from nonlocal sources. The "exotic" materials seemed to be limited to only the Batza Téna obsidian and an argillite that occurred in Component II.

Similarly, Graf and Goebel (2009) expanded their debitage analysis to include cores and tools to investigate whether elements of tool-stone selection occurred in the Dry Creek assemblages, another interpretation that Powers made in the original report. What they found again partially contradicted the original interpretation. There was little to no meaningful tool-stone selection in Component I, in as far as the production of expedient versus formal tools was concerned, but in Component II, side scrapers and burins were disproportionately manufactured on cherts, bifaces on basalt, and informal tools and microblades on cherts and obsidian.

Although the recent analyses of the site's lithic assemblages discussed here have led to some fine-tuning of Powers's team's interpretations of human behavior at the Dry Creek site, they have not altered the major functional argument of part 1 of this book. On the one hand, the lithic assemblages of both Component I and Component II reflect the production and maintenance of hunting equipment and the procurement and processing of large-mammal prey. On the other hand, the range of specific technological activities varies between the components.

Subsistence

Until the 1970s excavations at Dry Creek, no terminal Pleistocene Alaskan sites had yielded preserved faunal remains that could be used to interpret early hunter-gatherer subsistence in Beringia. Numerous paleontological discoveries had been made across Beringia indicating the presence of a "mammoth fauna" characterized chiefly by three large-mammal species: woolly mammoth (*Mammuthus primigenius*), steppe bison (*Bison priscus*), and horse (*Equus caballus*; Guthrie 1968, 1982). Across the Bering Land Bridge in Siberia, woolly rhinoceros (*Coelodonta antiquitatis*) was a fourth common megamammal species of the northern "mammoth steppe" (Vereshchagin and Baryshnikov 1982; Vereshchagin and Kuz'mina 1984). In Alaska, leading up to the excavations at Dry Creek, none of these had been discovered in an archaeological context, so the association of humans and the extinct fauna of the Pleistocene had not yet been proven with empirical evidence. Moreover, because of problems inherent in the conventional radiocarbon dating of bone prior to the "AMS revolution" of the 1980s, timings of the extinctions of these and other less frequently occurring Pleistocene fauna were not well understood.

Guthrie's analysis of the faunal remains from Dry Creek, however, provided the first link between Paleolithic humans and extinct megamammals in Alaska. The bison teeth recovered indicated the presence of the large form of northern bison, *Bison priscus*. With them, Guthrie interpreted that dwarfing of this species had not yet happened by the time of the Component II occupation and that the region's early people indeed hunted this large-bodied mammal of the mammoth steppe, as they did across the rest of the "Great Bison Belt" of northern Eurasia and the central North American Plains during the late Pleistocene. Guthrie further reasoned that because the bison teeth from Dry Creek represented males, bison hunting carried out from the site was opportunistic, not specialized herd-driving, which would have led to a preponderance of female bison teeth. In this respect, for Guthrie, the emerging pattern of bison hunting in central Alaska fit the Old World pattern as opposed to the Paleoindian Plains pattern. Since the early 1980s, the series of Paleolithic occupations recovered from Broken Mammoth, Swan Point, Upward Sun, and Gerstle River Quarry indeed indicate that steppe bison was the most regularly taken large-mammal species, especially after the onset of the Allerød interstadial (Potter et al. 2013). Moreover, no bison drive sites have yet been found in Beringia; instead, where well-preserved late Pleistocene faunal assemblages have been found, they repeatedly present diverse arrays of fauna (Potter 2007; Potter et al. 2013; Yesner 1996, 2001, 2007; Yesner et al. 2011). While at some sites bison might have been the targeted prey (e.g., Gerstle River Quarry), at others they appear to have been opportunistically hunted along with other prey (Potter et al. 2013). In these respects, the Beringian record of bison hunting still mirrors that of Siberia during the Upper Paleolithic (Guthrie 1990).

The discovery of wapiti teeth in the Component I deposit was unexpected and has since become a very important benchmark for our understanding of the late-glacial paleoecology of Beringia. Although common among Siberia's Upper Paleolithic sites (Goebel 1999, 2002), this species has been only rarely encountered in North America's Paleoindian record. In fact, at the time of Guthrie's analysis of the Dry Creek fauna, virtually no cases of early Paleoindian hunting of *Cervus canadensis* had been documented, and this is still the case today (e.g., Cannon and Meltzer 2004, 2008; Haynes 2002; Haynes and Hutson 2013). In Beringia, however, Dry Creek's unique record of wapiti hunting has been replicated at a number of sites—for example, Broken Mammoth and Mead from the terminal Pleistocene and earliest Holocene (Potter et al. 2013; Yesner 1996, 2001, 2007; Yesner et al. 2011). The importance of wapiti has not been lost on paleoecologists investigating terminal Pleistocene Beringia: Hoffecker and Elias (2007) recognized the importance of wapiti as a fossil indicator of the emerging shrub tundra during the Allerød interglacial of the region (but see Potter et al. 2013), and ancient DNA studies indicate that wapiti spread from Asia to America during end Pleistocene times, as the Bering Land Bridge and mammoth steppe waned (Meiri et al. 2013). The recent discovery that the Clovis antler rods from the Anzick burial, Montana, were from wapiti further suggests a strong ecological link between the first human explorers of Beringia and America and this cervid species (Rasmussen et al. 2014).

The presence of Dall sheep in both components suggested that this might have been the target species of the Dry Creek site's early human inhabitants, and Guthrie's chapter provides much insight into Dall sheep ethology and human hunting of this species (see chapter 6). Dry Creek, he infers, was used for sheep hunting in fall or early winter, this being the time that sheep are today found feeding on low-elevation mountain slopes near the Nenana River. Since the Dry Creek excavations, Dall sheep remains have been found in other early period sites (Potter et al. 2013), but with the exception of Broken Mammoth Cultural Zone 3 (Yesner 1996, 2001), these occur most frequently in mountain or near-mountain contexts like Dry Creek.

Some obvious faunal species of the mammoth steppe—horse and mammoth—were missing from Dry Creek, and if they had been hunted from the site, one would expect their presence in the faunal assemblage, despite the relatively poor preservation of remains. We now know that their absence is likely because by the time of the Component I occupation, these two important species of the mammoth steppe might have become locally extinct. Forty years after the completion of the Dry Creek excavation, only one archaeological site in Alaska has yielded remains of horse, Swan Point, which is also the oldest-known site, at ~14,000 cal yr BP. In fact, the directly dated horse bone from Swan Point is the youngest known evidence of Pleistocene horse from any context (archaeological or paleontological) in Alaska, at 11,950 ± 100 ^{14}C yr BP (13,800 cal yr BP). Mammoth remains, too, occur in the ~14,000 cal yr BP occupation at Swan Point, but this species appears to have persisted for some centuries in central Alaska, given the direct AMS ^{14}C dates of 11,540 ± 140 ^{14}C yr BP (13,370 cal yr BP) and 11,500 ± 160 ^{14}C yr BP (13,340 cal yr BP), respectively, on mammoth ivory from the Cultural Zone 4b of the Broken Mammoth site (Yesner 1996, 2001) and a paleontological find from near Galena (Guthrie 2006). The age of Dry Creek Component I reported here by Graf (chapter 7) overlaps with these youngest ages for mammoth, suggesting that the site's occupants might have seen mammoth during their lifetimes, but the animals would have become quite rare in the Alaskan landscape by that time, if not locally extinct.

Another interesting aspect of Dry Creek's subsistence record was Guthrie's identification and analysis of gastroliths, fossil gizzard stones found in the site's loess deposits (chapter 6). Since the excavations at Dry Creek, work at other Nenana Valley sites like Walker Road and Owl Ridge have yielded similar concentrations of rounded gravels; however, in these contexts, the rounded stones were interpreted to represent geological slopewash processes— their natural accumulation as the surface of the loess mantle was reworked by moving water during prehistoric times. We expect that this was the origin of the Dry Creek "gastrolithis," too, especially given that Guthrie did not find a good fit between the sizes of the stones from Dry Creek's sediments and actual gastroliths recovered experimentally from modern birds. Despite this, Guthrie was correct in assuming that central Alaska's terminal Pleistocene humans subsisted partly on birds. Remains of waterfowl (e.g., swans and geese), ptarmigan, and other birds have been found in the region's early components (Potter 2007; Potter et al. 2013; Yesner 1996, 2001), indicating that Beringia's early occupants were not big-game-hunting specialists.

Site Structure and Settlement Organization

The spatial analyses of the Dry Creek site (chapter 5) were accomplished before the development of the sophisticated computer-mapping software and statistics that we use today to analyze the structure of archaeological sites. Nonetheless, Hoffecker was able to map piece-plotted cores and tools as well as relative distributions of debitage for both components, calculating that roughly 70 percent of the archaeological debris came from readily discernible horizontal concentrations—3 in Component I and 14 in Component II. Some refitting was accomplished, too; however, these studies were primarily directed at reconstructing core-reduction processes, not the spatial organization of the site's occupation floors. Likewise, although rodent burrows and microfault features were mapped during the excavation, leading to the recognition that the components had been partially disturbed near the site's bluff edge, data important for measuring the effects of cryoturbation (e.g., trend and plunge of artifacts) were not gathered. The overall vertical distributions of artifacts, though, indicated that although any microstratigraphic patterning within the components had been obliterated, the components themselves were largely intact. These observations and interpretations were confirmed by Graf's recent geoarchaeological study (chapter 8). We agree with Hoffecker's initial assessment that there was likely little displacement of artifacts horizontally, so the artifact clusters identified in the spatial analysis reflected late Pleistocene human behavior.

The most important interpretation of the spatial analysis relates to the overall character of the site's 17 recognized artifact clusters. They present a series of subassemblages that appear to represent tool kits focused on the manufacture and repair of hunting weapons and initial butchery and processing of large-mammal carcasses. Within Component II, however, some fine-scale differences occur. As Hoffecker presents, five clusters (A, B, C, G, and N) are characterized by microblade cores, related microblade-core debitage, microblades, and burins, while the remaining nine clusters are characterized by bifacial points and point fragments as well as their preforms. Within the latter nine clusters, there are further differences in the proportions of high-quality and low-quality tool stones. Despite these differences, all the clusters seem to represent the same redundant pattern of preparing and repairing hunting implements, whether they were organic points with microblade insets or lithic bifacial points.

Hoffecker also identified two other patterns that were repeated throughout the site's clusters. Fire hearths were ephemeral and unprepared, and no evidence of site furniture or dwelling structures was identified.

Together, these observations suggested to Powers, Guthrie, and Hoffecker that the Dry Creek site was redundantly used as a short-term hunting camp by small special-task groups. This "spike camp" was used primarily as a hunting overlook, where a small party consisting of several adults (presumably males) could spot game, manufacture and repair hunting weapons, and bring carcasses for initial processing and retransport to a nearby base camp. Guthrie (chapter 6) further hypothesized that the season of use primarily was during the autumn and winter, when Dall sheep would have ranged close to the site. The bluff-edge spike-camp function of Dry Creek has since been replicated at numerous localities around the Nenana Valley and elsewhere in central Alaska. Finding the associated base camps, however, has not been so easy. Although Hoffecker (chapter 5) suggested that they could have been in the Tanana Lowlands north of the Nenana Valley, the nearby Walker Road site is one possibility. Its cultural component is close in age to Dry Creek's Component I, but its tool assemblage is more diverse, containing a variety of small end scrapers, gravers/drills, and bifacial knives, as well as some large scraper-planes and choppers (Goebel et al. 1996). Moreover, two substantial hearths were found there, centrally located in dense artifact clusters potentially representing light surface tents (Goebel and Powers 1989). Farther afield, as Hoffecker predicted, base camps are beginning to emerge along the middle Tanana River—for example, at the Upward Sun site, where remains of a dwelling and human infant burials have been found in a context synchronous to the newly dated hearth from Component II (chapter 8; Potter et al. 2011, 2013). Somewhat closer is the newly discovered McDonald Creek site (Goebel et al. 2015), which might eventually prove to represent an even earlier base camp.

Thus, to a significant extent, the conclusions presented by Powers, Guthrie, and Hoffecker on the context, chronology, and function of the Dry Creek site during the late Pleistocene remain valid and have been strengthened by new research.

Dry Creek and the Settlement of Beringia

Most of the debates generated by the 1983 Dry Creek site report presented as part 1 of this book (and subsequent publications on the site) revolve around the interpretation of the lower archaeological components and their wider significance for the peopling of Beringia and the Americas. Powers, Guthrie, and Hoffecker presented a detailed description of the artifacts and both their vertical (stratigraphic) and their horizontal distributions, and they noted significant variability in the technology and typology of the artifacts. They also discussed the potential sources of the variability in terms of function and cultural affiliation but left the resolution of the problem to subsequent research, noting that "the answer does not appear to inherently lie within the data from Dry Creek, or other known sites, but can only come from future sites."

The discussion of technological and typological variability at Dry Creek was confined to Component I and Component II—more specifically, to the technology and typology of the bifacial points in these two components. The uppermost component (IV) was assigned by the authors to the previously defined Northern Archaic Tradition on the basis of point typology (i.e., the presence of side-notched points; see Appendix A). The concentrations of artifacts containing evidence of microblade technology in Component II were assigned to another previously defined entity, the "Denali Complex" (or American Paleoarctic Tradition) on the basis of the wedge-shaped microcores, core parts, microblades, and multifaceted burins. The Component I assemblage, however, did not exhibit clear parallels with any known industries in Alaska, and the authors of the report did not assign this component to any previously defined archaeological entity. Its small triangular bifacial points were seemingly new to the region's record. Nor did Powers, Guthrie, and Hoffecker propose a new complex or industry on the basis of this single assemblage.

Compounding the problem of Component I's bifacial points and lack of microblades was that some of the concentrations of artifacts in Component II lacked diagnostic items of the Denali Complex but contained lanceolate bifacial point fragments similar to those of the Late Paleoindian Tradition of the Great Plains, temperate North America (specifically, the Hell Gap Complex). Although these bifacial point fragments were recovered from the same component as the microblades and related artifacts assigned to the Denali Complex, they had not been identified previously as diagnostic of this complex (e.g., West 1967, 1981). This spatial variability within Component II introduced the possibility that the Denali Complex contained a functionally and/or seasonally distinct lanceolate bifacial point technology and that the lowermost component also might represent a functional/seasonal variant of the same industry.

The idea was consistent with the interpretation of the Dry Creek site as a seasonally occupied large-mammal hunting camp.

In the late 1970s and early 1980s, few other archaeological sites in Alaska were known for Powers, Guthrie, and Hoffecker to draw comparisons. At the Healy Lake site in the upper Tanana basin, small bifacial points had been recovered from a level dating to as early as 13,000 cal yr BP in association with microblades (Cook 1969, 1996). The small bifacial points were eventually labeled Chindadn points, and they subsequently have turned up at other sites in eastern Beringia (Alaska/Yukon) and recently in northeast Asia (e.g., Easton et al. 2011; Goebel et al. 1996; Hoffecker 2001; Pitulko et al. 2014). While the authors of the Dry Creek report suggested the Healy Lake record could represent an example of points similar to those in Component I and that it appeared to represent a multiple-activity camp (i.e., less likely to contain a functional subset of the full spectrum of artifacts produced by its occupants), they also noted the possibility of postdepositional mixture of different industries in the site's relatively shallow eolian context. Thus in the early 1980s, Powers, Guthrie, and Hoffecker could only offer hypotheses for future testing to explain the assemblage variability in Dry Creek's early components.

During the past three decades, many new discoveries have been made on both sides of the Bering Strait that bear directly or indirectly on the problem of artifact variability at Dry Creek. Most of these discoveries pertain specifically to the problem of explaining the variability observed in the lowermost component. Because Component I was stratigraphically below Component II and had yielded at least one older radiocarbon date, it was viewed by some as a possible "premicroblade" (or "pre-Denali") industry in central Alaska (e.g., Haynes 1982). Thus considerable attention was devoted to finding and dating other sites containing assemblages of comparable age (i.e., antedating 13,000 cal yr BP). Conversely, as described previously and in chapter 7, there was discussion—during the 1980s and later—of the possibility that the apparent stratigraphic differences between Component II and Component I might be due to downward displacement of artifacts from the former to the latter by postdepositional disturbance processes (e.g., Dumond 2001:199; Thorson 2006).

The earliest discoveries of assemblages of comparable age to Component I occurred in the Nenana Valley, where many locations were found to contain deep eolian stratigraphy similar to that found at Dry Creek. Although both the Walker Road and the Moose Creek sites had been discovered and investigated by 1983, relatively little information was available from either (especially the former) at the time the Dry Creek report was written. A radiocarbon date obtained from Walker Road in 1984 suggested that the main (lower) component at this site was at least as old as Component I at Dry Creek (Powers and Hoffecker 1989). Later excavations yielded more supporting dates and a large assemblage containing teardrop-shaped bifacial points (i.e., Chindadn points) and end scrapers, but this assemblage lacked any evidence of microblade technology (Goebel et al. 1991, 1996). At Moose Creek, an early test unit produced two lanceolate point fragments and a possible microblade fragment from the lowermost sedimentary unit (Hoffecker 1985), but more extensive excavation in the 1990s distinguished two lower components in this unit, and their ages and contents were similar to Dry Creek's Components I and II (Pearson 1999, 2000). The lowermost component yielded Chindadn points (Pearson 1999:337). Lanceolate points were recovered, too, but during initial testing, so to this day their assignment to either of the site's two lower components remains problematic (Hoffecker 2011:171). Typologically, they resemble the lanceolate point fragments from Dry Creek's Component II (i.e., similar to point types of the late Paleoindian complexes of temperate North America) and not the bifacial point found in Component I (subsequently included in the category of Chindadn points). At Panguingue Creek, also in the Nenana Valley, yet another small assemblage of lanceolate points without microblades was recovered and dated to roughly the same time as Dry Creek's Component II, about 11,400 cal yr BP (Goebel and Bigelow 1992). In the neighboring Teklanika Valley, excavations at the Owl Ridge site during the 1980s had uncovered a fourth potential Nenana Complex occupation lacking microblades (Hoffecker et al. 1996; Phippen 1988). Thus by the start of the 1990s in the northern foothills of the Alaska Range, a repeating pattern had emerged: during the Allerød interstadial prior to 13,000 cal yr BP, small triangular and teardrop-shaped bifacial points predominated in the record, but subsequently during the Younger Dryas stadial and early Holocene, these were seemingly replaced by microblade technologies and lanceolate bifacial point technologies. Outside the Nenana Valley, in the nearby middle Tanana basin, a similar record began to emerge, with the reporting of a small assemblage similar to that of Component I at Dry Creek from the Chugwater site (Lively 1988, 1996). Although undated, it underlaid a Denali assemblage with early Holocene dates.

Equipped with this record, in a 1989 paper published in *American Antiquity*, Powers and Hoffecker defined the Nenana Complex on the basis of Component I and the other apparently pre-Denali assemblages with small bifacial points (and end scrapers) from the Nenana Valley. The lanceolate point fragments in Component II were grouped with the Denali Complex (Powers and Hoffecker 1989:272–77). The *American Antiquity* paper was followed by a systematic comparison of Nenana Complex and Clovis Complex assemblages (Goebel et al. 1991) and a January 1993 paper in *Science* that expanded the sites assigned to the Nenana Complex to include Chugwater in the Tanana basin (Hoffecker et al. 1993:49, figure 3). The characteristic stone artifacts of the Nenana Complex also were expanded to include side scrapers, planes, gravers, and *pièces esquillées* (now considered to represent bipolar cores). And, in addition to a postulated link between the Nenana and Clovis complexes, an explicit parallel was drawn between the two successive Alaskan industries (i.e., Nenana and Denali) and the two early migrations from Asia to Beringia (i.e., Amerind and Na-Dene) identified by Greenberg et al. (1986) in their "three-wave model" (Hoffecker et al. 1993:52). Hoffecker, Powers, and Goebel concluded that Beringia had been settled before 13,000 cal yr BP by people who made small bifacial points; the microblade industry represented a migration by a different group of people during the Younger Dryas cold event (which began after 13,000 cal yr BP). They noted that Kamchatka (southwest Beringia) had produced a record similar to that of central Alaska, where assemblages containing bifacial points (primarily stemmed bifacial points) had been found stratigraphically below microblade assemblages of Younger Dryas age (Dikov 1977, 1979). They conceded, however, that an "obvious source for the early Beringian complex" in northeast Asia was lacking (Hoffecker et al. 1993:52).

In the months following the *Science* paper, two major discoveries altered the landscape of the debate. During the summer of 1993, traces of microblade technology were unearthed from a level at the Swan Point site in the middle Tanana basin that antedated Component I at Dry Creek (Holmes 2001; Holmes et al. 1996). Although the small bifacial points considered diagnostic of the Nenana Complex were never found in this assemblage, it negated the status of the latter as a "premicroblade" industry in central Alaska. It also added weight to the argument that the Nenana assemblages were part of a larger Paleoarctic or Beringian Tradition (e.g., Holmes 2001; Meltzer 2001).

During the same year, new dates from the Mesa site in the Brooks Range suggested that lanceolate points similar to those of the Paleoindian Tradition in midlatitude North America were contemporaneous with the Nenana Complex and a more plausible source for the Clovis Complex (Kunz and Reanier 1994). As in the case of the Nenana Complex, this implied the presence of an industry (and the people who produced it) in Beringia before the arrival of microblade technology. And, as in the case of the Nenana Complex, an obvious source for the industry in northeast Asia (and in western Beringia) was lacking.

Additional radiocarbon dating of the Mesa site indicated, however, that most or all of its artifacts dated to the Younger Dryas (less than 13,000 cal yr BP) or later (Kunz et al. 2003). The assemblage, along with contemporaneous sites in northern Alaska yielding similar assemblages (e.g., Bever 2001; Reanier 1995, 1996), was assigned to the Mesa Complex of the Northern Paleoindian Tradition. The dating of the assemblages assigned to the Mesa Complex seemed to confirm that the late Paleoindian point types of northern Alaska (often alternatively called Plainview, Angostura, Agate Basin, or Hell Gap) were—as was long suspected (e.g., MacNeish 1963; Willey 1966:70–72; Wormington 1957)—derived from the Great Plains. The late Paleoindian point types of the plains were, in turn, derived from earlier Paleoindian industries of midlatitude North America (e.g., Clovis, Goshen, and Folsom). Paradoxically, in this model, the northern Paleoindian points reflect a migration into Beringia from midlatitude North America, rather than the reverse (e.g., Dumond 2001:202; Dumond 2011). Most recently, dating of northern Alaska's fluted-point complex to late Paleoindian times (Goebel et al. 2013; Smith et al. 2013; Young and Gilbert-Young 2007) has reinforced this notion. The expansion of steppic habitat and bison populations in eastern Beringia during the Younger Dryas (Shapiro et al. 2004) probably encouraged the northward movement of Plains bison hunters; not surprisingly, steppe bison remains are common in many of the Younger Dryas occupations in eastern Beringia (Hoffecker 2011:174).

A related issue that has remained unresolved, however, was the status of the lanceolate points in central Alaska, which—in some places at least—were associated with artifacts assigned to the Denali Complex. In the late 1990s, a Mesa Complex site (containing Hell Gap and Agate Basin points) was reported from southwestern Alaska (Ackerman 2001), and in 2005, one of the authors of the 1983 Dry Creek report suggested that the lanceolate bifacial points in Component II at Dry Creek (as well as the lanceolate point fragments from Moose Creek) might belong to the Mesa—not the Denali—Complex (Hoffecker 2005). If this is the case, then the apparent co-occurrence of diagnostic artifact forms from two separate industries conceivably reflects the occupation of some of the same site locations by different groups during the Younger Dryas. The pattern might extend into the Tanana basin, where the first Plains-like point forms (Agate Basin, Angostura) were discovered in the 1930s (Collins 1964:87; Rainey 1940). Alternatively, the lanceolate point and microblade industries might represent alternative hunting weaponry used by the same human groups. After all, both technologies continued temporally through the Holocene at numerous sites in the interior of Alaska, with large lanceolate points and microblade-inset points persisting as projectile weapons into the late prehistoric period (Dixon et al. 2005; Esdale 2008; Hare et al. 2012).

Continuing research at Swan Point during the first decade of the twenty-first century clarified the dating and character of its early microblade assemblage in the lowest occupation level (Holmes 2001; Hoffecker and Elias 2007:118–22). The occupation is firmly dated to roughly 14,000 cal yr BP, and the microblade technology exhibits a core preparation technique (Yubetsu) that is found in the Diuktai industry of northeast Asia. The earliest Diuktai level at Diuktai Cave on the Aldan River dates to about 15,000 cal yr BP and probably represents a movement of people from southern Siberia into the Lena basin after the Last Glacial Maximum (LGM; Graf 2013). The Swan Point assemblage might represent the subsequent movement of the same industry and people into Beringia as the late-glacial climate continued to warm (Gómez Coutouly 2012).

Thus, three decades after original production of the Dry Creek report, archaeological proxies for two migrations into eastern Beringia had been documented: (1) a movement of people making wedge-shaped microblade cores from northeast Asia into Beringia roughly 14,000 cal yr BP (i.e., Diuktai) and (2) a movement of people making lanceolate (and fluted) points from temperate North America into eastern (but not western) Beringia during the Younger Dryas cold interval (~12,000 cal yr BP; i.e., Mesa). Component II at Dry Creek, dating to the Younger Dryas, contained traces of at least one of these events (a younger microblade assemblage) and possibly both (spatially segregated debris concentrations containing lanceolate points similar to those of the Mesa Complex). Component I and the other assemblages assigned to the Nenana Complex remained highly controversial—regarded by some as an activity facies of the microblade industry (Bever 2001; West 1996a; Holmes 2001) and by others as a separate complex and possibly representative of a separate movement of people into Beringia (Clark and Gottardt 1999; Dixon 1999, 2001; Dumond 2011; Hoffecker 2001).

At the beginning of the twenty-first century, some important developments occurred in western Beringia that potentially relate to the early record at Dry Creek. The first of a group of sites near the mouth of the Yana River dating to roughly 30,000 cal yr BP was reported (Pitulko et al. 2004). The Yana River sites later yielded a large number of radiocarbon dates unequivocally demonstrating that the settlement of northwestern Beringia had taken place before the beginning of the LGM (Pitulko et al. 2013). Beringia was either occupied continuously from 30,000 cal yr BP to the time of the early components at Swan Point and Dry Creek, or it was abandoned during the LGM and subsequently reoccupied. Although the Yana River assemblages do not exhibit significant diagnostic commonalities with any of Beringia's late-glacial industries, their discovery significantly altered the broader context of Beringian archaeology and potentially the significance of the Nenana Complex.

Another development in western Beringia was renewed research at Ushki in Kamchatka (southwest Beringia), which provided a more reliable chronology of the lowest horizons at a group of sites originally excavated in the 1960s and 1970s (Dikov 1977, 1979). New dates from Ushki confirmed the presence of an industry containing small stemmed points dating to about 13,000 cal yr BP, coeval with Component I at Dry Creek, underlying a microblade assemblage dating to roughly the same age as Component II (Goebel et al. 2003, 2010). The Ushki sites appeared to be long-term occupations (with traces of dwelling structures) and less likely than the Nenana Complex sites to represent a limited-activity locus (i.e., the absence of microblade technology is less easily explained as a product of site function bias). The repeated patterns at Ushki and in the Nenana Valley suggest the presence of industries other than Diuktai in Beringia before the beginning of the Younger Dryas. However, the relationship of Ushki and Nenana remains unclear, as do their relationships to Asian industries farther south in the Japanese archipelago or farther west in Siberia (e.g., Goebel 2004; Nagai 2007).

In 2007, an international team of geneticists introduced a model suggesting that a human population had been present throughout the LGM in Beringia (Tamm et al. 2007; see also González-José et al. 2008; Mulligan and Kitchen 2013; Mulligan et al. 2008). The original hypothesis ("Beringian Standstill") was based on the analysis of mitochondrial DNA, and more recent studies of nuclear DNA seem to support it, although with a much shorter incubation period in Beringia (Raghavan et al. 2015). Although there is not yet compelling archaeological evidence for a longstanding population that inhabited Beringia through the LGM, paleoecological data indicate that central Beringia supported a mesic tundra environment with some woody plants (i.e., shrubby willow and possibly birch; Brubaker et al. 2005; Elias and Crocker 2008), a potential habitat for humans to have persisted. Meiri et al. (2013) similarly propose, based on ancient DNA evidence, that wapiti (*Cervus* sp.) persisted in western Beringia through the LGM. If a small human population was concentrated in a refugium in either central or northwestern Beringia during the LGM, its sites would possess low archaeological visibility today (i.e., because of the predicted small size of the human population as well as the possible inundation of its refugium by rising sea levels at the end of the Pleistocene; Hoffecker et al. 2014).

The Beringian Standstill model predicts that people were already present in Beringia at the time that the microblade industry arrived at Swan Point, about 14,000 cal yr BP, thus offering a possible explanation for why the Nenana Complex (and early Ushki Complex) might lack any obvious source in post-LGM Asia. In other words, the occupants of Dry Creek during the time of Component I might represent the descendants of people who had occupied a Beringian refugium between > 25,000 and 15,000 cal yr BP. Recently, new field research at Berelekh in northwestern Beringia yielded obvious Chindadn points in a context dating to ~14,900–13,500 cal yr BP (Pitulko et al. 2014), indicating a shared Beringian cultural trait on both sides of the land bridge. And, as described in chapter 7, Graf et al. (2015) have reported new dates of ~13,500 cal yr BP on Component I at Dry Creek. These developments indicate that the Nenana Complex (*sensu lato*, including sites like Berelekh) is either contemporaneous to or slightly older than the Diuktai-like microblade industry at Swan Point and is present in both western and eastern Beringia before the Younger Dryas. In the years to come, as we continue to learn more about the archaeology of Beringia predating 14,000 cal yr BP, the relationship of the Nenana Complex to the Yubetsu industry at Swan Point, as well as the pre-LGM industry at Yana, will become clear. At the same time, behavioral studies like those discussed earlier in this chapter (e.g., Goebel 2011; Gore and Graf, in press; Graf and Goebel 2009; Potter 2011; Rasic 2011; Wygal 2011; Yesner et al. 2011) need to be expanded to discern whether the technological and formal differences observed in Beringia's late Pleistocene industries are products of site-specific variation in tool-stone procurement, technological activities, subsistence pursuits, seasonality, or occupation duration. In other words, developing a clear understanding of the peopling of Beringia is going to require more than just finding and excavating earlier archaeological sites in Alaska.

Dry Creek and the Peopling of the Western Hemisphere

Discovery, excavation, and analysis of the Dry Creek site have had a major impact on the continuing debate over when and how Native Americans arrived in the Western Hemisphere. Undeniably, Dry Creek yielded the first compelling evidence for the presence of a Beringian industry predating 13,000 cal yr BP in Alaska, the first plausible source for the earliest known artifacts in midlatitude North America. In the early 1980s, the latter was represented by the Clovis Complex (Paleoindian Tradition) and dated to roughly the same age as Dry Creek's Component I. The potential link between the Component I assemblage and the early Paleoindian complexes of North America was recognized by several researchers shortly after the conclusion of the excavations at Dry Creek (e.g., Greenberg et al. 1986; Haynes 1982; Haynes 1987) and further explored in the years that followed (Goebel et al. 1991; Hoffecker et al. 1993). The significance of Component I was magnified by the discovery and dating of similar assemblages in other central Alaskan sites (as described previously). But the issue of Clovis origins has remained controversial and without clear resolution, because the link between the Nenana Complex and the early Paleoindian industry or industries is problematic. Despite general technological and typological similarities between the two (Goebel 2004; Goebel et al. 1991), a common diagnostic point form—which probably would have resolved the debate—remains unknown. Fluted points have not been found in any of central Alaska's late-glacial sites, while Chindadn points have not been found in any Paleoindian sites of temperate North America. Moreover, repeated discoveries of potential pre-Clovis sites south of the Canadian ice sheets (e.g., Gilbert et al. 2008; Goebel et al. 2008b; Waters et al. 2011) suggest that to find the archaeological source of the first Americans in Beringia, we need to dig deeper than the layers preserved at Dry Creek. In other words, comparisons of Nenana and Clovis might no longer be relevant, despite the new ages reported here by Graf et al. (chapter 8) suggesting that Component I at Dry Creek potentially predates the oft-cited age of Clovis by as much as three or four centuries (Waters and Stafford 2007).

Besides the Nenana-Clovis relationship developed in the 1990s, throughout the twentieth century, a number of other theories have been put forth to explain a Beringian origin for the first Americans. From the 1930s onward, archaeologists were confronted with a major dilemma in their search for material evidence of the peopling of the Western Hemisphere. The surviving remnants of the land connection between Asia and North America (i.e., far northeast Asia and Alaska/Yukon) yielded traces of two early industries—both eventually dated to the final millennia of the Pleistocene—neither one of which indicated a movement of people from Asia to North America across the Bering Land Bridge. One of these industries is represented by a microblade technology with clear roots in Asia (i.e., American Paleoarctic Tradition) but unknown in midlatitude North America or farther south (Nelson 1937). The other industry is characterized by the production of lanceolate points that closely resemble artifacts of the later Paleoindian complexes of the plains (Rainey 1940). At the time that Dry Creek was excavated (1974–77), neither Beringian industry was widely regarded as a proxy for the people who initially settled the Western Hemisphere.

In the years after the completion of the Dry Creek site report (1983), there were several different approaches to resolving the dilemma presented by the apparent lack of an Asian and/or Beringian archaeological proxy for the first Native Americans. One of them—pursued from multiple perspectives—entailed reinterpretation of the archaeological record to derive the earliest Paleoindian complexes of midlatitude North America from one of the two identified industries in eastern Beringia. In 1993, a radiocarbon date (13,500 cal yr BP) obtained on wood charcoal from a hearth at the Mesa site in the central Brooks Range suggested that the source of the early Paleoindian assemblages might indeed lie in Beringia (Kunz and Reanier 1994). Mesa was one of several localities in northern Alaska that had yielded lanceolate points (including fluted forms) similar to those of the later Paleoindian complexes. As described previously, however, additional dating of the hearths at Mesa confirmed that the site was younger—both the lanceolate and the fluted points of this industry (Northern Paleoindian Tradition) date to the Younger Dryas cold episode or later (~12,800–11,300 cal yr BP; Goebel et al. 2013; Kunz et al. 2003; Smith et al. 2013).

Another approach to resolving the dilemma posed by the Beringian archaeological record was to argue that the microblade industry was a plausible source for the early Paleoindian assemblages after all. This approach also reflected developments in the dating of early Alaskan sites, including Dry Creek, where the radiocarbon dating of Component II in the late 1970s had provided important support for the terminal Pleistocene age of the microblade industry (American Paleoarctic Tradition; Thorson and Hamilton 1977; Powers and Hamilton 1978). The dating of the early microblade assemblages in central and northern Alaska had been problematic until the Dry Creek radiocarbon chronology was reported (e.g., Anderson 1970a; West 1975).

During the 1990s, the discovery and dating of a microblade assemblage in the lowest level at Swan Point in the Tanana Valley suggested that this industry antedated Component I at Dry Creek (as well as other assemblages assigned to the Nenana Complex; Holmes 2001). This supported the argument—originally based on the co-occurrence of Chindadn points and microblades at Healy Lake in the upper Tanana Valley (Cook 1969, 1996)—that the Nenana assemblages were simply activity facies of a larger industry characterized by production of both (e.g., Meltzer 2001). Accordingly, the definition of the American Paleoarctic Tradition was expanded to include the assemblages assigned by others to the Nenana Complex (e.g., West 1996a). Moreover, Component II at Dry Creek—assigned to the microblade industry—also contained lanceolate point fragments that might actually represent the Mesa Complex (Hoffecker 2005, 2011; Hoffecker and Elias 2007). With the American Paleoarctic Tradition broadened to encompass greater time depth and at least two, and possibly three, distinct sets of artifacts, the contrast with the Clovis Complex was less stark (West 1996a:553–56).

A radically different approach to the problem was proposed in the late 1990s by researchers who despaired of identifying an archaeological link between Asia and North America. Building on a conference paper delivered in October 1999, Stanford and Bradley (2012) suggested that similarities between the lithic technology of the Clovis Complex and that of the Solutrean Upper Paleolithic industry reflected a migration across the North Atlantic during the LGM (i.e., when the Solutrean industry was present in southwest Europe). The "Solutrean Hypothesis" has been critiqued by other archaeologists (e.g., Straus 2000; Straus et al. 2005; O'Brien et al. 2014) but, more important, has been significantly weakened by the genetic evidence (including ancient DNA), which now indicates beyond a reasonable doubt that all Native Americans are derived from the late Pleistocene and Holocene populations of Asia (e.g., Raff and Bolnick 2015; Raghavan et al. 2014; Raghavan et al. 2015; Rasmussen et al. 2014).

Despite important new discoveries in western and eastern Beringia during the past two decades (many of which are described in the preceding section), it is new research in midlatitude North America (and South America) that has had the largest impact on the debate over the peopling of the Western Hemisphere. Especially significant are new dates on the earliest sites in both North and South America and genetics research, including the analysis of ancient DNA extracted from dated human remains. Radiocarbon dates on a number of recently discovered and/or dated archaeological sites indicate that people dispersed widely across North and South America by 14,500 cal yr BP (Goebel et al. 2008b; Waters and Stafford 2013). The presence of people in midlatitude North America antedates the earliest dated diagnostic Clovis artifacts by more than a thousand calendar years. The earliest firmly dated sites also antedate the oldest dated microblade assemblage in Beringia (i.e., Swan Point) by at least a few centuries.

Analysis of Native American genetics indicates a rapid dispersal out of Beringia during the late glacial, perhaps no more than 15,000 cal yr BP (Raghavan et al. 2015). The Beringian source population for the dispersal appears to have contained significant diversity (as many as 16 maternal lineages are identified in the Beringian source population), and there is little evidence of a genetic "bottleneck" in Native American population history (Achilli et al. 2013; O'Rourke and Raff 2010). The distributions of both living groups and ancient DNA suggest that much of the initial population dispersal was concentrated along the Pacific coast and adjoining western regions of North America, although there also is some indication of movement in the interior "ice-free corridor" (Perego et al. 2009). Ancient DNA from the Anzick burial in Montana suggests that at least some of the people who made the Clovis artifacts were derived from one of the groups that spread south along the Pacific coast (Rasmussen et al. 2014).

An archaeological link between the Beringian industries and the Clovis Complex remains to be found, despite the general similarities between the Nenana and Clovis assemblages (Goebel et al. 1991; Hoffecker et al. 1993). However, another potential link that deserves exploration is between Beringia's early industries and the stemmed-point assemblages of the western United States. New dates from Paisley Caves (Oregon) suggest that stemmed points in the intermountain west of North America are as old (> 13,000 cal yr BP) as the oldest diagnostic Clovis artifacts (Jenkins et al. 2012). Stemmed points also have been found at a series of sites on the Channel Islands off the southern coast of California, in slightly younger occupations that reflect an established maritime economy (Erlandson 2013; Erlandson et al. 2011), although the Channel Island stemmed points are stylistically and metrically different from those in the interior of the Great Basin. The age and context of these artifacts suggest a relationship with southwest Beringia (i.e., the stemmed point assemblages at Ushki on Kamchatka) and a movement of people along the northwest coast of North America (e.g., Erlandson 2013). According to the Beringian Standstill theory, these coastal migrants descended from the population that occupied Beringia during the LGM (Tamm et al. 2007). Significantly, two of the maternal lineages associated with this hypothesized population have been identified in ancient DNA from Paisley Caves (Gilbert et al. 2008; Jenkins et al. 2012).

Like the early Ushki Culture, the Nenana Complex—which was defined on the basis of Component I at Dry Creek and several similar assemblages from the Nenana Valley—also conceivably represents a descendant group of the hypothesized "Standstill" population. The differences in material culture between the Ushki and Nenana industries might reflect the diversity of the Beringian population inferred from the genetics of living people. Perhaps, in contrast to the people who made the stemmed points, the makers of the Nenana Complex played a very limited role in the settlement of the Western Hemisphere. The only evidence for a movement from interior Beringia into midlatitude North America (i.e., the route that the makers of the interior Nenana Complex presumably would follow) is based on the genetics of living people (Perego et al. 2009). An even more limited role may be assumed for the makers of the Beringian microblade industry—the Denali Complex artifacts in Component II at Dry Creek—which eventually spread into northwest North America and might ultimately be identified as speakers of the Na-Dene languages (Greenberg et al. 1986). Tying specific projectile technologies to different late Pleistocene populations, however, is a dangerous occupation and one that requires careful consideration of the repercussions of variable behaviors. In other words, the stark differences between the stemmed and Clovis complexes of temperate North America and their counterparts in Beringia could be the result of adaptation to new environmental opportunities and challenges or cultural drift from centuries of long-distance expansion between the arctic and temperate latitudes. Many more "Dry Creeks" will need to be intensively studied across Beringia before a clear understanding of the peopling of the Americas emerges.

Conclusions

Most of the original conclusions presented in the 1983 Dry Creek report—part 1 of this book—stand as accurate and sound. The work accomplished at the site in the 1970s has withstood the test of time, and the interpretations made by Powers, Guthrie, and Hoffecker have presaged many of the questions and controversies that have driven Beringian archaeology during the last three decades. Here we offer a set of updated conclusions based on our understanding of the site and the archaeological and paleoecological records of Beringia forty years after Powers's backfilling of the Dry Creek excavation in 1977.

1. Unequivocally, there are two stratigraphically separate cultural components represented in the geological deposits of the Dry Creek site, and both of them are intact geoarchaeologically.
2. Component I at Dry Creek is now firmly dated to 13,600–13,300 cal yr BP. It represents the earliest known human occupation of the Nenana Valley. This occupation coincides with significant warming of the Allerød interstadial, locally expressed by the spread of birch shrubs and replacement of the open tundra-steppe of the full glacial with a new biome, the shrub tundra.
3. The Younger Dryas at Dry Creek and in the surrounding foothills of the central Alaska Range, 12,900–11,700 cal yr BP, was a time of increased wind intensity as well as a cooler and drier climate. Vegetation appears to have reverted back toward the earlier tundra-steppe conditions of the full glacial, with an expansion of *Artemisia* and retraction of shrub birch, leading to a good habitat for bison.
4. Component II is more difficult to date than Component I, largely because it was found to contain so many horizontally segregated concentrations of artifacts that

could be of significantly variable ages. For certain, some of its concentrations date to about 10,700 cal yr BP, but others might date to centuries before or after this time, with a potential chronological range of 12,700–10,200 cal yr BP. Component II's association with a well-developed paleosol, however, suggests that its human occupations occurred after the Younger Dryas, during a period of significant early Holocene warming.

5. Archaeologically, Component I's assemblage of small triangular bifacial points is now known to represent a unique set of assemblages that have since been found in numerous contexts in the foothills of the central Alaska Range and called the Nenana Complex. At Walker Road, Moose Creek, and Owl Ridge, they occur in terminal Pleistocene loess deposits consistently radiocarbon dated to before 13,000 cal yr BP. The origins of the Nenana Complex are still not well understood but might relate to a population of Beringians who persisted in the Bering Land Bridge region through the LGM.

6. Component II's classic microblade industry is a derivative ultimately of the northeast Siberian Diuktai Complex, now known to be earliest expressed in central Alaska at the Swan Point site ca. 14,000 cal yr BP. Component II's microcore technology is much more variable than the antecedent Swan Point assemblage, suggesting some local adaptation of Denali Complex technology in response to central Alaska's variable and often less-than-ideal tool-stone resources.

7. Component II's lanceolate-point concentrations might represent the same people as the component's microblade concentrations, with the alternative possibility being that Dry Creek's lanceolate points reflect the Mesa Complex. It might be, too, that the variability inherent in the artifact clusters of Component II might simply reflect variable weaponries maintained in the tool kits of a single group who returned to Dry Creek at variable times of the year to hunt different animals.

8. The lack of the two keynote species of Beringia's tundra-steppe ecosystem—mammoth and horse—in the Dry Creek faunal assemblage could be the product of poor preservation, but based on the emerging record of these taxa elsewhere in central Alaska, it more likely means that these species had become extinct or were nearing extinction and were rarely encountered by human hunters by the time of the Component I occupation.

9. Dry Creek's faunal assemblage, despite being depauperate and poorly preserved, represents a dynamic and changing ecosystem—first an Allerød-aged shrub-birch community that promoted expansion of a wapiti habitat, and subsequently the possible Younger Dryas–aged resurgence of open steppe-like communities encouraging the local expansion of steppe bison.

10. Throughout the late glacial, Dry Creek served as a fall-winter spike camp from which central Alaskan hunters sought out Dall sheep and other large-animal prey. Although many other temporary hunting lookouts like Dry Creek have been found in the foothills of the central Alaska Range and dating to before 11,000 cal yr BP, none has presented a clear case of a linked base camp. Instead, the Dry Creek inhabitants' base camps might have been much farther removed, in the Tanana lowlands or even middle Tanana Valley.

The importance of Dry Creek in our understanding of the peopling of Beringia has not been diminished with time. Despite being the first site in Alaska to break the 13,000-cal-yr-BP "barrier," today its lower component remains among the oldest occupations in central Alaska and certainly the oldest in the Nenana Valley. As such, its continued consideration in the framework of early Beringian prehistory is warranted. As even older sites are searched for and investigated, the problems laid out in the 1970s Dry Creek research design and the 1980s Dry Creek report are still central to our ongoing quest to explain the peopling of Beringia and the Americas.

Through the exercise of preparing this final version of the Dry Creek report, several new goals have become clear to us that relate to developing a comprehensive understanding of the Nenana Valley's paleoecological and archaeological records for the late Pleistocene. First, we need to achieve a better understanding of the Nenana River basin's late Pleistocene glacial history through continued application of new chronometric techniques. Second, continued palynological analyses of the valley's lakes are needed to develop a fine-scale record of biotic change during the late Pleistocene and early Holocene, and this needs to be more firmly correlated to potential climate signals preserved in the region's loess profiles. When did loess deposition begin in the valley, and what truly is the significance of Sand 1 at Dry Creek and other nearby sites? Answering these questions will ultimately provide a stronger framework for understanding human adaptive change 13,500–10,000 cal yr BP. Third, we need to obtain clear chronological control over as many of Component II's variable artifact concentrations as possible. Are the microblade clusters and lanceolate-point clusters synchronous, or do they represent successive occupations tied to different climatic chronozones? Solving this problem can only be achieved by renewed excavations of still-preserved portions of the component's clusters, as Graf et al. (2015; chapter 8) have shown is possible. Fourth, new molecular methods of zooarchaeological analysis need to be applied to the existing collection of fragmented faunal remains to squeeze as much information as possible out of the site's meager faunal assemblage. At the same time, more detailed analyses of Component II should be carried out to shed light on the proximate causes of the variability noted by Hoffecker (chapter 5) and whether the differences could ultimately be the product of different populations or just the same population behaving variably. Such studies should also include Component I so that we can clearly gauge how different its assemblage is from the Component II lanceolate-point clusters. To be fully successful, these technological studies of Dry Creek's early components will require detailed information on the origins of the lithic raw materials used to construct the variable tool forms they left behind. Without precise knowledge of the lithic landscape around Dry Creek and farther afield in the foothills of the Alaska Range, we will never be able to achieve full understanding of the site's impressive record of late Pleistocene human activity.

Certainly the big future discoveries in Beringian archaeology are going to occur far from the Dry Creek site. Continued surveys of the Tanana basin likely will eventually uncover the long-sought-after base camps of the Nenana Valley's earliest human visitors. Moreover, testing of the current genomics-based peopling models and the search for antecedents of the Nenana Complex require us to turn our attention increasingly from upland settings like the Nenana Valley to the lowlands of central Beringia. Nonetheless, the archaeological and paleoecological frameworks developed by the original team of researchers who excavated the Dry Creek site, refined by the new studies reviewed here, will continue to play an important role in the science of Beringia.

REFERENCES CITED

Abbott, M. B., B. P. Funney, M. E. Edwards, and K. R. Kelts
    2000   Lake Level Reconstruction and Paleohydrology of Birch Lake, Central Alaska, Based on Seismic Reflection Profiles and Core Transects. *Quaternary Research* 53:154–66.

Abramova, Z. A.
    1967   O Vkladyshevykh Orudiiakh v Paleolita Yeniseia. *Kratkie Soobshcheniia Instituta Arkheologii* 111:12–18.
    1973   *Concerning the Cultural Contacts between Asia and America in the Late Paleolithic.* Paper prepared for the 9th International Congress of Anthropological and Ethnological Sciences, Chicago.
    1979a   K Voprosu o Vozraste Aldanskogo Paleolita. *Sovetskaia Arkheologiia* (4):5–14.
    1979b   *Paleolit Eniseia: Kokorevkaia Kul'tura.* Nauka, Novosibirsk.

Achilli, A., U. A. Perego, H. Lancioni, A. Olivieri, F. Gandini, B. H. Kashani, V. Battaglia, V. Grugni, N. Angerhofer, M. P. Rogers, and R. J. Herrera
    2013   Reconciling Migration Models to the Americas with the Variation of North American Native Mitogenomes. *Proceedings of the National Academy of Sciences* 110:14308–13.

Ackerman, Robert E.
    2001   Spein Mountain: A Mesa Complex Site in Southwestern Alaska. *Arctic Anthropology* 38(2):81–97.
    2011   Microblade Assemblages in Southwestern Alaska: An Early Holocene Adaptation. In *From the Yenisei to the Yukon: Interpreting Lithic Assemblage Variability in Late Pleistocene/Early Holocene Beringia*, edited by T. Goebel and I. Buvit, pp. 255–69. Texas A&M University Press, College Station.

Agenbroad, L. D.
    1978   *The Hudson-Meng Site.* University Press of America, Washington, DC.

Ager, T. A.
    1975   *Quaternary Environmental History of the Tanana Valley, Alaska.* Institute of Polar Studies, Report No. 54. Ohio State University, Columbus.
    1983   Holocene Vegetational History of Alaska. In *Late Quaternary Environments of the United States*, vol. 1, *The Holocene*, edited by Henry E. Wright, pp. 128–41. University of Minnesota Press, Minneapolis.

Aigner, Jean S.
- 1970 The Unifacial Core and Blade Site on Anangula Island, Aleutians. *Arctic Anthropology* 7(2):59–88.
- 1978 *The Lithic Remains from Anangula: An 8500 Year Old Aleut Coastal Village*. Urgeschichtliche Materialhefte, No. 3. Tubingen.

Alley, R. B.
- 2000 The Younger Dryas Cold Interval as Viewed from Central Greenland. *Quaternary Science Reviews* 19:213–26.

Anderson, Douglas D.
- 1968a A Stone Age Campsite at the Gateway to America. *Scientific American* 218(6):24–33.
- 1968b *Early Notched Point and Related Assemblages in the Western American Arctic*. Paper presented at the Annual Meeting of the American Anthropological Association, Seattle.
- 1970a *Akmak: An Early Archeological Assemblage from Onion Portage, Northwest Alaska*. Acta Arctica, No. 16. University of Arizona Press, Tucson. Copenhagen.
- 1970b Microblade Traditions in Northwestern Alaska. *Arctic Anthropology* 7(2):2–16.
- 1978 *Continuity and Change in the Prehistoric Record from North Alaska*. Alaska Native Culture and History, Senri Ethnological Studies No. 4. National Museum of Ethnology, Osaka, Japan.

Anderson, Patricia M., Mary E. Edwards, and Linda B. Brubaker
- 2004 Results and Paleoclimate Implications of 35 Years of Paleoecological Research in Alaska. *Development in Quaternary Science* 1:427–40.

Bacon, Glenn, and Charles Holmes
- 1980 *Archaeological Survey and Inventory of Cultural Resources at Fort Greely, Alaska 1979*. Final report prepared by Alaskarctic for the Corps of Engineers, Alaska District.

Ballenger, Jesse A. M., Vance T. Holliday, Andrew L. Lowler, William T. Reitze, Mary M. Prasciunas, D. Shane Miller, and Jason D. Windingstad
- 2011 Evidence for Younger Dryas Global Climate Oscillation and Human Response in the American Southwest. *Quaternary International* 242:502–19.

Behrensmeyer, Anna K.
- 1978 Taphonomic and Ecologic Information from Bone Weathering. *Paleobiology* 4(2):150-62.

Bell, R. H. V.
- 1969 The Use of the Herb Layer by Grazing Ungulates in the Serengeti. In *Animal Populations in Relation to Their Food Resources*, edited by A. Watson, pp. 111–24. Blackwell, Oxford.

Bever, Michael R.
- 2001 An Overview of Alaskan Late Pleistocene Archaeology: Historical Themes and Current Perspectives. *Journal of World Prehistory* 15(2):125–91.
- 2006 Too Little, Too Late? The Radiocarbon Chronology of Alaska and the Peopling of the New World. *American Antiquity* 71(4):595–620.

Bigelow, Nancy H.
   1992 *Analysis of Late Quaternary Soils and Sediments in the Nenana Valley, Central Alaska*. MA thesis, Department of Anthropology, University of Alaska Fairbanks.

Bigelow, Nancy H., Jim Begét, and Roger Powers
   1990 Latest Pleistocene Increase in Wind Intensity Recorded in Eolian Sediments from Central Alaska. *Quaternary Research* 34:160–68.

Bigelow, Nancy H., and Mary E. Edwards.
   2001 A 14,000 Yr Paleoenvironmental Record from Windmill Lake, Central Alaska: Lateglacial and Holocene Vegetation in the Alaska Range. *Quaternary Science Reviews* 20:203–15.

Bigelow, Nancy H., and W. Roger Powers
   1994 New AMS Dates from the Dry Creek Paleoindian Site, Central Alaska. *Current Research in the Pleistocene* 11:114–16.
   2001 Climate, Vegetation, and Archaeology 14,000–9000 Cal Yr BP in Central Alaska. *Arctic Anthropology* 38:171–95.

Binford, L. R.
   1978 Dimensional Analysis of Behavior and Site Structure: Learning from an Eskimo Hunting Stand. *American Antiquity* 43(3):330–61.
   1981 *In Pursuit of the Past: Decoding the Archaeological Record*. Thames and Hudson, New York.
   1983 Long Term Land Use Patterns: Some Implications for Archaeology. In *Lulu Linear Punctuated: Essays in Honor of George Irving Quimby*, edited by R. C. Dunnell and D. K. Grayson, pp. 27–53. Museum of Anthropology Anthropological Papers No. 72. University of Michigan, Ann Arbor.

Blott, Simon J., and Kenneth Pye
   2012 Particle Size Scales and Classification of Sediment Types Based on Particle Size Distributions: Review and Recommended Procedures. *Sedimentology* 59:2017-96.

Bordes, F.
   1961 *Typologie du Paleolithique Ancien et Moyen*, 2nd ed. Memoire No. 1. Publications de l'Institute de Prehistoire de l'Universite de Bordeaux, Bordeaux.

Bottema, S.
   1975 The Use of Gastroliths in Archaeology. In *Archaeozoological Studies*, edited by A. T. Calson, pp. 397–406. Elsevier, New York.

Bowers, Peter M.
   1978 *Research Summary: 1977 Investigations of the Carlo Creek Site, Central Alaska*. Report submitted to the University of Alaska Museum, Fairbanks.
   1980 *The Carlo Creek Site: Geology and Archaeology of an Early Holocene Site in the Central Alaskan Range*. Occasional Paper No. 27. Cooperative Park Studies Unit, University of Alaska Fairbanks.

Bowers, Peter M., R. Bonnichsen, and D. M. Hoch
   1983   Flake Dispersal Experiments: Noncultural Transformations of the Archaeological Record. *American Antiquity* 48(3):553–72.

Bowers, Peter M., and Joshua D. Reuther
   2008   AMS Re-dating of the Carlo Creek Site, Nenana Valley, Central Alaska. *Current Research in the Pleistocene* 25:58–60.

Brain, C. K.
   1976   Some Principles in the Interpretation of Bone Accumulations Associated with Man. In *Human Origins*, edited by G. L. Isaac and E. R. McCown, pp. 97–116. W. A. Benjamin, Menlo Park.
   1981   *The Hunters or the Hunted? An Introduction to African Cave Taphonomy*. University of Chicago Press, Chicago.

Brauer, A., G. H. Haug, P. Dulski, D. M. Sigman, and J. F. W. Negendank
   2008   An Abrupt Wind Shift in Western Europe at the Onset of the Younger Dryas Cold Period. *Nature Geoscience* 1:520–23.

Broecker, W. S., G. H. Denton, L. R. Edwards, H. Cheng, R. B. Alley, and A. E. Putnam
   2010   Putting the Younger Dryas Cold Event into Context. *Quaternary Science Reviews* 29:1078–81.

Bronk Ramsey, Christopher
   2009   Bayesian Analysis of Radiocarbon Dates. *Radiocarbon* 51(1):337–60.
   2013   OxCal v.4.2.4. https://c14.arch.ox.ac.uk/embed.php?file=.

Brubaker, L. B., P. M. Anderson, M. E. Edwards, and A. V. Lozhkin
   2005   Beringia as a Glacial Refugium for Boreal Trees and Shrubs: New Perspectives from Mapped Pollen Data. *Journal of Biogeography* 32:833–48.

Burt, Rebecca, and Soil Survey Staff
   2014   *Kellogg Soil Survey Laboratory Methods Manual*. Soil Survey Investigations Report No. 42, Version 5.0. US Department of Agriculture, Natural Resources Conservation Service.

Butler, B. R.
   1978   *A Guide to Understanding Idaho Archaeology*. Idaho Museum of Natural History, Pocatello.

Butzer, K. W.
   1982   *Archaeology as Human Ecology: Method and Theory for a Contextual Approach*. Cambridge University Press, Cambridge.

Cahen, D., and L. H. Keeley
   1980   Not Less than Two, Not More than Three. *World Archaeology* 12(2):166–80.

Campbell, Iain B., and Gordon G. C. Claridge
   1987   *Antarctica: Soils, Weathering Processes, and Environment*. Developments in Soil Science, vol. 16. Elsevier, Amsterdam.

Campbell, John M.
   1961   The Tuktu Complex of Anaktuvuk Pass. *Anthropological Papers of the University of Alaska* 9(2):61–80.

Cannon, Michael D., and David J. Meltzer
   2004   Early Paleoindian Foraging: Examining the Faunal Evidence for Large Mammal Specialization and Regional Variability in Prey Choice. *Quaternary Science Reviews* 23:1955–87.
   2008   Explaining Variability in Early Paleoindian Foraging. *Quaternary International* 191:5–17.

Chard, C. S.
   1974   *Northeast Asia in Prehistory*. University of Wisconsin Press, Madison.

Cinq-Mars, Jacques
   1979   Blue Fish Cave I: A Late Pleistocene Eastern Beringian Cave Deposit in the Northern Yukon. *Canadian Journal of Archaeology* 3:1–32.

Clark, Donald W.
   1978   *Discussion on "Clovis Culture."* Abstract of the Fifth Biennial Meeting of the American Quaternary Association, Edmonton.

Clark, Donald W., and Ruth M. Gotthardt
   1999   *Microblade Complexes and Traditions in the Interior Northwest as Seen from the Kelly Creek Site, West-Central Yukon*. Occasional Papers in Archaeology No. 6. Heritage Branch, Government of the Yukon, Hude Hudan Series.

Clark, J. D. G.
   1954   *Excavations at Starr Carr*. Cambridge University Press, Cambridge.

Coffman, Samuel C.
   2011   *Archaeology at Teklanika West (HEA-001): An Upland Archaeological Site, Central Alaska*. MA thesis, Department of Anthropology, University of Alaska Fairbanks.

Coffman, Samuel C., and Ben A. Potter
   2011   Recent Excavations at Teklanika West: A Late-Pleistocene Multicomponent Site in Denali National Park and Preserve, Central Alaska. *Current Research in the Pleistocene* 28:29–31.

Colinvaux, Paul A.
   1980   Vegetation of the Bering Land Bridge Revisited. *Quaternary Review of Archaeology* 1:2–15.

Colinvaux, Paul A., and Frederick H. West
   1984   The Beringian Ecosystem. *Quarterly Review of Archaeology* 5(3):10–16.

Collins, Henry B., Jr.
   1964   The Arctic and Subarctic. In *Prehistoric Man in the New World*, edited by J. D. Jennings and E. Norbeck, pp. 85–114. University of Chicago Press, Chicago.

Cook, John P.
- 1968 Some Microblade Cores from the Western Boreal Forest. *Arctic Anthropology* 5(1):121–27.
- 1969 *The Early Prehistory of Healy Lake*. PhD dissertation, University of Wisconsin, Madison.
- 1995 Characterization and Distribution of Obsidian in Alaska. *Arctic Anthropology* 32(1):92–100.
- 1996 Healy Lake. In *American Beginnings: The Prehistory and Paleoecology of Beringia*, edited by F. H. West, pp. 323–27. University of Chicago Press, Chicago.

Cook, John P., and Robert A. McKennan
- 1970 *The Village Site at Healy Lake, Alaska: An Interim Report*. Paper presented at the 35th Annual Meeting of the Society for American Archeology, Mexico City.

Couey, Faye Morrison
- 1950 *Rocky Mountain Bighorn Sheep of Montana*. Helena: Montana Fish and Game Commission.

Crabtree, D. E.
- 1967 Notes on Experiments in Flintknapping: 4. Tools Used for Making Flaked Stone Artifacts. *Tebiwa* 10(1):60–73.

Cwynar, Les C.
- 1982 A Late-Quaternary Vegetation History from Hanging Lake, Northern Yukon. *Ecological Monographs* 52(1):1–24.

Cwynar, Les C., and J. C. Ritchie
- 1980 Arctic Steppe-Tundra: A Yukon Perspective. *Science* 208:1375–77.

Dacey, M. F.
- 1973 Statistical Tests of Spatial Association in the Location of Tools Types. *American Antiquity* 38(3):320–28.

Davis, C. W., D. C. Linck, K. M. Schoenberg, and H. M. Shields
- 1981 *Slogging, Humping and Mucking through the NPR-A: An Archaeological Interlude*. Occasional Paper No. 25, Cooperative Park Studies Unit, University of Alaska Fairbanks.

Davis, Stanley D.
- 1996 Hidden Falls. In *American Beginnings: The Prehistory and Paleoecology of Beringia*, edited by F. H. West, pp. 413–23. University of Chicago Press, Chicago.

Del Bene, Terry A.
- 1981 *The Anangula Lithic Technological System: An Appraisal of Eastern Aleutian Technology circa 8250–8750 B.P.* PhD dissertation, University of Connecticut.

Dennell, R. S.
- 1983 *European Economic Prehistory: A New Approach*. Academic Press, London.

de Vernal, A., C. Hillaire-Marcel, and G. Bilodeau
- 1996 Reduced Meltwater Outflow from the Laurentide Ice Margin during the Younger Dryas. *Nature* 381:774–77.

Dikov, N. N.
- 1977 *Arkeologicheskie Pamiatniki Kamchatki, Chukotki i Verkhnei Kolymy.* Nauka, Moscow.
- 1978 Ancestors of Palaeoindians and Proto-Eskimo-Aleuts in the Palaeolithic of Kamchatka. In *Early Man in America from a Circum-Pacific Perspective*, edited by Alan L. Bryan. Occasional Papers No. 1 of the Department of Anthropology, University of Alberta, Edmonton.
- 1979 *Drevnie Kul'tury Severo-Vostochnoi Azii.* Moscow, Nauka.

Dixon, E. James
- 1975 The Gallagher Flint Station, an Early Man Site on the North Slope, Arctic Alaska, and Its Role in Relation to the Bering Land Bridge. *Arctic Anthropology* 7(1):68–75.
- 1976 The Pleistocene Prehistory of Arctic North America. In *Habitats Humains Antérieurs à l'Holocène en Amérique*, edited by J. B. Griffen, pp. 168–98. Proceedings of the 9th International Congress of Anthropological Sciences, Nice.
- 1999 *Bones, Boats, and Bison: Archeology and the First Colonization of Western North America.* University of New Mexico Press, Albuquerque.
- 2001 Human Colonization of the Americas: Timing, Technology and Process. *Quaternary Science Reviews* 20:277–99.

Dixon, E. James, William F. Manley, and Craig M. Lee.
- 2005 The Emerging Archaeology of Glaciers and Ice Patches: Examples from Alaska's Wrangell–St. Elias National Park and Preserve. *American Antiquity* 70:129–43.

Dortch, Jason M.
- 2006 *Defining the Timing of Glaciation in the Central Alaska Range.* MA thesis, Department of Geology, University of California, Riverside.

Dortch, Jason M., Lewis A. Owen, Mark W. Caffee, Dewen Li, and Thomas V. Lowell
- 2010 Beryllium-10 Surface Exposure Dating of Glacial Successions in the Central Alaska Range. *Journal of Quaternary Science* 25:1259–69.

Dumond, Don E.
- 1977 *The Eskimos and Aleuts.* Thames and Hudson, London.
- 1978 *A Chronology of Native Alaskan Subsistence Systems.* Alaska Native Culture and History, Senri Ethnological Studies No. 4. National Museum of Ethnology, Osaka, Japan.
- 1980 The Archaeology of Alaska and the Peopling of America. *Science* 209(4460):984–91.
- 2001 The Archaeology of Eastern Beringia: Some Contrasts and Connections. *Arctic Anthropology* 38(2):196–205.
- 2011 Technology, Typology, and Subsistence: A Partly Contrarian Look at the Peopling of Beringia. In *From the Yenisei to the Yukon: Interpreting Lithic Assemblage Variability in Late Pleistocene/Early Holocene Beringia*, edited by T. Goebel and I. Buvit, pp. 345–61. Texas A&M University Press, College Station.

Easton, N. A., G. R. Mackay, P. B. Young, P. Schnurr, and D. R. Yesner
- 2011 Chindadn in Canada? Emergence Evidence of the Pleistocene Transition in Southeast Beringia as Revealed by the Little John Site, Yukon. In *From the Yenisei to the Yukon: Interpreting Lithic Assemblage Variability in Late Pleistocene/Early Holocene Beringia*, edited by T. Goebel and I. Buvit, pp. 389–407. Texas A&M University Press, College Station.

Edwards, M. E., C. J. Mock, B. P. Finney, V. A. Barber, and P. J. Bartlein
   2001   Potential Analogues for Paleoclimatic Variations in Eastern Interior Alaska during the Past 14,000 Years: Atmospheric-Circulation Controls of Regional Temperature and Moisture Responses. *Quaternary Science Reviews* 20:189–202.

Elias, Scott A., and Barnaby Crocker
   2008   The Bering Land Bridge: A Moisture Barrier to the Dispersal of Steppe-Tundra Biota? *Quaternary Science Reviews* 27:2473–83.

Erlandson, Jon M.
   2013   After Clovis-First Collapsed: Reimagining the Peopling of the Americas. In *Paleoamerican Odyssey*, edited by Kelly E. Graf, Caroline V. Ketron, and Michael R. Waters, pp. 127–32. Center for the Study of the First Americans, Texas A&M University, College Station.

Erlandson, Jon M., T. C. Rick, T. J. Braje, M. Casperson, B. Culleton, B. Fulfrost, T. Garcia, D. A. Guthrie, N. Jew, D. J. Kennett, and M. L. Moss
   2011   Paleoindian Seafaring, Maritime Technologies, and Coastal Foraging on California's Channel Islands. *Science* 331:1181–85.

Esdale, Julie A.
   2008   A Current Synthesis of the Northern Archaic. *Arctic Anthropology* 45(2):3–38.

Ewers, J. C.
   1955   *The Horse in Blackfoot Indian Culture*. Bureau of American Ethnology Bulletin No. 159. Washington, DC.

Feder, Kenneth L.
   2014   *The Past in Perspective: An Introduction to Human Prehistory*, 6th ed. Oxford University Press, New York.

Flannigan, Thomas H.
   2002   *Use-Wear Analysis of Experimental Stone Tools, and a Sample of Lithics from Component I of the Walker Road Site*. MA thesis, Department of Anthropology, University of Alaska Fairbanks.

Freeman, L. G.
   1978   The Analysis of Some Occupation Floor Distributions from Earlier and Middle Paleolithic Sites in Spain. In *Views of the Past: Essays in the Old World Prehistory and Paleoanthropology*, edited by L. G. Freeman, pp. 57–116. The Hague, Mouton.

Frison, G. C.
   1968   A Functional Analysis of Certain Chipped Stone Tools. *American Antiquity* 33(2):149–55.
   1974   *The Casper Site*. Academic Press, New York.
   1978   *Prehistoric Hunters of the High Plains*. Academic Press, New York.
   1986   Human Artifacts, Mammoth Procurement, and Pleistocene Extinctions as Viewed from the Colby Site. In *The Colby Mammoth Site Taphonomy and Archaeology of a Clovis Kill in Northern Wyoming*, edited by G. C. Frison and L. C. Todd, pp. 91–114. University of New Mexico Press, Albuquerque.

Frison, G. C., and L. C. Todd (editors)
   1986   *The Colby Mammoth Site Taphonomy and Archaeology of a Clovis Kill in Northern Wyoming*. University of New Mexico Press, Albuquerque.

Fu, Qiaomei, Alissa Mittnik, Philip L. F. Johnson, Kirsten Bos, Martina Lari, Ruth Bollogino, Chengkai Sun, Liane Giemsch, Ralf Schmitz, Joachim Burger, Anna Maria Ronchitelli, Fabio Martini, Renata G. Cremonesi, Jiří Svoboda, Peter Bauer, David Caramelli, Sergi Castellano, David Reich, Svante Pääbo, and Johannes Krause
   2013   A Revised Timescale for Human Evolution Based on Ancient Mitochondrial Genomes. *Current Biology* 23(7):553–59.

Gaines, Edmund P., Kate S. Yeske, Scott J. Shirar, William C. Johnson, and James F. Kunesh
   2011   Pleistocene Archaeology of the Tanana Flats, Eastern Beringia. *Current Research in the Pleistocene* 28:42–44.

Gal, Robert, and Edwin S. Hall
   1982   Provisional Culture History. Archaeological Investigations by the U.S. Geological Survey and the Bureau of Land Management in the National Petroleum Preserve in Alaska. *Anthropological Papers of the University of Alaska* 20(1–2):3–5.

Geist, V.
   1971   *Mountain Sheep, a Study in Behavior and Evolution*. University of Chicago Press, Chicago.
   1978   *Life Strategies, Human Evolution, and Environmental Design*. Springer-Verlag, New York.

Gifford-Gonzalez, D. P., D. B. Damrosch, D. R. Damrosch, J. Pryor, and R. L. Thunen
   1985   The Third Dimension and Vertical Dispersal. *American Antiquity* 50(4):803–18.

Gilbert, M. T. P., D. L. Jenkins, A. Götherstrom, N. Naveran, J. J. Sanchez, M. Hofreiter, P. F. Thomsen, J. Binladen, T. F. Higham, R. M. Yohe, R. Parr, L. Scott Cummings, and E. Willerslev
   2008   DNA from Pre-Clovis Human Coprolites in Oregon, North America. *Science* 320:786–89.

Goebel, Ted
   1990   *Early Paleoindian Technology in Beringia*. MA thesis, Department of Anthropology, University of Alaska Fairbanks.
   1996   Recent Research at Teklanika West: Site Stratigraphy and Dating. In *American Beginnings: The Prehistory and Paleoecology of Beringia*, edited by F. H. West, pp. 341–43. University of Chicago Press, Chicago.
   1999   Pleistocene Human Colonization of Siberia and the Peopling of the Americas: An Ecological Approach. *Evolutionary Anthropology* 8:208–27.
   2002   The "Microblade Adaptation" and Recolonization of Siberia during the Late Upper Pleistocene. In *Thinking Small: Global Perspectives on Microlithization*, edited by Robert G. Elston and Steven L. Kuhn, pp. 117–131. Archeological Papers of the American Anthropological Association, No. 12. American Anthropological Association, Arlington.
   2004   The Search for a Clovis Progenitor in Sub-Arctic Siberia. In *Entering America: Northeast Asia and Beringia before the Last Glacial Maximum*, edited by David B. Madsen, pp. 311–56. University of Utah Press, Salt Lake City.

    2011   What Is the Nenana Complex? Raw Material Procurement and Technological Organization at Walker Road, Central Alaska. In *From the Yenisei to the Yukon: Interpreting Lithic Assemblage Variability in Late Pleistocene/Early Holocene Beringia*, edited by T. Goebel and I. Buvit, pp. 199–214. Texas A&M University Press, College Station.

Goebel, Ted, and Nancy H. Bigelow
    1992   The Denali Complex at Panguingue Creek, Central Alaska. *Current Research in the Pleistocene* 9:15–17.
    1996   Panguingue Creek. In *American Beginnings: The Prehistory and Paleoecology of Beringia*, edited by F. H. West, pp. 366–70. University of Chicago Press, Chicago.

Goebel, Ted, and Ian Buvit
    2011   Introducing the Archaeological Record of Beringia. In *From the Yenisei to the Yukon: Interpreting Lithic Assemblage Variability in Late Pleistocene/Early Holocene Beringia*, edited by T. Goebel and I. Buvit, pp. 1–30. Texas A&M University Press, College Station.

Goebel, Ted, Bryan Hockett, Kenneth D. Adams, David Rhode, and Kelly Graf
    2011   Climate, Environment, and Humans in North America's Great Basin during the Younger Dryas, 12,900–11,600 Calendar Years Ago. *Quaternary International* 242:479–501.

Goebel, T., and W. R. Powers
    1989   A Possible Paleoindian Dwelling in the Nenana Valley: Spatial Analysis at Walker Road. Paper presented at the 16th Annual Meeting of the Alaska Anthropological Association, Anchorage.

Goebel, Ted, W. Roger Powers, and Nancy H. Bigelow
    1991   The Nenana Complex of Alaska and Clovis Origins. In *Clovis: Origins and Adaptations*, edited by R. Bonnichsen and K. L. Turnmire, pp. 49–79. Center for the Study of the First Americans, Oregon State University Press, Corvallis.

Goebel, Ted, W. Roger Powers, Nancy H. Bigelow, and Andrew S. Higgs
    1996   Walker Road. In *American Beginnings: The Prehistory and Paleoecology of Beringia*, edited by F. H. West, pp. 356–63. University of Chicago Press, Chicago.

Goebel, Ted, Sergei B. Slobodin, and Michael R. Waters
    2010   New Dates from Ushki-1, Kamchatka, Confirm 13,000 Cal BP Age for Earliest Paleolithic Occupation. *Journal of Archaeological Science* 37(10):2640–49.

Goebel, Ted, H. L. Smith, L. DiPietro, M. R. Waters, B. Hockett, K. E. Graf, R. Gal, S. B. Slobodin, R. J. Speakman, S. G. Driese, and D. Rhode
    2013   Serpentine Hot Springs, Alaska: Results of Excavations and Implications for the Age and Significance of Northern Fluted Points. *Journal of Archaeological Science* 40:4222–33.

Goebel, Ted, Robert J. Speakman, and Joshua D. Reuther
    2008a   Results of Geochemical Analysis of Obsidian Artifacts from the Walker Road Site. *Current Research in the Pleistocene* 25:88–90.

Goebel, Ted, Michael R. Waters, and Margarita Dikova
    2003   The Archaeology of Ushki Lake, Kamchatka, and the Pleistocene Peopling of the Americas. *Science* 301:501–5.

Goebel, Ted, Michael R. Waters, and Mikhail N. Meshcherin
 2000 Masterov Kluich and the Early Upper Paleolithic of the Transbaikal, Siberia. *Asian Perspectives* 39(1):47–70.

Goebel, Ted, Michael R. Waters, and Dennis H. O'Rourke
 2008b The Late Pleistocene Dispersal of Modern Humans in the Americas. *Science* 319:1497–1502.

Gómez Coutouly, Yan Axel
 2011 Identifying Pressure Flaking Modes at Diuktai Cave: A Case Study of the Siberian Upper Paleolithic Microblade Tradition. In *From the Yenisei to the Yukon*, edited by T. Goebel and I. Buvit, pp. 75–90. Texas A&M University Press, College Station.
 2012 Pressure Microblade Industries in Pleistocene-Holocene Interior Alaska: Current Data and Discussions. In *The Emergence of Pressure Blade Making*, edited by P. M. Desrosiers, pp. 347–74. Springer-Verlag, New York.

González-José, R., M. C. Bortolini, F. R. Santos, and S. L. Bonatto
 2008 The Peopling of America: Craniofacial Shape Variation on a Continental Scale and Its Interpretation from an Interdisciplinary View. *American Journal of Physical Anthropology* 137:175–87.

Gore, Angela K., and Kelly E. Graf
 In press Human Response to Late Pleistocene and Early Holocene Environmental Change in Central Alaska. In *Lithic Technological Organization and Paleoenvironmental Change: Global and Diachronic Perspectives*, edited by Erick Robinson and Frederic Sellet.

Gould, R. A.
 1980 *Living Archaeology*. Cambridge University Press, Cambridge.

Graf, Kelly E.
 2008 *Uncharted Territory: Late Pleistocene Hunter-Gatherer Dispersals in the Siberian Mammoth Steppe*. Unpublished PhD dissertation. University of Nevada Reno.
 2009 "The Good, the Bad, and the Ugly": Evaluating the Radiocarbon Chronology of the Middle and Late Upper Paleolithic in the Enisei River Valley, South-Central Siberia. *Journal of Archaeological Science* 36:694–707.
 2010 Hunter-Gatherer Dispersals in the Mammoth-Steppe: Technological Provisioning and Land-Use in the Enisei River Valley, South-Central Siberia. *Journal of Archaeological Science* 37(1):210–23.
 2013 Siberian Odyssey. In *Paleoamerican Odyssey*, edited by Kelly E. Graf, Caroline V. Ketron, and Michael R. Waters, pp. 65–80. Center for the Study of the First Americans, Texas A&M University, College Station.

Graf, Kelly E., and Nancy H. Bigelow
 2011 Human Response to Climate during the Younger Dryas Chronozone in Central Alaska. *Quaternary International* 242(2):434–51.

Graf, Kelly E., John Blong, and Ted Goebel
 2010 A Concave-Based Projectile Point from New Excavations at the Owl Ridge Site, Central Alaska. *Current Research in the Pleistocene* 27:88–90.

Graf, Kelly E., Lyndsay M. DiPietro, Kathryn E. Krasinski, Angela K. Gore, Heather L. Smith, Brendan J. Culleton, Douglas J. Kennett, and David Rhode
    2015   Dry Creek Revisited: New Excavations, Radiocarbon Dates, and Site Formation Inform on the Peopling of Eastern Beringia. *American Antiquity* 80:671–94.

Graf, Kelly E., and Ted Goebel
    2009   Upper Paleolithic Toolstone Procurement and Selection across Beringia. In *Lithic Materials and Paleolithic Societies*, edited by Brian Adams and Brooke S. Blades, pp. 55–77. Blackwell, West Sussex.

Greenberg, Joseph H., Christy G. Turner, and Steven L. Zegura
    1986   The Settlement of the Americas: A Comparison of the Linguistic, Dental, and Genetic Evidence. *Current Anthropology* 27:477–97.

Guthrie, R. D.
    1966   The Extinct Wapiti of Alaska and the Yukon Territory. *Canadian Journal of Zoology* 44:45–57.
    1968   Paleoecology of the Large Mammal Community in Interior Alaska during the Late Pleistocene. *American Midland Naturalist* 79:246–63.
    1980   Bison and Man in North America. *Canadian Journal of Anthropology* 1:55–73.
    1982   Mammals of the Mammoth Steppe as Paleoenvironmental Indicators. *In Paleoecology of Beringia*, edited by D. M. Hopkins, J. V. Matthews Jr., C. E. Schweger, and S. B. Young, pp. 307–26. Academic Press, New York.
    1983   Paleoecology of the Site and Its Implications for Early Hunters. In *Dry Creek: Archaeology and Paleoecology of a Late Pleistocene Alaskan Hunting Camp*, edited by W. Roger Powers, R. Dale Guthrie, and John F. Hoffecker, pp. 209–87. US National Park Service, Washington, DC.
    1984a   Mosaics, Allelochemics, and Nutrients: An Ecological Theory of Late Pleistocene Extinctions. In *Quaternary Extinctions*, edited by P. S. Margin and R. G. Klein, pp. 259–98. University of Arizona Press, Tucson.
    1984b   Alaskan Megabucks, Megabulls, and Megarams: The Issue of Pleistocene Gigantism. *Carnegie Museum of Natural History Special Report* 8:482–510.
    1985   Woolly Arguments against the Mammoth Steppe—a New Look at the Palynological Data. *Quarterly Review of Archaeology* 6:9–14.
    1990   *Frozen Fauna of the Mammoth Steppe*. University of Chicago Press, Chicago.
    2006   New Carbon Dates Link Climatic Change with Human Colonization and Pleistocene Extinctions. *Nature* 441:207–9.

Hahn, J.
    1977   Besiedlung und Sedimentation der Pra-Dorset-Station Umingmak ID, Banks Island, N.W.T. *Polarforschung* 47:26–37.

Hajdas, I., G. Bonani, P. Boden, D. M. Peteet, and D. H. Mann
    1998   Cold Reversal on Kodiak Island, Alaska, Correlated with the European Younger Dryas by Using Variations of Atmospheric ^{14}C Content. *Geology* 26:1047–50.

Hamilton, Thomas D.
    1970   Geological Relations of the Akmak Assemblage, Onion Portage, Alaska. *Acta Arctica* 16:71–80.

Hammatt, H.
　1970　A Paleo-Indian Butchering Kit. *American Antiquity* 35(2):141–52.

Hansen, Robert M., and M. J. Morris
　1968　Movement of Rocks by Northern Pocket Gophers. *Journal of Mammalogy* 49:391–99.

Hare, P. G., S. Greer, R. Gotthardt, R. Farnell, V. Bowyer, C. Schweger, and D. Strand
　2004　Ethnographic and Archaeological Investigations of Alpine Ice Patches in Southwest Yukon, Canada. *Arctic* 57(3):260–72.

Hare, P. G., C. D. Thomas, T. N. Topper, and R. M. Gotthardt
　2012　The Archaeology of Yukon Ice Patches: New Artifacts, Observations, and Insights. *Arctic* 65(S1):118–35.

Harington, C. R.
　1978　Quaternary Vertebrate Faunas of Canada and Alaska and Their Suggested Chronological Sequence. *Sylloques* 15:1–105.

Hayashi, Kensaku
　1968　The Fukui Microblade Technology and Its Relationships in Northeast Asia and North America. *Arctic Anthropology* 5:128–90.

Haynes, C. Vance
　1978　The Clovis Culture. In *Abstracts of the Fifth Biennial Meeting of the American Quaternary Association*, pp. 129–35. Edmonton.
　1982　Were Clovis Progenitors in Beringia? In *Paleoecology of Beringia*, edited by D. M. Hopkins, J. V. Matthews Jr., C. E. Schweger, and S. B. Young, pp. 383–98. Academic Press, New York.
　1987　Clovis Origins Update. *The Kiva* 52(2):83–93.

Haynes, Gary A.
　1984　Tooth Wear Rate in Northern Bison. *Journal of Mammalogy* 65(3):487–91.
　2002　*The Early Settlement of North America: The Clovis Era*. Cambridge University Press, Cambridge.

Haynes, Gary A., and Jarod M. Hutson
　2013　Clovis-Era Subsistence: Regional Variability, Continental Patterning. In *Paleoamerican Odyssey*, edited by Kelly E. Graf, Caroline V. Ketron, and Michael R. Waters, pp. 293–310. Center for the Study of the First Americans, Texas A&M University, College Station.

Heimer, W. E.
　1973　*Dall Sheep Movements and Mineral Lick Use*. Final Report Job No. 6.1R. Alaska Department of Fish and Game, Juneau.

Heintzman, P. D., D. Froese, J. W. Ives, A. E. R. Soares, G. D. Zazula, B. Letts, T. D. Andrews, J. C. Driver, E. Hall, P. G. Hare, C. N. Jass, G. MacKay, J. R. Southon, M. Stiller, R. Woywitka, M. A. Suchad, and B. Shapiro

2016 Bison Phylogeography Constrains Dispersal and Viability of the Ice-Free Corridor in Western Canada. *Proceedings of the National Academy of Sciences* 113(29):8057-8063.

Henn, Winfield
- 1975 *Current Research on Eskaleut Prehistory in the Ugashik River Drainage, Alaska Peninsula.* Paper presented at the Annual Meeting of the American Anthropological Association, San Francisco.
- 1978 Archaeology of the Alaska Peninsula: The Ugashik Drainage 1973–1975. University of Oregon Anthropological Papers No. 14. Department of Anthropology, University of Oregon, Eugene.

Hesse, A.
- 1973 Essai Sur les Distributions Spatiales des Vestiges en Prehistoire et en Archeology. In *L'Homme Hier et Aujourd'hui*, edited by M. Sauter, pp. 551–63. Cujas, Paris.

Hester, J. J.
- 1972 *Blackwater Locality No. 1: A Stratified Early Man Site in Eastern New Mexico.* Ranchos de Taos, Fort Burgwin Research Center, New Mexico.

Hester, T., D. Gilbrow, and A. Albee
- 1973 A Functional Analysis of "Clear Fork" Artifacts from the Rio Grande Plain, Texas. *American Antiquity* 38(1):90–96.

Higgs, Andrew S.
- 1992 *Technological and Spatial Considerations of the Walker Road Site: Implications from a Lithic Refit Study.* MA thesis, Department of Anthropology, University of Alaska Fairbanks.
- 1994 Lithic Refits at Walker Road: Continuing Studies into the Nenana Complex of Central Alaska. *Current Research in the Pleistocene* 11:132–34.

Hoefs, M. E. G.
- 1974 Food Selection by Dall Sheep *Ovis dalli*. In *Behavior of Ungulates and Its Relation to Management*, edited by V. Geist and F. Walther, pp. 759–86. International Union for Conservation of Nature and Natural Resources, New Series No. 24. Morges, Switzerland.

Hoffecker, John F.
- 1978 *On the Potential of the North Alaska Range for Archaeological Sites of Pleistocene Age.* Report to the National Geographic Society and the National Park Service.
- 1980 *Archaeological Field Research 1980: The North Alaska Range Early Man Project.* Report to the National Geographic Society and the National Park Service.
- 1982 *The Moose Creek Site: An Early Man Occupation in Central Alaska.* Preliminary report to the National Park Service and the National Geographic Society.
- 1983 Human Activity at the Dry Creek Site: A Synthesis of the Artifactual, Spatial and Environmental Data. In *Dry Creek: Archaeology and Paleoecology of a Late Pleistocene Alaskan Hunting Camp*, edited by W. Roger Powers, R. Dale Guthrie, and John F. Hoffecker, pp. 182–208. US National Park Service, Washington, DC.
- 1985 The Moose Creek Site. *National Geographic Society Research Reports* 19:33–48.

1988 Applied Geomorphology and Archaeological Survey Strategy for Sites of Pleistocene Age: An Example from Central Alaska. *Journal of Archaeological Science* 15:683–713.

1996 Moose Creek. In *American Beginnings: The Prehistory and Paleoecology of Beringia*, edited by F. H. West, pp. 363–65. University of Chicago Press, Chicago.

2001 Late Pleistocene and Early Holocene Sites in the Nenana River Valley, Central Alaska. *Arctic Anthropology* 38(2):139–53.

2005 Incredible Journey: Plains Bison Hunters in the Arctic. *The Review of Archaeology* 26(2):18–23.

2011 Assemblage Variability in Beringia: The Mesa Factor. In *From the Yenisei to the Yukon: Interpreting Lithic Assemblage Variability in Late Pleistocene/Early Holocene Beringia*, edited by T. Goebel and I. Buvit, pp. 165–78. Texas A&M University Press, College Station.

Hoffecker, John F., and Scott A. Elias
2007 *Human Ecology of Beringia*. Columbia University Press, New York.

Hoffecker, John F., Scott A. Elias, and Dennis H. O'Rourke.
2014 Out of Beringia? *Science* 343:979–80.

Hoffecker, John F., W. Roger Powers, and Nancy H. Bigelow
1996a Dry Creek. In *American Beginnings: The Prehistory and Paleoecology of Beringia*, edited by F. H. West, pp. 343–52. University of Chicago Press, Chicago.

Hoffecker, John F., W. Roger Powers, and Ted Goebel
1993 The Colonization of Beringia and the Peopling of the New World. *Science* 259:46–53.

Hoffecker, John F., W. Roger Powers, and Peter G. Phippen
1996b Owl Ridge. In *American Beginnings: The Prehistory and Paleoecology of Beringia*, edited by F. H. West, pp. 353–55. University of Chicago Press, Chicago.

Holmes, Charles E.
2001 Tanana River Valley Archaeology circa 14,000 to 9000 B.P. *Arctic Anthropology* 38(2):154–70.

2011 The Beringian and Transitional Periods in Alaska: Technology of the East Beringian Tradition as Viewed from Swan Point. In *From the Yenisei to the Yukon*, edited by T. Goebel and I. Buvit, pp. 179–91. Texas A&M University Press, College Station.

Holmes, Charles E., and Glenn Bacon
1982 *Holocene Bison in Central Alaska: A Possible Explanation for Technological Conservatism*. Paper presented at the 9th Annual Meeting of the Alaska Anthropological Association, Fairbanks.

Holmes, Charles E., Joshua D. Reuther, and Peter Bowers
2010 *The Eroadaway Site: Early Holocene Lithic Technological Variability in the Central Alaska Range*. Paper presented at the 37th Annual Meeting of the Alaska Anthropological Association, Anchorage, Alaska.

Holmes, Charles E., R. VanderHoek, and T. E. Dilley
 1996 Swan Point. In *American Beginnings: The Prehistory and Paleoecology of Beringia*, edited by F. H. West, pp. 319–22. University of Chicago Press, Chicago.

Hopkins, David M., John V. Matthews, Charles E. Schweger, and Steven B. Young (editors)
 1982 *Paleoecology of Beringia*. Academic Press, New York.

Horejsi, B.
 1976 Suckling and Feeding Behavior in Relation to Lamb Survival in Big Horn Sheep (*Ovis canadensis* Shaw). PhD dissertation, University of Calgary, Alberta.

Hoskins, C. M., R. D. Guthrie, and B. L. D. Hoffman
 1970 Pleistocene and Recent Bird Gastroliths from Interior Alaska. *Arctic* 23:14–23.

Hu, Feng Sheng, Linda B. Brubaker, and Patricia M. Anderson
 1993 A 12 000 Year Record of Vegetation Change and Soil Development from Wien Lake, Central Alaska. *Canadian Journal of Botany* 71:1133–42.

Inizan, Marie-Louise, Michèle Reduron-Ballinger, Hélène Roche, and Jacques Tixier
 1999 *Technology and Terminology of Knapped Stone*. Préhistorie de le Pierre Taillée, Tome 5. CREP, Nanterre.

Isaac, G. L.
 1971 The Diet of Early Man: Aspects of Archaeological Evidence from Lower and Middle Pleistocene Sites in Africa. *World Archaeology* 2:278–99.

Jarman, P. J.
 1974 The Social Organization of Antelope in Relation to Their Ecology. *Behavior* 48:215–66.

Jenkins, D. L., L. G. Davis, T. W. Stafford, P. F. Campos, B. Hockett, G. T. Jones, L. S. Cummings, C. Yost, T. J. Connolly, R. M. Yohe, and S. C. Gibbons
 2012 Clovis Age Western Stemmed Projectile Points and Human Coprolites at the Paisley Caves. *Science* 337:223–28.

Jochim, Michael A.
 1976 *Hunter-Gatherer Subsistence and Settlement: A Predictive Model*. Academic Press, New York.

Johnson, Donald L.
 1989 Subsurface Stone Lines, Stone Zones, Artifact-Manuport Layers, and Biomantles Produced by Bioturbation via Pocket Gophers (*Thomomys bottae*). *American Antiquity* 54:370–89.

Johnson, Donald L., and Kenneth L. Hansen
 1974 The Effects of Frost-Heaving on Objects in Soils. *Plains Anthropologist* 19(64):81–98.

Johnson, Donald. L., Daniel R. Muhs, and Michael L. Barnhardt
- 1977 The Effects of Frost-Heaving on Objects in Soils, II: Laboratory Experiments. *Plains Anthropologist* 22(76):133–47.

Johnson, Donald L., and Donna Watson-Stegner
- 1990 The Soil-Evolution Model as a Framework for Evaluating Pedoturbation in Archaeological Site Formation. In *Archaeological Geology of North America*, Centennial Special Vol. 4, edited by Norman P. Lasca and Jack Donahue, pp. 541–60. Geological Society of America, Boulder.

Keeley, Lawrence H.
- 1980 *Experimental Determination of Stone Tools Uses: A Microwear Approach*. University of Chicago Press, Chicago.

Kehoe, J. F.
- 1973 *The Gull Lake Site: A Prehistoric Bison Drive Site in Southwestern Saskatchewan*. Publications in Anthropology and History, No. 1. Milwaukee Public Museum.

Klein, D. R.
- 1965 Ecology of Deer Range in Alaska. *Ecological Monographs* 35:259–84.
- 1970 Tundra Ranges of the Boreal Forests. *Journal of Range Management* 23:8–14.

Klein, R. G.
- 1973 *Ice-Age Hunters of the Ukraine*. University of Chicago Press, Chicago.
- 1980 The Interpretation of Mammalian Faunas from Stone-Age Archeological Sites, with Special Reference to Sites in the Southern Cape Province, South Africa. In *Fossils in the Making*, edited by A. K. Behrensmeyer and A. P. Hill, pp. 223–46. University of Chicago Press, Chicago.
- 2009 *The Human Career: Human Biological and Cultural Origins*, 3rd ed. University of Chicago Press, Chicago.

Kobayashi, T.
- 1970 Microblade Industries in the Japanese Archipelago. *Arctic Anthropology* 7(2):38–58.

Kokorowski, H. D., P. M. Anderson, C. J. Mock, and A. V. Lozhkin
- 2008 A Re-evaluation and Spatial Analysis of Evidence for a Younger Dryas Climatic Reversal in Beringia. *Quaternary Science Reviews* 27:1710–22.

Krasinski, Kathryn E.
- 2010 *Broken Bones and Cutmarks: Taphonomic Analyses and Implications for the Peopling of North America*. PhD dissertation, Department of Anthropology, University of Nevada, Reno.

Krasinski, Kathryn E., and David R. Yesner
- 2008 Late Pleistocene/Early Holocene Site Structure in Beringia: A Case Study from the Broken Mammoth Site, Interior Alaska. *Alaska Journal of Anthropology* 6(1–2):27–42.

Kunz, M. L., M. Bever, and C. Adkins
- 2003 *The Mesa Site: Paleoindians above the Arctic Circle*. BLM-Alaska Open File Report No. 86. US Department of the Interior, Bureau of Land Management, Anchorage.

Kunz, M. L., and R. E. Reanier
　1994　Paleoindians in Beringia: Evidence from Arctic Alaska. *Science* 263:660–62.

Kurtén, B., and E. Anderson
　1972　The Sediments and Fauna of Jaguar Cave, II: The Fauna. *Tebiwa* 15:21–39.

Larsen, H.
　1968　*Trail Creek: Final Report on Excavations of Two Caves on the Seward Peninsula*. Acta Arctica No. 15.

Laws, R. M., I. C. S. Parker, and R. C. B. Johnson
　1975　*Elephants and Habitats*. Oxford University Press, Oxford.

Lee, C., and T. Goebel
　2016　The Slotted Antler Points from Trail Creek Caves, Alaska: New Information on Their Age and Technology. *PaleoAmerica* 2:40–47.

Lee, R. B., and I. Devore (editors)
　1968　*Man the Hunter*. Aldine, Chicago.

Leroi-Gourhan, A., and M. Brezillon
　1966　L'habitation Magdelemienne No. 1 de Pincevent (Seine-et-Marne). *Gallia Prehistoire* 9:263–385.

Levenson, H.
　1979　Sciurid Growth Rates: Some Corrections and Additions. *Journal of Mammalogy* 60:232–34.

Lewis, M., and W. Clark
　1893　*History of the Expedition under the Command of Lewis and Clark 1804–5–6*, new ed. Elliot Coves, New York.

Lewis, R. O.
　1978　Use of Opal Phytoliths in Paleo-Environmental Reconstruction at the Hudson-Meng Site. In *The Hudson-Meng Site: An Alberta Kill in the Nebraska High Plains*, edited by Larry D. Agenbroad, pp. 211–15. University Press of America, Washington, DC.

Lively, R. A.
　1988　*A Study of the Effectiveness of a Small Scale Probabilistic Sampling Design at an Interior Alaska Site, Chugwater (FAI-035)*. Report on file with the US Army Corps of Engineers, Alaska District, Anchorage.
　1996　Chugwater. In *American Beginnings: The Prehistory and Paleoecology of Beringia*, edited by F. H. West, pp. 308–11. University of Chicago Press, Chicago.

Lloyd, Andrea H., Mary E. Edwards, Bruce P. Finney, Jason A. Lynch, Valerie Barber, and Nancy H. Bigelow
　2006　Holocene Development of the Alaskan Boreal Forest. In *Alaska's Changing Boreal Forest*, edited by F. Stuart Chapin III, Mark W. Oswood, Keith Van Cleve, Leslie A. Viereck, and David L. Verbyla, pp. 62–78. Oxford University Press, New York.

Lothrop, Jonathan C., Paige E. Newby, Arthur E. Speiss, and James W. Bradley
- 2011 Paleoindians and the Younger Dryas in the New England-Maritimes Region. *Quaternary International* 242:546–69.

MacNeish, Richard S.
- 1956 The Engigstciak Site on the Yukon Arctic Coast. *Anthropological Papers of the University of Alaska* 4(2):91–111.
- 1963 The Early Peopling of the New World: As Seen from the Southwestern Yukon. *Anthropological Papers of the University of Alaska* 19(2):93–106.

Martin, Paul S.
- 1982 The Pattern and Meaning of Holarctic Mammoth Extinction. In *Paleoecology of Beringia*, edited by D. Hopkins, J. Matthews, C. Schweger, and S. Young, pp. 399–408. Academic Press, New York.

Mason, Owen K., Peter M. Bowers, and David M. Hopkins
- 2001 The Early Holocene Milankovitch Thermal Maximum and Humans: Adverse Conditions for the Denali Complex of Eastern Beringia. *Quaternary Science Reviews* 20:525–48.

Maxwell, Howard E.
- 1987 *Archeology of Panguingue Creek*. MA thesis, Department of Anthropology, University of Alaska Fairbanks.

McDonald, Jerry N.
- 1981 *North American Bison: Their Classification and Evolution*. University of California Press, Berkeley.

McHugh, T.
- 1972 *The Time of the Buffalo*. Knopf, New York.

McNaughton, S. J., J. L. Tarrants, M. M. McNaughton, and R. D. Davis
- 1985 Silica as a Defense against Herbivory and a Growth Promotor in African Grasses. *Ecology* 66(2):528–35.

Medvedev, G. I.
- 1968 K Voprosu ob Iznachal'nykh Formakh Nakonechnikov Strel v Priangar'e. *Trudy Irkutskogo Gosudarstvennogo Universiteta Im. A. A. Zhdanova* 55(1). Irkutsk.

Meiri, Meirav, Adrian M. Lister, Matthew J. Collins, Noreen Tuross, Ted Goebel, Simon Blockley, Grant D. Zazula, Nienke van Doorn, R. Dale Guthrie, Gennady G. Boeskorov, Gennady F. Baryshnikov, Andrei Sher, and Ian Barnes
- 2013 Faunal Record Identifies Bering Isthmus Conditions as Constraint to End-Pleistocene Migration to the New World. *Proceedings of the Royal Society B* 281:2013–2167.

Meltzer, David J.
- 2001 Late Pleistocene Cultural and Technological Diversity of Beringia: A View from Down Under. *Arctic Anthropology* 38(2):206–13.

Miquelle, D.
   1985   *Food Habits and Range Conditions of Bison and Sympatric Ungulates on the Upper Chita River, Wrangell-Saint Elias National Park and Preserve.* Alaska Region Research Resources Management Report No. AR-8. National Park Service, Anchorage.

Mobley, Charles M.
   1982   The Landmark Gap Trail Site, Tangle Lakes, Alaska: Another Perspective on the Amphitheater Mountain Complex. *Arctic Anthropology* 19(1):81–102.
   1991   *The Campus Site: A Prehistoric Camp at Fairbanks, Alaska.* University of Alaska Press, Fairbanks.

Mochanov, IU. A.
   1973   *Early Migrations to America in Light of a Study of the Dyuktai Paleolithic Culture in Northeast Asia.* Paper prepared for the 9th International Congress of Anthropological and Ethnological Sciences, Chicago.
   1977   *Drevneishie Etapy Zaseleniia Chelovekom Severovostochnoi Azii.* Nauka, Novosibirsk.
   1978   Stratigraphy and Absolute Chronology of the Paleolithic of Northeast Asia, According to the Work of 1963–1973. In *Early Man in America from a Circum-Pacific Perspective*, edited by Alan L. Bryan, pp. 54–66. Occasional Papers No. 1 of the Department of Anthropology. University of Alberta, Edmonton.

Morlan, R. E.
   1965   *The Wedge-Shaped Microblade Core Has Been Given a Great Deal of Attention in the Northern Archeological Literature.* Unpublished manuscript.
   1970   Wedge-Shaped Core Technology in Northern North America. *Arctic Anthropology* 7(2):17–37.
   1974   Gladstone: An Analysis of Horizontal Distributions. *Arctic Anthropology* 11 (supplement):82–93.
   1977   Fluted Point Makers and the Extinction of the Arctic-Steppe Biome in Eastern Beringia. *Canadian Journal of Archaeology* 1:95–108.
   1978   Technological Characteristics of Some Wedge-Shaped Cores in Northwestern North America and Northeast Asia. *Asian Perspectives* 19(1):96–106.

Morlan, R. E., and J. Cinq-Mars
   1982   Ancient Beringians: Human Occupation in the Late Pleistocene of Alaska and the Yukon Territory. In *Paleoecology of Beringia*, edited by D. M. Hopkins, J. V. Matthews Jr., C. E. Schweger, and S. B. Young, pp. 353–81. Academic Press, New York.

Muhs, Daniel, R., and James R. Budahn
   2006   Geochemical Evidence for the Origin of Late Quaternary Loess in Central Alaska. *Canadian Journal of Earth Science* 43:323-37.

Muhs, Daniel R., James R. Budahn, John P. McGeehin, E. Arthur Bettis III, Gary Skipp, James B. Paces, and Elisabeth A. Wheeler
   2013   Loess Origin, Transport, and Deposition Over the Past 10,000 Years, Wrangell-St. Elias National Park, Alaska. *Aeolian Research* 11:85-99.

Müller Beck, H.
  1967   On Migration of Hunters across the Bering Land Bridge in the Upper Pleistocene. In *Bering Land Bridge*, edited by D. M. Hopkins, pp. 373–408. Stanford University Press, Stanford.

Mulligan, Connie J., and Andrew Kitchen
  2013   Three-Stage Colonization Model for the Peopling of the Americas. In *Paleoamerican Odyssey*, edited by Kelly E. Graf, Caroline V. Ketron, and Michael R. Waters, pp. 171–82. Center for the Study of the First Americans, Texas A&M University, College Station.

Mulligan, Connie J., Andrew Kitchen, and Michael M. Miyamoto.
  2008   Updated Three-Stage Model for the Peopling of the Americas. *PLoS One* 3:e3199.

Murie, A.
  1944   *The Wolves of Mount McKinley*. Fauna of the National Parks of the United States, Fauna Series No. 5. US Government Printing Office, Washington, DC.

Murie, O. J.
  1951   *The Elk of North America*. Stackpole, Harrisburg.

Murphy, E. G.
  1974   *An Age Structure and a Reevaluation of the Population Dynamics of Dall Sheep (Ovis dalli dalli)*. MA thesis, University of Alaska Fairbanks.

Nagai, Kenji
  2007   Flake Scar Patterns of Japanese Tanged Points: Toward an Understanding of Technological Variability during the Incipient Jomon. *Anthropological Science* 115:223–26.

Nelson, N. C.
  1935   Early Migration of Man to America. *Natural History* 35(4):356.
  1937   Notes on Cultural Relations between Asia and America. *American Antiquity* 2(4):267–72.

Norden, H. C., I. M. Cowan, and A. T. Wood
  1968   Nutritional Requirements of Black Tailed Deer in Captivity. In *Comparative Nutrition of Wild Animals*, edited by M. A. Crawford, pp. 89–96. Symposia of the Zoological Society of London, No. 21. Zoological Society of London.

O'Brien, Michael J., M. T. Boulanger, M. Collard, B. Buchanan, L. Tarle, L. G. Straus, and M. I. Eren
  2014   On Thin Ice: Problems with Stanford and Bradley's Proposed Solutrean Colonisation of North America. *Antiquity* 88:606–13.

Odell, George. H.
  2004   *Lithic Analysis*. Manuals in Archaeological Method, Theory and Technique. Springer, New York.

Odess, Daniel, and Scott Shirar
  2007   New Evidence of Microblade Technology in the Nenana Complex Type Site at Dry Creek, Central Alaska. *Current Research in the Pleistocene* 24:129–31.

Ognev, S. I.
   1963   *Mammals of the U.S.S.R. and Adjacent Countries*, vol. 5, Rodents. Translated by A. Birron and Z. S. Cole. Israel Program for Scientific Translations, Jerusalem.

O'Rourke, Dennis H., and Jennifer A. Raff
   2010   The Human Genetic History of the Americas: The Final Frontier. *Current Biology* 20:R202–7.

Pearson, G. A.
   1999   Early Occupations and Cultural Sequence at Moose Creek: A Late Pleistocene Site in Central Alaska. *Arctic* 52(4):332–45.
   2000   Late Pleistocene and Holocene Microblade Industries at the Moose Creek Site. *Current Research in the Pleistocene* 17:64–65.

Peden, D. G., G. M. Van Dyne, R. W. Rice, and R. M. Hausen
   1974   The Trophic Ecology of *Bison bison* on Short Grass Prairie. *Journal of Applied Ecology* 11:489–98.

Perego, U. A., A. Achilli, N. Angerhofer, M. Accetturo, M. Pala, A. Olivieri, B. H. Kashani, K. H. Ritchie, R. Scozzari, Q. P Kong, and N. M. Myres
   2009   Distinctive Paleo-Indian Migration Routes from Beringia Marked by Two Rare mtDNA Haplogroups. *Current Biology* 19:1–8.

Petocz, R. G.
   1973   *Marco Polo Sheep of the Afghan Pamirs*. Unpublished report FAO/UNDP.

Péwé, T. L.
   1975   *Quaternary Geology of Alaska*. US Geological Survey Professional Paper No. 835. Washington, DC.

Phippen, Peter
   1988   *Pleistocene-Holocene Boundary Archaeology in the North Alaska Range Foothills: A View from Owl Ridge*. MA thesis, Department of Anthropology, University of Alaska Fairbanks.

Pike, R. L., and M. L. Brown
   1967   *Nutrition: An Integrated Approach*. Wiley and Sons, New York.

Pitulko, Vladimir V., Aleksandr E. Basilyan, and Elena Y. Pavlova
   2014   The Berelekh Mammoth "Graveyard": New Chronological and Stratigraphical Data from the 2009 Field Season. *Geoarchaeology* 29(4):277–99.

Pitulko, Vladimir V., Pavel A. Nikolskiy, Aleksandr Basilyan, and Elena Pavlova
   2013   Human Habitation in Arctic Western Beringia Prior to the LGM. In *Paleoamerican Odyssey*, edited by Kelly E. Graf, Caroline V. Ketron, and Michael R. Waters, pp. 13–44. Center for the Study of the First Americans, Texas A&M University, College Station.

Pitulko, Vladimir V., Pavel A. Nikolsky, Yu. E. Girya, A. E. Basilyan, V. E. Tumskoy, S. A. Koulakov, S. N. Astakhov, E. Yu Pavlova, and M. A. Anisimov
   2004   The Yana RHS Site: Humans in the Arctic before the Last Glacial Maximum. *Science* 303:52–56.

Plaskett, D. C.
  1977 *The Nenana River Gorge Site: A Late Prehistoric Athabaskan Campsite in Central Alaska.* MA thesis, Department of Anthropology, University of Alaska Fairbanks.

Pontti, Elizabeth B.
  1997 *Defining the Denali Complex: A Comparative Study of Lithic Assemblages from Panguingue Creek and Dry Creek, Central Alaska.* MA thesis, Department of Anthropology, University of Alaska Fairbanks.

Potter, Ben A.
  2005 *Site Structure and Organization in Central Alaska: Archaeological Investigations at Gerstle River.* PhD dissertation, Department of Anthropology, University of Alaska Fairbanks.
  2007 Models of Faunal Processing and Economy in Early Holocene Interior Alaska. *Environmental Archaeology* 12(1):3–23.
  2008 Radiocarbon Chronology of Central Alaska: Technological Continuity and Economic Change. *Radiocarbon* 50(2):181–204.
  2011 Late Pleistocene and Early Holocene Assemblage Variability in Central Alaska. In *From the Yenisei to the Yukon: Interpreting Lithic Assemblage Variability in Late Pleistocene/Early Holocene Beringia*, edited by T. Goebel and I. Buvit, pp. 215–33. Texas A&M University Press, College Station.

Potter, Ben A., Charles E. Holmes, and David R. Yesner
  2013 Technology and Economy among the Earliest Prehistoric Foragers in Interior Eastern Beringia. In *Paleoamerican Odyssey*, edited by Kelly E. Graf, Caroline V. Ketron, and Michael R. Waters, pp. 81–103. Center for the Study of the First Americans, Texas A&M University, College Station.

Powers, M. C.
  1953 A New Roundness Scale for Sedimentary Particles. *Journal of Sedimentary Petrology* 23:117–19.

Powers, William R.
  1973 Palaeolithic Man in Northeast Asia. *Arctic Anthropology* 10(2):1–106.
  1978 Perspectives on Early Man. *Abstracts of the Fifth Biennial Meeting of the American Quaternary Association.* Edmonton.

Powers, William R., F. E. Goebel, and Nancy H. Bigelow
  1990 Late Pleistocene Occupation at Walker Road: New Data on the Central Alaskan Nenana Complex. *Current Research in the Pleistocene* 7:40–43.

Powers, William R., R. Dale Guthrie, and John F. Hoffecker
  1983 *Dry Creek: Archaeology and Paleoecology of a Late Pleistocene Alaskan Hunting Camp.* US National Park Service, Washington, DC.

Powers, William R., and T. D. Hamilton
  1978 Dry Creek: A Late Pleistocene Human Occupation in Central Alaska. In *Early Man in America from a Circum-Pacific Perspective*, edited by Alan L. Bryan, pp. 72–77. Occasional Paper No. 1 of the Department of Anthropology, University of Alberta, Edmonton.

Powers, William R., and John F. Hoffecker
    1989   Late Pleistocene Settlement in the Nenana Valley, Central Alaska. *American Antiquity* 54(2):263–87.

Powers, William R., and Howard Evan Maxwell
    1986   *Lithic Remains from Panguingue Creek: An Early Holocene Site in the Northern Foothills of the Alaska Range*. Report submitted to the Alaska Historical Commission, Anchorage.

Raff, Jennifer A., and Deborah A. Bolnick
    2015   Does Mitochondrial X Indicate Ancient Trans-Atlantic Migration to the Americas? A Critical Re-evaluation. *PaleoAmerica* 1:297–304.

Raghavan, Maanasa, Pontus Skoglund, Kelly E. Graf, Mait Metspalu, Anders Albrechtsen, Ida Moltke, Simon Rasmussen, Thomas W. Stafford Jr., Ludovic Orlando, Ene Metspalu, Monika Karmin, Kristiina Tambets, Siiri Rootsi, Reedik Mägi, Paula F. Campos, Elena Balanovska, Oleg Balanovsky, Elza Khusnutdinova, Sergey Litvinov, Ludmila P. Osipova, Sardana A. Federova, Mikhail I. Voevoda, Michael DeGiorgio, Thomas Sicheritz-Ponten, Søren Brunak, Svetlana Demeshchenko, Toomas Kivisild, Richard Villems, Rasmus Nielsen, Mattias Jakobsson, and Eske Willerslev
    2014   Upper Paleolithic Siberian Genome Reveals Dual Ancestry of Native Americans. *Nature* 505(7481):87–91.

Raghavan, Maanasa, Matthias Steinrücken, Kelley Harris, Stephan Schiffels, Simon Rasmussen, Michael DeGiorgio, Anders Albrechtsen, Cristina Valdiosera, Maria Ávila-Arcos, Anna Sapfo Malaspinas, Anders Eriksson, Ida Moltke, Mait Metspalu, Julian R. Homburger, Jeff Wall, Omar E. Cornejo, J. Victor Moreno-Mayar, Thorfinn Korneliussen, Tracey Pierre, Morten Rasmussen, Paula F. Campos, Peter De Barros, Morten E. Allentoft, John Lindo, Ene Metspalu, Ricardo Rodríguez-Varela, Josefina Mansilla, Celeste Henrickson, Andaine Seguin-Orlando, Helena Malmström, Thomas Stafford, Suyash S. Shringarpure, Andres Moreno-Estrada, Monika Karmin, Kristiina Tambets, Anders Bergström, Yali Xue, Vera Warmuth, Andrew D. Friend, Joy Singarayer, Paul Valdes, Francois Balloux, Ilan Leboreiro, Jose Luis Vera, Hector Rangel-Villalobos, Davide Pettener, Donata Luiselli, Loren G. Davis, Evelyne Heyer, Christoph P. E. Zollikofer, Marcia S. Ponce de León, Colin I. Smith, Vaughan Grimes, Kelly-Anne Pike, Michael Deal, Benjamin T. Fuller, Bernardo Arriaza, Vivien Standen, Maria F. Luz, Francois Ricaut, Niede Guidon, Ludmila Osipova, Mikhail I. Voevoda, Olga L. Posukh, Oleg Balanovsky, Maria Lavryashina, Yuri Bogunov, Elza Khusnutdinova, Marina Gubina, Elena Balanovska, Sardana Fedorova, Sergey Litvinov, Boris Malyarchuk, Miroslava Derenko, M. J. Mosher, David Archer, Jerome Cybulski, Barbara Petzelt, Joycelynn Mitchell, Rosita Worl, Paul J. Norman, Peter Parham, Brian M. Kemp, Toomas Kivisild, Chris Tyler-Smith, Manjinder S. Sandhu, Michael Crawford, Richard Villems, David Glenn Smith, Michael R. Waters, Ted Goebel, John R. Johnson, Ripan S. Malhi, Mattias Jakobsson, David J. Meltzer, Andrea Manica, Richard Durbin, Carlos D. Bustamante, Yun S. Song, Rasmus Nielsen, and Eske Willerslev
    2015   Genomic Evidence for the Pleistocene and Recent Population History of Native Americans. *Science* 349:aab3884.

Rainey, Froelich
    1940   Archaeological Investigation in Central Alaska. *American Antiquity* 5:299–308.

Rapp, George Robert, and Christopher L. Hill
    1998   *Geoarchaeology: The Earth-Science Approach to Archaeological Interpretation*. Yale University Press, New Haven.

Rasic, Jeffrey T.
   2011   Functional Variability in the Late Pleistocene Archaeological Record of Eastern Beringia: A Model of Late Pleistocene Land Use and Technology from Northwest Alaska. In *From the Yenisei to the Yukon: Interpreting Lithic Assemblage Variability in Late Pleistocene/Early Holocene Beringia*, edited by T. Goebel and I. Buvit, pp. 128–64. Texas A&M University Press, College Station.

Rasmussen, Morten, Sarah L. Anzick, Michael R. Waters, Pontus Skoglund, Michael DeGeorgio, Thomas W. Stafford Jr., Simon Rasmussen, Ida Moltke, Anders Albrechtsen, Shane M. Doyle, G. David Poznik, Valborg Gundmundstottir, Rachita Yadav, Anna-Sapfo Malaspinas, Samuel Stockton White V, Morten E. Allentoft, Omar E. Cornejo, Kristiina Tambets, Anders Eriksson, Peter D. Heintzman, Monika Karmin, Thorfinn Sand Korneliussen, David J. Meltzer, Tracey L. Pierre, Jesper Stenderup, Lauri Saag, Vera M. Warmuth, Margarida C. Lopes, Ripan S. Malhi, Søren Brunak, Thomas Sicheritz-Ponten, Ian Barnes, Matthew Collins, Ludovic Orlando, Francois Balloux, Andrea Manica, Ramneek Gupta, Mait Metspalu, Carlos D. Bustamante, Mattias Jakobsson, Rasmus Nielsen, and Eske Willerslev
   2014   The Genome of a Late Pleistocene Human from a Clovis Burial Site in Western Montana. *Nature* 506(7487):225–29.

Reanier, R. E.
   1995   The Antiquity of Paleoindian Materials in Northern Alaska. *Arctic Anthropology* 32(1):1–50.
   1996   Putu and Bedwell. In *American Beginnings: The Prehistory and Paleoecology of Beringia*, edited by F. H. West, pp. 505–11. University of Chicago Press, Chicago.

Reher, C. A.
   1974   Population Study of the Casper Site Bison. In *The Casper Site: A Hell Gap Bison Kill on the Great Plain*, edited by G. C. Frison, pp. 113–24. Academic Press, New York.

Reimer, Paula. J., M. G. L. Baillie, Edourd Bard, Alex Bayliss, J. Warren Beck, C. Bertrand, Paul G. Blackwell, C. E. Buck, G. Burr, K. B. Cutler, P. E. Damon, R. Lawrence Edwards, Richard G. Fairbanks, M. Friedrich, T. P. Guilderson, K. A. Hughen, B. Kromer, F. G. McCormac, S. Manning, C. Bronk Ramsey, Ron W. Reimer, S. Remmele, John R. Southon, M. Stuiver, S. Talamo, F. W. Taylor, Johannes van der Plicht, and C. E. Weyhenmeyer
   2004   The IntCal05 Data Set. *Radiocarbon* 46:1029–58.

Reimer, Paula J., Edouard Bard, Alex Bayliss, J. Warren Beck, Paul G. Blackwell, Christopher Bronk Ramsey, Caitlin E. Buck, Hai Cheng, R. Lawrence Edwards, Michael Friedrich, Pieter M. Grootes, Thomas P. Guilderson, Haflidi Haflidason, Irka Hajdas, Christine Hatté, Timothy J. Heaton, Dirk L. Hoffmann, Alan G. Hogg, Konrad A. Hughen, K. Felix Kaiser, Bernd Kromer, Stuart W. Manning, Mu Niu, Ron W. Reimer, David A. Richards, E. Marian Scott, John R. Southon, Richard A. Staff, Christian S. M. Turney, and Johannes van der Plicht
   2013   Intcal13 and Marine13 Radiocarbon Age Calibration Curves 0–50,000 Years Cal BP. *Radiocarbon* 55(4):1869–87.

Reitz, Elizabeth J. and Elizabeth S. Wing
   1999   *Zooarchaeology*. Cambridge University Press, Cambridge.

Reuther, Joshua D., Natalya S. Slobodina, Jeffrey T. Rasic, John P. Cook, and Robert J. Speakman
   2011   Gaining Momentum: Late Pleistocene and Early Holocene Archaeological Obsidian Source Studies in Interior and Northeastern Beringia. In *From the Yenisei to the Yukon:*

Interpreting Lithic Assemblage Variability in Late Pleistocene/Early Holocene Beringia, edited by T. Goebel and I. Buvit, pp. 270–86. Texas A&M University Press, College Station.

Ritchie, J. C.
    1984   *Past and Present Vegetation of the Far Northwest of Canada*. University of Toronto Press, Toronto.

Ritchie, J. C., and L. C. Cwynar
    1982   The Late Quaternary Vegetation of the North Yukon. In *Paleoecology of Beringia*, edited by D. M. Hopkins, J. V. Matthews Jr., C. E. Schweger, and S. B. Young, pp. 113–26. Academic Press, New York.

Ritter, Dale F.
    1982   Complex River Terrace Development in the Nenana Valley near Healy, Alaska. *Geological Society of America Bulletin* 93:346–56.

Ritter, Dale F., and N. W. Ten Brink
    1986   Alluvial Fan Development and the Glacial-Glaciofluvial Cycle, Nenana Valley, Alaska. *Journal of Geology* 94:613–25.

Roseneau, D. G.
    1977   Northeast by East: Review of Alaskan Arctic Gas Research. *Kutchin Caribou Fence Studies* 2(2).

Sadek-Koores, H.
    1966   *Jaguar Cave: An Early Man Site in the Beaverhead Mountains of Idaho*. PhD dissertation, Harvard University.

Sattler, Robert A., Thomas E. Gillispie, Norman A. Easton, and Michael Grooms
    2011   Linda's Point: Results from a New Terminal-Pleistocene Human Occupation at Healy Lake, Alaska. *Current Research in the Pleistocene* 28:75–78.

Schweger, Charles E.
    1985   Geoarchaeology of Northern Regions: Lessons from Cryoturbation at Onion Portage, Alaska. In *Archaeological Sediments in Context*, edited by J. K. Stein and W. R. Farrand, pp. 127–41. Center for the Study of Early Man, University of Maine, Orono.

Schweger, Charles E., John V. Matthews Jr., David M. Hopkins, and Steven B. Young
    1982   Beringian Paleoecology—a Synthesis. In *Paleoecology of Beringia*, edited by David M. Hopkins, John V. Matthews, Charles E. Schweger, and Steven B. Young, pp. 425–44. Academic Press, New York.

Semenov, S. A.
    1964   *Prehistoric Technology*. Translated by M. W. Thompson. Gary, Adams and MacKay, London.

Shapiro, Beth, Alexei J. Drummond, Andrew Rambaut, Michael C. Wilson, Paul E. Matheus, Andrei V. Sher, Oliver G. Pybus, M. Thomas P. Gilbert, Ian Barnes, Jonas Binladen, Eske Willerslev, Anders J. Hansen, Gennady F. Baryshnikov, James A. Burns, Sergei Davydov, Jonathan C. Driver, Duane G. Froese, C. Richard Harington, Grant Keddie, Pavel Kosintsev, Michael L. Kunz, Larry D. Martin, Robert O. Stephenson, John Storer, Richard Tedford, Sergei Zimov, and Alan Cooper
    2004   The Rise and Fall of the Beringian Steppe Bison. *Science* 306:1561–65.

Shipman, P.
    1981   *Life History of a Fossil: An Introduction to Taphonomy and Paleoecology*. Harvard University Press, Cambridge.

Skarland, Ivar, and Charles J. Keim
    1958   Archaeological Discoveries on the Denali Highway, Alaska. *Anthropological Papers of the University of Alaska* 6(2):79–88.

Skinner, M. P., and O. C. Kaisen
    1947   The Fossil Bison of Alaska and Preliminary Revision of the Genus. *Bulletin of the American Museum of Natural History* 89:125–256.

Smith, Dwight R.
    1954   *The Bighorn Sheep in Idaho: Its Status, Life History and Management*. No. 3352. Boise: Idaho Department of Fish and Game.

Smith, Heather L., Jeffrey T. Rasic, and Ted Goebel
    2013   Biface Traditions of Northern Alaska and Their Role in the Peopling of the Americas. In *Paleoamerican Odyssey*, edited by Kelly E. Graf, Caroline V. Ketron, and Michael R. Waters, pp. 105–23. Center for the Study of the First Americans, Texas A&M University, College Station.

Smith, Timothy Alan
    1977   *Obsidian Hydration as an Independent Dating Technique*. MA thesis, Department of Anthropology, University of Alaska Fairbanks.
    1981   *Spatial Analysis of the Dry Creek Archaeological Site*. KPS-NGS Early Man Studies Program. Progress report to National Park Service and National Geographic Society.
    1985   Spatial Analysis of the Dry Creek Archeological Site. *National Geographic Society Research Reports* 19:6–11.

Speiss, A. E.
    1979   *Reindeer and Caribou Hunters: An Archaeological Study*. Academic Press, New York.

Stanford, Dennis J., and Bruce A. Bradley
    2012   *Across Atlantic Ice: The Origin of America's Clovis Culture*. University of California Press, Berkeley.

Stiner, Mary C.
    1994   *Honor among Thieves: A Zooarchaeological Study of Neandertal Ecology*. Princeton University Press, Princeton.

Straus, Lawrence G.
- 1977 Of Deerslayers and Mountain Men: Paleolithic Faunal Exploitation in Cantabrian Spain. In *For Theory Building in Archaeology*, edited by L. R. Binford, pp. 41–76. Academic Press, New York.
- 2000 Solutrean Settlement of North America? A Review of Reality. *American Antiquity* 65:219–26.

Straus, Lawrence G., and Ted Goebel
- 2011 Humans and Younger Dryas: Dead End, Short Detour, or Open Road to the Holocene? *Quaternary International* 242:259–61.

Straus, Lawrence G., David J. Meltzer, and Ted Goebel
- 2005 Ice Age Atlantis? Exploring the Solutrean-Clovis "Connection." *World Archaeology* 37:507–32.

Stuiver, Minze, and Henry A. Polach
- 1977 Discussion: Reporting of C-14 Data. *Radiocarbon* 19(3):355–63.

Sudgen, Lawson G.
- 1961 *The California Bighorn in British Columbia, with Particular Reference to the Churn Creek Herd*. Victoria: British Columbia Department of Recreation and Conservation.

Summerfield, B. L.
- 1974 *Population Dynamics and Movement Patterns of Dall Sheep in the Atigun Canyon Area Brooks Range, Alaska*. MA thesis, Department of Biology, University of Alaska Fairbanks.
- 1975 *Population Dynamics and Seasonal Movement Patterns of Dall Sheep in the Atigun Canyon Area, Brooks Range, Alaska*. MS thesis, University of Alaska Fairbanks.

Swift, Michael J., O. W. Heal, and J. M. Anderson
- 1979 Decomposition in Terrestrial Ecosystems. In *Studies in Ecology*, vol. 5, edited by D. J. Anderson, P. Greig-Smith, and Frank A. Pitelka, p. 372. Blackwell Scientific Publications, Oxford.

Tamm, Erika, Toomas Kivisild, Maere Reidla, Mait Metspalu, David Glenn Smith, Connie J. Mulligan, Claudio M. Bravi, Olga Rickards, Cristina Martinez-Labarga, Elsa K. Khusnutdinova, Sardana A. Federova, Maria V. Golubenko, Vadim A. Stepanov, Marina A. Gubina, Sergey I. Zhadanov, Ludmila P. Ossipova, Larisa Damba, Mikhail I. Voevoda, Jose E. Dipierri, Richard Villems, and Ripan S. Malhi.
- 2007 Beringian Standstill and Spread of Native American Founders. *PLoS One* 2(9):E829.

Ten Brink, N., and C. F. Waythomas
- 1985 Late Wisconsin Glacial Chronology of the North-Central Alaska Range: A Regional Synthesis and Its Implications for Early Human Settlements. *National Geographic Society Research Reports* 19:15–32.

Thorson, Robert M.
- 1986 Late Cenozoic Glaciation of the Northern Nenana River Valley. In *Glaciation in Alaska: The Geologic Record*, edited by Thomas D. Hamilton, Katherine M. Reed, and Robert M. Thorson, pp. 99–121. Alaska Geological Society, Anchorage.

1990   Geologic Contexts of Archaeological Sites in Beringia. In *Archaeological Geology of North America*, Centennial Special Vol. 4, edited by Norman P. Lasca and Jack Donahue, pp. 399–420. Geological Society of America, Boulder.

2006   Artifact Mixing at the Dry Creek Site, Interior Alaska. *Anthropological Papers of the University of Alaska, New Series* 4(1):1–10.

Thorson, Robert M., and G. Bender
1985   Eolian Deflation by Ancient Katabatic Winds: A Late Quaternary Example from the North Alaska Range. *Geological Society of America Bulletin* 96:702–9.

Thorson, Robert M., and Thomas D. Hamilton
1977   Geology of the Dry Creek Site: A Stratified Early Man Site in Interior Alaska. *Quaternary Research* 7(2):149–76.

Tinner, Willy, Feng Shen Hu, Ruth Beer, Petra Boltshauser-Kaltenrieder, Brigitte Scheurer, and Urs Krähenbühl
2006   Postglacial Vegetational and Fire History Pollen, Plant Macrofossil and Charcoal Records from Two Alaskan Lakes. *Vegetation History and Archaeobotany* 15(4):279-93.

Tixier, J.
1974   Glossary for the Depiction of Stone Tools. *Newsletter of Lithic Technology: Special Publication Number 1*. Washington State University, Pullman.

Vanstone, James
1974   *Athapaskan Adaptations*. Aldine, Chicago.

Van Vliet-Lanoë, Bridget
2010   Frost Action. In *Interpretation of Micromorphological Features of Soils and Regoliths*, edited by George Stoops, Vera Marcelino, and Florias Mees, pp. 81–108. Elsevier, Amsterdam.

Vereshchagin, N. K.
1967   *The Mammals of the Caucasus: A History of the Evolution of the Fauna*. Israel Program for Scientific Translations, Jerusalem.

Vereshchagin, N. K., and G. F. Baryshnikov
1982   Paleocology of the Mammoth Fauna in the Eurasian Arctic. In *Paleoecology of Beringia*, edited by D. M. Hopkins, J. V. Matthews Jr., C. E. Schweger, and S. B. Young, pp. 267–79. Academic Press, New York.

Vereshchagin, N. K., and I. E. Kuz'mina
1984   Late Pleistocene Mammal Fauna of Siberia. In *Late Quaternary Environments of the Soviet Union*, edited by H. E. Wright and C. Barnowsky, pp. 219–22. University of Minnesota Press, Minneapolis.

Voevodskij, M. V.
1952   Paleoliticheskaia Stoianka Rabochii Rov (Chulatovo II). *Uchenye Zapiski Moskovskogo Gosudarstvennogo Universiteta* 158:101–32.

Wahrhaftig, C.
- 1958 *Quaternary Geology of the Nenana River Valley and Adjacent Parts of the Alaska Range.* US Geological Survey Professional Paper No. 482. US Government Printing Office, Washington, DC.
- 1965 *Physiographic Divisions of Alaska.* US Geological Survey Professional Paper No. 482. US Government Printing Office, Washington, DC.

Ward, G. K., and S. R. Wilson
- 1978 Procedures for Comparing and Combining Radiocarbon Age Determinations: A Critique. *Archaeometry* 20(1):19–31.

Washburn, A. L.
- 1979 *Geocryology: A Survey of Periglacial Processes and Environments.* Edwards Arnold, Norwich.

Waters, Michael R.
- 1992 *Principles of Geoarchaeology: A North American Perspective.* University of Arizona Press, Tucson.

Waters, Michael R., S. L. Forman, T. A. Jennings, L. C. Nordt, S. G. Driese, J. M. Feinberg, J. L. Keene, J. Halligan, A. Lindquist, J. Pierson, and C. T. Hallmark
- 2011 The Buttermilk Creek Complex and the Origins of Clovis at the Debra L. Friedkin Site, Texas. *Science* 331:1599–1603.

Waters, Michael. R., and Thomas W. Stafford Jr.
- 2007 Redefining the Age of Clovis: Implications for the Peopling of the Americas. *Science* 315(5815):1122–26.
- 2013 The First Americans: A Review of the Evidence for the Late-Pleistocene Peopling of the Americas. In *Paleoamerican Odyssey*, edited by Kelly E. Graf, Caroline V. Ketron, and Michael R. Waters, pp. 541–60. Center for the Study of the First Americans, Texas A&M University, College Station.

Weeden, R. B.
- 1964 Spatial Separation of Sexes in Rock and Willow Ptarmigan in Winter. *Auk* 81:534–41.

West, Frederick Hadleigh
- 1967 The Donnelly Ridge Site and the Definition of an Early Core and Blade Complex in Central Alaska. *American Antiquity* 32(2):360–82.
- 1975 Dating the Denali Complex. *Arctic Anthropology* 12(1):76–81.
- 1980 Late Paleolithic Cultures in Alaska. In *Early Native Americans: Prehistoric Demography, Economy, and Technology*, edited by David L. Browman, pp. 161–87. The Hague, Mouton.
- 1981 *Archaeology of Beringia.* Columbia University Press, New York.
- 1996a The Archaeological Evidence. In *American Beginnings: The Prehistory and Paleoecology of Beringia*, edited by F. H. West, pp. 537–60. University of Chicago Press, Chicago.
- 1996b Donnelly Ridge. In *American Beginnings: The Prehistory and Paleoecology of Beringia*, edited by F. H. West, pp. 302–7. University of Chicago Press, Chicago.

West, Frederick H., Brian S. Robinson, and Mary Lou Curran
- 1996a Phipps Site. In *American Beginnings: The Prehistory and Paleoecology of Beringia*, edited by F. H. West, pp. 381–85. University of Chicago Press, Chicago.

West, Frederick H., Brian S. Robinson, and R. Greg Dixon.
- 1996b Sparks Point. In *American Beginnings: The Prehistory and Paleoecology of Beringia*, edited by F. H. West, pp. 394–98. University of Chicago Press, Chicago.

Whallon, R.
- 1973a Spatial Analysis of Occupation Floors. In *The Explanation of Culture Change*, edited by C. Renfrew, pp. 115–30. Duckworth, London.
- 1973b Spatial Analysis of Occupation Floors I: Application of Dimensional Analysis of Variance. *American Antiquity* 38(2):266–78.
- 1974 Spatial Analysis of Occupation Floors II: Application of Dimensional Analysis of Variance. *American Antiquity* 39(1):16–34.

Wheat, J. B.
- 1972 *The Olsen-Chubbuck Site: A Paleoindian Bison Kill*. Society for American Archaeology Memoir No. 26. Washington, DC.

Whitten, K. R.
- 1975 *Habitat Relationships and Population Dynamics of Dall Sheep (Ovis dalli dalli) in Mt. McKinley National Park*. MA thesis, Department of Biology, University of Alaska Fairbanks.

Willey, G. R.
- 1966 *An Introduction to American Archaeology*, vol. 1, North and Middle America. Prentice-Hall, New York.

Wilmsen, E. N.
- 1968 A Functional Analysis of Flaked Stone Artifacts. *American Antiquity* 33(3):156–61.
- 1970 Lithic Analysis and Cultural Inference: A Paleo-Indian Case. *Anthropological Papers of the University of Arizona* No. 16. University of Arizona Press, Tucson.
- 1974 *Lindenmeier: A Pleistocene Hunting Society*. Harper and Row, New York.

Wilmsen, E. N., and F. H. H. Roberts
- 1984 Lindenmeier, 1934–1974: Concluding Report on Investigations. *Smithsonian Contributions to Anthropology* (reprint ed.):24.

Wilson, M.
- 1975 *Holocene Fossil Bison from Wyoming and Adjacent Areas*. MA thesis, Department of Anthropology, University of Wyoming, Laramie.

Wood, J. J.
- 1978 Optimal Location in Settlement Space: A Model for Describing Location Strategies. *American Antiquity* 43(2):258–70.

Wood, W. R., and D. L. Johnson
- 1978 A Survey of Disturbance Processes in Archaeological Site Formation. In *Advances in Archaeological Method and Theory*, vol. 1, edited by M. Schiffer, pp. 315–81. Academic Press, New York.

Wormington, H. Marie
- 1957 *Ancient Man in North America*. Denver Museum of Natural History, Denver.

Wygal, Brian T.
- 2010 Prehistoric Upland Tool Production in the Central Alaska Range. *Alaska Journal of Anthropology* 8(1):107–19.
- 2011 The Microblade/Non-microblade Dichotomy: Climatic Implications, Toolkit Variability, and the Role of Tiny Tools in Eastern Beringia. In *From the Yenisei to the Yukon: Interpreting Lithic Assemblage Variability in Late Pleistocene/Early Holocene Beringia*, edited by T. Goebel and I. Buvit, pp. 234–54. Texas A&M University Press, College Station.

Yellen, John E.
- 1977 *Archaeological Approaches to the Present*. Academic Press, New York.

Yellen, John E., and Henry Harpending
- 1972 Hunter-Gatherer Populations and Archaeological Inference. *World Archaeology* 4(2):244–53.

Yesner, David R.
- 1994 Subsistence Diversity and Hunter-Gatherer Strategies in Late Pleistocene/Early Holocene Beringia: Evidence from the Broken Mammoth Site, Big Delta, Alaska. *Current Research in the Pleistocene* 11:154-56.
- 1996 Human Adaptation at the Pleistocene-Holocene Boundary (Circa 13,000 to 8,000 BP) in Eastern Beringia. In *Humans at the End of the Ice Age*, edited by Lawrence G. Straus, Berit V. Eriksen, Jon M. Erlandson, and David R. Yesner, pp. 255–76. Plenum Press, New York.
- 2001 Human Dispersal into Interior Alaska: Antecedent Conditions, Mode of Colonization, and Adaptations. *Quaternary Science Reviews* 20(1):315–27.
- 2007 Faunal Extinction, Hunter-Gatherer Foraging Strategies, and Subsistence Diversity among Eastern Beringian Paleoindians. In *Foragers of the Terminal Pleistocene in North America*, edited by R. B. Walker and B. N. Driskell, pp. 15–31. University of Nebraska Press, Lincoln.

Yesner, David R., K. J. Crossen, and N. A. Easton
- 2011 Geoarchaeological and Zooarchaeological Correlates of Early Beringian Artifact Assemblages. In *From the Yenisei to the Yukon: Interpreting Lithic Assemblage Variability in Late Pleistocene/Early Holocene Beringia*, edited by T. Goebel and I. Buvit, pp. 308–22. Texas A&M University Press, College Station.

Yesner, David R., Charles E. Holmes, and Kristine J. Crossen
- 1992 Archaeology and Paleoecology of the Broken Mammoth Site, Central Tanana Valley, Interior Alaska, USA. *Current Research in the Pleistocene* 9(1):12.

Yoshizaki, M.
- 1961 Shirataki Iseki to Hakkaido no Mukoki Bunka. *Minzokugako Kenkyu* 36(1):13–23.

Young, Christopher, and Sabra Gilbert-Young
- 2007 A Fluted Projectile-Point Base from Bering Land Bridge National Preserve, Northwest Alaska. *Current Research in the Pleistocene* 24:154–56.

Younie, Angela M., and Thomas E. Gillispie
- 2016 Lithic Technology at Linda's Point, Healy Lake, Alaska. *Arctic* 69(1):79–98.

CONTRIBUTORS

Brendan J. Culleton, Human Paleoecology and Isotope Geochemistry Lab, Department of Anthropology, Penn State University, University Park, PA 16802

Lyndsay M. DiPietro, Department of Geology, Baylor University, Waco, TX 76798

Ted Goebel, Center for the Study of the First Americans, Department of Anthropology, Texas A&M University, College Station, TX 77843

Angela K. Gore, Center for the Study of the First Americans, Department of Anthropology, Texas A&M University, College Station, TX 77843

Kelly E. Graf, Center for the Study of the First Americans, Department of Anthropology, Texas A&M University, College Station, TX 77843

R. Dale Guthrie, Department of Biology, University of Alaska Fairbanks, Fairbanks, AK 99775

John F. Hoffecker, Institute of Arctic and Alpine Research, University of Colorado, Boulder, CO 80303

Douglas J. Kennett, Human Paleoecology and Isotope Geochemistry Lab, Department of Anthropology, Penn State University, University Park, PA 16802

Kathryn E. Krasinski, Department of Sociology and Anthropology, Fordham University, Rose Hill Campus, Bronx, NY 10458

W. Roger Powers, deceased

Heather L. Smith, Department of Anthropology, Eastern New Mexico University, Portales, NM 88130

INDEX

Page numbers with *t* indicate tables; those with *f* indicate images.

aberrant microblade cores, 49, 57–58, 59*f*
acid/base/acid (ABA) pretreatment, 224
activity areas, 208–10, 209*t*
Afontova Gora III, 68
Akmak
   Complex, 105, 200
   microblades, 67
Alaska
   Consortium of Zooarchaeology
     in Anchorage, 226
   Department of Fish and Game, 168
   Interior, 226
   megafauna of, 6
   Peninsula, 105, 200
   Pleistocene, 165
   Range, 9, 161, 171, 177, 181, 188,
     198–99, 263, 265, 267, 279
Alberta, 165
Aldan River, 281
Aleutians, 200
America
   Great Plains of, 165, 180
   Paleoarctic Tradition of, 284
American Antiquity, 279
AMS analysis, 225, 262, 265–69, 274
Anangula, 200
angle burins, 69–70, 70*f*
anvil stones, 48, 101–3
Anzick burial, 285
Appalachians, 172
archaeological investigations, 13–22
   1973, 13–15, 14–15*f*
   1974, 15–18, 17–18*f*
   1976, 19–20
   1977, 20–22, 21–22*f*
artifacts, 22*t*, 204–8, 205–6*f*
   UA76-155-443, 255–56

asymmetric triangular knives, 76, 76*f*
Athabascans, 170–71, 184
Aurignacoid blade technology, 20

Batza Téna source, 272
Beringia, 7, 157, 166, 172, 186, 190, 194,
   198, 220, 271, 273, 282, 283–85
   settlement in, 277–82
   Standstill, 282
   Tradition, 106, 199
   wapiti, 173
Bering Land Bridge, 273–74, 283
Beryllium-10 dating, 262
Beta Analytic, Inc., 224
biface base, 206–7
   fragments, 84–85, 84*f*
   knives, 40–41, 40*f*
   midsection, 85, 86*f*
   miscellaneous, 85–86, 87*f*
   point bases, 39–40
   projectile point, 38–39
   tip, 40, 85, 86*f*
   tools, 38–41, 39–40*f*, 204–6
Bigelow, Nancy, 262
big-game hunting in North America, 180–82
bioliths, 186–90, 187*f*, 189*f*
bird gastroliths, 22
bison, 6, 16, 19, 154, 156, 166–67, 195
   antiquus, 168–69
   bison, 170, 177
   ethnographic studies of, 181
   occidentalis, 166
   priscus, 133, 136, 142
Blackwater Draw, 170
bladelike flake, 99
   tools, 99–101, 100–101*f*, 100*t*
Blue-fish Caves, 197

blue grouse, 186
boulder spall tools, 207–8
Brink, Norman Ten, 25–26
Broken Mammoth, 220–21, 263, 274–75
Brooks Range, 280, 284
brown chert, 37, 39, 40, 62, 64, 65–67t, 69, 71, 72t, 80, 86
Browne Glaciation, 25
buffalo procurement complex, 181–82
burins, 68–72, 69–73f, 72t
    angle, 70, 70f
    core-, 71, 72f
    dihedral, 69–70, 70f
    ordinary, 69–70
    scrapers, 45
    on snaps, 69, 69f
    spalls, 71–72, 72t, 73f
    transverse, 70–71, 71f
by-products, 49–58

California, 285
Calmes, Mary, 190
calories, 178–79
Campus
    cores, 67
    method, 272
    site, 5, 181, 183
Capra, 157
carbohydrate supplements, 179
caribou, 175
Carlo
    Creek, 163, 199, 262–63
    Readvance, 36
Casper site, 74, 200
Caucasus, 165
Cervus, 154, 156, 172–73, 195
chalcedony, 64–65
Channel Islands, 285
Chindadn Complex, 105, 196, 278–79, 282–83
Chloridae phytoliths, 3
chronology, 268–70
Chugwater site, 279
Chulatovo II, 108
Clovis Complex, 193, 198, 220, 279–80, 284–86
    technology, 7
Cluster A, 117f, 134, 134t
Cluster B, 118f, 134–35, 134t
Cluster C, 119f, 135–36, 135t
Cluster D, 120f, 136, 136t
Cluster E, 121f, 136–37, 137t
Cluster F, 122f, 137, 138t
Cluster G, 123f, 137–38, 139t
Cluster H, 124f, 138–40, 140t
Cluster I, 125f, 140, 141t
Cluster J, 126f, 140–42, 141t
Cluster K, 127f, 142, 142t
Cluster L, 128f, 142–43, 143t
Cluster M, 129f, 143–45, 144–45t
Cluster N, 130f, 145, 146t
Cluster X, 112–31, 113f, 131t
Cluster Y, 131–32, 114f, 131t
Cluster Z, 132, 115f, 132t
cobble, 91f
    cores, 46–48, 47–48f
    with lateral working edge, 88
    tools, 90
    with working edges on end and side, 88–90
Cody knives, 3
Colby site, 75
Component I, 38–48, 112–33
    bifacial tools, 38–41, 39–40f
    Cluster X, 112–31, 113f, 131t
    Cluster Y, 131–32, 114f, 131t
    Cluster Z, 132, 115f, 132t
    miscellaneous artifacts, 46–48, 47–48f
    occupation floors, 112–33, 111f, 113–30f, 131–32t
    research on implications of, 195–99
    scrapers, 41–45, 42–44f
Component II, 49–103, 116f, 133–46
    burins, 68–72, 69–73f, 72t
    by-products, 49–58
    Cluster A, 117f, 134, 134t
    Cluster B, 118f, 134–35, 134t
    Cluster C, 119f, 135–36, 135t
    Cluster D, 120f, 136, 136t
    Cluster E, 121f, 136–37, 137t
    Cluster F, 122f, 137, 138t
    Cluster G, 123f, 137–38, 139t
    Cluster H, 124f, 138–40, 140t
    Cluster I, 125f, 140, 141t
    Cluster J, 126f, 140–42, 141t
    Cluster K, 127f, 142, 142t
    Cluster L, 128f, 142–43, 143t
    Cluster M, 129f, 143–45, 144–45t
    Cluster N, 130f, 145, 146t
    core preforms, 58, 59–60f
    core tablets, 58–68, 62–65f, 65–67t
    core technology, 96–101, 98–100f, 101t

heavy percussion flaked implements, 86–90
knives, 76–86, 76f, 78–79f, 81–84f, 86–87f
large bifacial tools, miscellaneous, 90
microblades, 49–58, 58f
miscarried microcore preforms, 58
occupation floors, 133–46, 134–46t
projectile point, 73–75, 73f, 75f
research on implications of, 199–200
scrapers, 90–96, 92f, 94–95f, 97f
wedge-shaped cores, 49–58, 50–51t, 53–56f
Component IV, 203–15
   activity areas, 208–10, 209t
   artifacts, 204–8, 205–6f
   Northern Archaic Tradition, 212–13
   paleoecology and regional relationships in, 210–11
contrasts of occupation floors, 149
convergent side scrapers, 96, 97f
Cook, John, 268
core, 46–48, 47–48f
   microblades, 64–68
   preforms, 58, 59–60f
   tablets, 58–68, 62–65f, 65–67t
   wedge-shaped core parts, miscellaneous, 64
core-burin, 71
core technology, 96–101, 98–100f, 101t
   anvil stones, 101–3
   bladelike flake, 99–101
   hammerstones, 101
   miscellany, 101–3
   subprismatic cores, 96–99
   unshaped flake tools, 101
Coutouly, Gómez, 272
cows, 167
Cryepts, 32

Dall sheep, 156–65, 158f, 159t, 161–62f, 175, 274–76
dating
   cultural features of, 239–45
   dihedral burins, 69–70, 70f
   geology and, 34–36, 35f
Del Bene, Dr. T. A., 38, 73, 136, 138, 141
Delta River, 181, 190
deltoid biface, 83
Denali
   Complex, 5, 7, 105, 181, 183, 193, 199–200, 220–21, 277, 280, 286
   National Park and Preserve, 161, 262–63
Department of Anthropology at Texas A&M University, 224

Desert Research Institute, 224
Desna River, 108
dihedral burins, 69–70
discoidal biface, 83–84
Diuktai
   Cave, 281
   Culture, 7, 196, 200
   industry, 281–82
   Tradition, 105, 272–73
Division of Parks of the State of Alaska, 19
Donnelly Ridge, 67, 268
double end scrapers, 45, 95f
Dry Creek site
   archaeological investigations at, history of, 13–22, 14–15f, 17–18f, 21–22f
   chronology of, 177–78
   Clovis technology in, 7
   Component IV at, 203–15
   discovery of, 4
   Diuktai Culture and, 7
   extinction at, hunting of, 6–7
   formation of, 226–30
   geoarchaeology at (*see* geochronology)
   geochronology of, 34–36, 219–60, 222f, 227f, 231–38f, 238t, 240–41f, 242t, 243–44f, 245t, 247t, 248f, 249t, 251t, 252–53f, 253–55t
   geology of, 25–36, 28–32f
   Glaciation, 25
   lithic technology of, 37–106
   location of, 9
   North, early archaeology and place in, 193–201
   occupation floors at, 107–52
   paleoecology of, and implications for early hunters, 153–92
   regional setting of, 9–13, 10–12f
   research hypothesis of, 4–5
   research philosophy of, multidisciplinary approach to, 3–4
   retrospective on, 261–88, 262–63t, 264f
      Beringia, settlement in, 277–82
      site setting, Quaternary geomorphology, chronology, and environments, 264–67
      temporal context, geoarchaeology and chronology, 268–70
      terminal Pleistocene hunter-gatherer behavior, archaeological components, 271–77
      Western Hemisphere, people in, 283–86
   settlement pattern of, 5–6

Dry Creek site (*continued*)
    speculations concerning function of, 149–52
    synthesis at, 4

Early Man in Alaska Program, 4, 20
Eastern Hemisphere, 180
elliptical knives, 77–80, 77*f*
end scrapers, 42–45, 44*f*, 207
    flat, 207
    steep, 207
Epi-Levallois, 96
Equus, 16, 168, 181, 190, 273
Eskimos, 182, 184–85
Eurasia, 108, 153, 157, 171, 194
    large game hunting activity in, 179
    Paleolithic hunter's diet, 165
    Paleolithic sites, 181
excavation methods and analytical
    procedures, 223–26
    extinction, hunting of, 6–7
    faunal analysis, 226
    geological laboratory analysis, 223–24
    lithic analysis, 225–26
    radiocarbon analysis, 224–25
    sample, 223

Far North, 155, 166, 178, 183–84, 186
Far Northwest, 197
faunal analysis, 226, 250–55
flat end scrapers, 207
flying squirrels, 176
fossil ungulates, 154–73, 158*f*,
    159*t*, 162–63*f*, 168*t*, 169*f*
frost, 108–9

Gallagher Flint Station, 200
gastroliths, 186–90, 187*f*, 189*f*, 275
geoarchaeology, 268–70. *See
    also* geochronology
geobiochemical weathering, 109
geochronology, 34–36, 219–60, 222*f*, 227*f*
    Artifact UA76-155-443,
        reanalysis of, 255–56
    excavation methods and analytical
        procedures, 223–26
    2011 results on, 226–55, 227*f*, 230–38*f*,
        239*t*, 240–41*f*, 242*t*, 243–44*f*, 245*t*, 247*t*,
        248*f*, 249*t*, 251*t*, 252–53*f*, 253–55*t*
geology, 25–36, 28–32*f*, 226–30
    context of occupation floors, 108–10
    dating and, 34–36, 35*f*

laboratory analysis, 223–24
    regional, 25
Gerstle River Quarry, 274
get-by survival strategy, 155
Gillispie, Tom, 262
gizzard stones, 186–87
Goebel, Ted, 262
Golden Valley Electrical Association, 35
Great Basin, 285
Great Bison Belt of North
    America, 194, 199, 274
Great Plains, 7, 74, 105, 154, 164–66, 170–71,
    177–78, 181–82, 191, 193–94, 277
Gromov, 68
Group K obsidian, 272
Guthrie, R. Dale, 16

Hadleigh-West, Frederick, 177
Hamilton, Thomas D., 13, 19, 268
hammerstone, 101, 103*f*
Haskett points, 75, 200
Healy
    Age, 9, 12, 35
    Creek, 196
    Glaciation, 25–26, 264
    Lake, 105, 196, 268, 270, 278, 284
Healy–McKinley Park, 156
heavy percussion flaked implements,
    86–90, 89*f*. *See also* cobble
Hell Gap, 7, 74–75, 200
herb zone, 267
High Plains, 192
Holarctic, 172
Holmes, Charles, 13
Holocene, 16, 49, 173, 178, 180, 190
    bison, 177
    dwarfing, 154–56
    fauna, 6
    packet of sediments, 266
horse, 6, 16, 165, 176, 180,
    190–91, 197, 273, 275
Human Paleoecology and Isotope
    Geochemistry Lab, 224
HydroMU dispersion, 223

Idaho, 163, 200
Illinois, 172
Indian Mountain, 37
inductively coupled plasma atomic emission
    spectrometry (ICP-MS/ICP-AES), 224
industrial accidents, 64

Inland Eskimo, 182
Inner Range, 9, 161
IntCal13 Northern Hemisphere atmospheric curve, 225
Intermontane Plateau, 9

Jaguar Cave, 163

Kamchatka Peninsula, 200, 281
Kansas, 172
knives, 76–86, 78–79f, 81–84f, 86–87f. *See also* biface base
    asymmetric triangular, 76, 76f
    base fragments of bifaces, 84–85
    elliptical, 77–80, 79f
    oblong, 76
    oval, 80, 81f
    ovate, 80–81, 82f
    slightly stemmed, 77
    small spatulate, 77, 78f
Kobuk River, 105
Kokorevo
    I, 68, 151, 170
    II, 68
Koyukuk River, 37
krotovinas, 108–9
Kukhtui III, 197
Kutchin, 173

lanceolate bifaces, 81–83, 83f
large flake tools, 90
Last Glacial Maximum (LGM), 281–82, 284–85
Late Paleoindian Tradition, 277
Leica DFC 450 camera attachment, 224
Lignite Creek, 12
limited activity sites, 149–50
Lindenmeier, 105t, 151
lithic analysis, 225–26
lithic artifact assemblages, 246–50
    natural site formation and, 250
    technological organization and behavioral site formation, 246–50
lithic technology, 37–106
    Component I, 38–48
    Component II, 49–103
Little Panguingue Creek, 199
Llano Estacado, 166
loess, 27–33, 36, 108–9, 219, 221

Main Range, 161
Malvern Mastersizer2000, 223

Mammalogy Department of the University of Alaska Museum of the North, 226
mammoth steppe, 174, 266
Mammuthus, 16, 170, 190, 273
marginal habitats, 167
Maxwell, Howard, 262
McCalpin, James, 19–20, 26, 190
McDonald Creek, 277
mean daily caloric requirements, 179
    analysis and site chronology of, 177–78
    megafauna, reconstruction of late glacial, 173–78, 173t, 174f
    ungulate community, modern, 175–76
    ungulate winter range, in Nenana Valley, 176–77
Menlo Park, California, 20
Mercer, Beryl, 176
Mesa
    Northern Paleoindian Tradition, Complex of 280
    site, 280, 284
microblades, 49–58, 55–56f, 58f, 62–64f, 64–68, 65–66t
micromorphology, 234
mineral licks, 160
miscarried microcore preforms, 58
miscellaneous tools, 46–48, 47–48f. *See also specific types of*
    anvil stones, 48
    biface, 85–86, 87f
    cobble cores, 46–48
    cobble tools, 90
    large bifacial, 90
    large flake tools, 90
    quadrilateral uniface, 46
    split boulder, 48
    split cobble, 46
    unshaped flake, 46
    wedge-shaped core parts, 64
miscellany, 101–3
modern ungulate community, 175–76
Mongolia, 49
Montana, 166, 285
moose, 175
Moose Creek, 105, 109, 149–50, 196, 262, 198, 220–21, 262, 266, 270, 278
mountain sheep, 157–59
Mount Healy, 12, 163
Mousteroid technology, 20
Munsell colors, 26

Na-Dene languages, 286
National Geographic Society, 19–20
National Park Service, 20
National Science Foundation, 15, 19
Native Americans, 283, 285
natural site formation, 250
Nenana
   Canyon, 163
   Complex, 220, 256, 279–83, 285–86
   Glacier, 9
   Gorge, 191
   River, 9–12, 32–33, 37, 150, 190, 199
     bison and, 168
     Dall sheep and, 160–61
   Valley, 12, 33, 150, 152, 186, 196–97,
     199, 261, 265–66, 285
     Dall sheep and, 160–61
     glacial episodes in, 25
     Holocene alluvium in, 147
     Pleistocene geology of, 25, 34–36
     ungulate winter range in, 176–77
     Wisconsinan Glaciation and, 156
New World, 49, 153, 156–57,
   165, 173, 179–80, 201
nonmicroblade, 221
North
   early archaeology and place in, 193–201
   research on implications of, 194–200
North Alaska Range Early
   Man Project, 25, 261
North American Great Plains, 197, 200, 274
Northern Archaic Tradition, 212–13, 277
northern grazing genera, 170
Northern Hemisphere, 265
number of identified specimens (NISP), 226
Nunamiut hunting-pouch contents, 181
nutritional considerations, 178–80

oblong knives, 76
occupation floors, 107–52, 113–30f, 131–32t
   analysis of, 146–49
   Component I, 112–33, 113–30f, 131–32t
   Component II, 133–46, 134–46t
   contrasts of, 149
   geological context of, 108–10
   methodology of studying, 110–12, 111f
Old World, 49, 180, 274
Olympus BX-51 research microscope, 224
Onion Portage, 105
Orb model of hunting camp
   settlement, 182–86

ordinary burins, 69–70
Outer Range, 9, 12, 161, 191
oval knives, 80, 80f
ovate knives, 80–81, 81f
Ovis, 154, 195
Owl Ridge, 220–21, 262–63, 266, 270

Pacific Mountain System, 9
Paisley Caves, 285
paleoecology
   bioliths as, 186–90, 187f, 189f
   conclusions on, 190–92
   fossil ungulates, 154–73, 158f,
     159t, 162–63f, 168t, 169f
   gastroliths as, 186–90
   implications for early hunters,
     153–92, 210–11
   megafauna, reconstruction of late
     glacial, 173–78, 173t, 174f
   nutritional considerations, 178–80
   Orb model of hunting camp
     settlement, 182–86
   phytoliths as, 186–90
   significance of, 153–54
Paleoindians, 153, 176, 181
   archaeological sites, 173
   evidence, 6
   fish, 178
   Plains, 274
   Tradition, 280
Paleolithic, 176
   age, 108
   diet, 165
   evidence, 6
paleosols, 27–28, 33, 36, 108–9, 145, 219
palynological vacuum, 266
Panguingue Creek, 262, 266, 272
Parks Highway, 9
Pastureland, 155
Pearson, Georges, 262
Pennsylvania State University, 224
Phippen, Peter, 262
physiological fuel value, 178–79
phytoliths, 189–90
Plains tribes, 171
Plano, 74–75
Pleistocene, 198
   age, 4, 13, 16
   bison, 167, 170
   fauna, 6–7
   gigantism, 178

mammoth steppe, 155–56
sites, 150
ungulates, 155–57, 164
point bases, 39–40
Powers, W. R., 219
Powers Era, 262
Powers roundedness scale, 188
premicroblade industry, 278–80
projectile point, 38–39, 73–75, 73f, 75f
bases, 74–75, 75f
tips, 75
ptarmigan, 187–89

quadrilateral uniface, 46
Quaternary
geomorphology, 264–67
soils, 3–4

radiocarbon analysis, 35t, 224–25
Rangifer, 172, 175, 199. *See also* caribou
red-back voles, 176
red squirrels, 176
regional geology, 25
research
Component I, 195–99
Component II, 199–200
hypothesis of Dry Creek site, 4–5
on North, early archaeology and place in, 194–200
philosophy of Dry Creek site, multidisciplinary approach to, 3–4
residential camp, 183
results of 2011 study, 226–55, 227f, 230f, 232–38f, 239t, 240–41f, 242t, 243–44f, 245t, 247t, 248f, 249t, 251t, 252–53f, 253–55t
dating of cultural features, 239–45
faunal analysis, 250–55, 253f
geology, stratigraphy and site formation, 226–30
lithic artifact assemblages, 246–50
micromorphology, 234
sedimentology, 230–33
soil geochemistry, 234–37
soil micromorphology and formation of Loess, 237–39
retouched flakes, 207
Riley Creek Glaciation, 12f, 25, 36
Rocky Mountains, 172, 199
rodent burrowing, 108–9
ruffed grouse, 187

sage grouse, 186
salt licks, 159–60
sample excavation methods and analytical procedures, 223
scrapers, 41–45, 42–44f, 90–96, 92f, 94–95f, 97f
convergent side, 96, 97f
double end, 45, 95f
end, 42–45, 44f
side, 41–42, 43f, 93–95, 94f
spokeshaves, 92f, 93
transverse, 41, 42f, 92–93, 92f
sedimentology, 230–33
Seward Peninsula, 197
sharp-tailed grouse, 187
Siberian-American Paleoarctic Tradition, 106
Siberian Diuktai Culture, 197
side scrapers, 41–42, 43f, 93–95, 94f
Silva Ranger clinometer compasses, 223
site. *See also* Dry Creek site
setting, 264–67
slightly stemmed knives, 77
small spatulate knives, 77, 77f
Smith, Tim, 208
Smithsonian Institution, 13, 34
Snake River Plain of Idaho, 75
soil
geochemistry of, 234–37
loess, micromorphology and formation of, 237–39
Sokkia EDM total station, 223
Solutré, 180
Solutrean Upper Paleolithic industry, 284
Sosnovskii, 68
South Siberia
Paleolithic age of, 68
spalls, 72t, 73f
special purpose locations, 183
spike camp, 182–85, 276
split boulder, 48
split cobble tools, 46
spokeshaves, 92f, 93
spruce grouse, 186
"Standstill" population, 285–86
Star-Carr, 172–73
steep end scrapers, 207
steppe bison, 165–71, 168t, 169f, 273–74
stratigraphy, 26–34, 28–32f, 222f, 226–30, 227f
Stuckenrath, Robert, 13, 34
subprismatic cores, 96–99

subsistence, 273–75
Sugar Loaf Mountain, 12
Swan Point, 220, 263, 273–75, 281–82, 284

Tanana
   basin, 259, 278, 279–80, 288
   River, 9–12, 277
   Uplands, 175, 177
   Valley, 175, 196, 220, 263, 284
Tanana-Kuskokwim lowlands, 12
Tangle Lakes, 5, 268
technologies, 271–73
   organization and behavioral
      site formation, 246–50
Teklanika
   River, 67
   Valley, 177, 261, 279
   West, 262–63, 266, 268, 270
temporal context, 268–70
terminal Pleistocene hunter-gatherer behavior
   archaeological components, 271–77
Texas, 165
Texas A&M Soil Characterization
   Laboratory, 224
Thermo Finnegan Flash 1112 Series, 224
thermoluminescence (TL) date, 269
Trail Creek Cave, 68, 197
trail food bones, 185
   transverse burins, 70–71, 71f
transverse scrapers, 41, 42f, 92–93, 92f
Tuktu core, 33, 212

Ugashik Narrows Phase, 105, 200
ungulate community, modern, 175–76
Union Grid, 15
United States Geological
   Survey (USGS), 19–20
University of Alaska, 16, 19, 273
University of Alaska Museum of
   the North, 221, 225, 255
University of California, Irvine, 224
University of Connecticut, 38

unshaped flake tools, 46, 101
Upward Sun, 274
USDA-NRCS, 224
Ushki, 281
   Culture, 285–86
   Lake, 7, 200
Usibelli, Joseph, 177, 196
   coal mine, 268
USSR, 108

Wahrhaftig, C., 25
Walker Road, 149, 220–21, 262,
   265–66, 270, 275–76, 278
wapiti, 171–73, 173t, 174f
Waythomas, C., 269
wedge-shaped cores, 49–58,
   50–51t, 53–56f, 64
   microblade cores, 50–58, 55–56f, 59f
West, Frederick, 268
Wiki Peak, 272
Windmill Lake, 267
Wisconsinan Glaciation, 25, 105,
   157, 166–68, 177, 265
W. M. Keck Carbon Cycle Accelerator
   Mass Spectrometry facility, 224
woolly mammoth, 273
woolly rhinoceros, 273
Wyoming, 157

X-ray fluorescence (pXRF) technologies, 272

Yakutia, 200
Yana River, 281
Yanert Fork, 267
Yenisei River, 68, 151
Younger Dryas, 265–66
Younger Younger Dryas, 266–67,
   269, 279–80, 282, 284
Yubetsu method, 272–73, 281
Yukon
   River, 198
   Territory, 173–74, 177, 197